Lecture Notes in Mathematics

Edited by A. Dold and B. Eckmann

Subseries: Mathematisches Institut der Universität und Max-Planck-Institut für Mathematik, Bonn – vol. 10

Adviser: F. Hirzebruch

1290

G. Wüstholz (Ed.)

Diophantine Approximation and Transcendence Theory

Seminar, Bonn (FRG)
May – June 1985

Springer-Verlag
Berlin Heidelberg New York London Paris Tokyo

Editor

Gisbert Wüstholz
Mathematik, ETH Zentrum
8092 Zürich, Switzerland

Mathematics Subject Classification (1980): 10B10, 10B45, 10F05, 10F35, 10F37, 12C15, 14D99, 14K99

ISBN 3-540-18597-6 Springer-Verlag Berlin Heidelberg New York
ISBN 0-387-18597-6 Springer-Verlag New York Berlin Heidelberg

Printed in Germany

Printing and binding: Druckhaus Beltz, Hemsbach/Bergstr.
2146/3140-543210

INTRODUCTION

In 1985 the traditional Arbeitstagung at Bonn was cancelled. Instead a number of workshops were organized during Spring and Summer 1985 by various organizers. One of these workshops was on number theory with special emphasis on diophantine problems and transcendence. It took place at the Max-Planck-Institut für Mathematik at Bonn in May - June 1985. A great number of leading mathematicians in the subject were invited for a certain period to discuss mathematics and problems. It seems that a very fruitful atmosphere was created which is reflected by quite a number of joint papers which were written during this time at Bonn or at least initiated there.

We are very happy to present in this volume a selection of papers that grew out of this workshop at Bonn. It consists entirely of research papers which cover various important aspects of the field and each of them presents a fundamental contribution to the subject.

In the first contribution by Colliot-Thélène, Kanevsky and Sansuc an effective algorithm for the calculation of the Manin obstruction for the Hasse principle for diagonal cubic surfaces is given.

Then in the next article by Masser on small values of heights on families of abelian varieties, an effective lower bound for the variation of the Néron-Tate height in families of abelian varieties is given.

In two further papers, one by Brownawell and another by Brownawell and Tubbs, questions on large transcendence degree are studied. The authors obtain very precise lower bounds for the transcendence degree of fields generated by values of elliptic or exponential functions.

In the next two contributions multiplicity estimates for group varieties are studied in the most simplest case. Here extremely sharp results are given and applied to Baker's theory of linear forms in logarithms.

In the last paper, by E. Bombieri, the number of solutions in integers in number fields of the so-called Thue equation is bounded effectively. This generalizes in one direction the earlier work of Bombieri and Schmidt on the Thue equation.

This workshop profited very much from the friendly atmosphere at the MPI. We are very grateful to its Director F. Hirzebruch for giving us the opportunity to organize this workshop and for keeping everything as unbureaucratic as possible. We have also to thank Ms. A. Franz for doing the very careful typing of most of the contributions at the Bergische - Universität - Gesamthochschule Wuppertal.

Zurich 1987

G. Wüstholz

TABLE OF CONTENTS

ARITHMÉTIQUE DES SURFACES CUBIQUES DIAGONALES

par

Jean-Louis Colliot-Thélène
Dimitri Kanevsky
Jean-Jacques Sansuc

Etant donnée une surface cubique V d'équation homogène diagonale:

(1) $ax^3 + by^3 + cz^3 + dt^3 = 0 \qquad a, b, c, d \in \mathbb{Q}^*$

peut-on décider si l'ensemble $V(\mathbb{Q})$ de ses points rationnels est non vide? Autrement dit, peut-on décider s'il existe une solution non triviale $(x,y,z,t) \in \mathbb{Q}^4$ de l'équation (1)?

Une première condition, évidente, est que (1) possède des solutions non triviales dans chaque complété p-adique $\mathbb{Q}_p$ de $\mathbb{Q}$ (le complété réel $\mathbb{R}$ ne pose pas de problème pour une surface cubique). Vérifier si ces *conditions locales* $V(\mathbb{Q}_p) \neq \emptyset$ sont satisfaites ne requiert, comme il est bien connu, qu'un nombre fini de vérifications faciles, cf. §4, pas nº0.

En 1949, Mordell conjecturait [15] que ce test suffisait pour assurer l'existence d'un point rationnel sur une surface cubique définie sur $\mathbb{Q}$. Dans le cas particulier des surfaces cubiques diagonales (1) telles que l'un des quotients ab/cd soit le cube d'un nombre rationnel, ceci fut établi par Selmer [17] en 1953, et de plus Selmer vérifia aussi que toutes les surfaces (1) avec a,b,c,d entiers et $|abcd|<500$ ont un point rationnel dès que les conditions locales sont vérifiées. En d'autres termes, dans tous ces cas, le *principe de Hasse* est vérifié. Mais on sait, depuis 1962, que le test local ne suffit pas pour les surfaces cubiques générales (Swinnerton-Dyer [21], Mordell [16]), et, en 1966, Cassels et Guy [4] donnèrent le premier exemple de surface cubique diagonale ne satisfaisant pas le principe de Hasse:

(2) $5x^3 + 9y^3 + 10z^3 + 12t^3 = 0$.

Leur ordinateur leur avait fourni une liste de quelques 250 équations du type (1) avec a,b,c,d entiers (petits), ayant des solutions dans chaque complété de $\mathbb{Q}$, mais n'ayant pas de solution rationnelle de petite hauteur (i.e. avec x,y,z,t entiers et petits en valeur absolue). Ce n'est qu'en 1978 que fut traité un second exemple (Bremner [3]), lui aussi pris dans la liste de Cassels et Guy:

(3) $x^3 + 4y^3 + 10z^3 + 25t^3 = 0$.

Chez Cassels et Guy comme chez Bremner, l'analyse qui mène à la conclusion $V(\mathbb{Q})=\emptyset$ est très élaborée. Elle nécessite la connaissance précise des groupes de classes et d'unités de diverses extensions cubiques et bicubiques de $\mathbb{Q}$, et elle ne semble pas se prêter à une généralisation systématique qui mènerait à un algorithme (au moins conjectural) permettant de décider si l'équation (1) a une solution rationnelle non triviale.

En 1970, Manin [13] proposa une interprétation générale de divers contre-exemples au principe de Hasse figurant dans la littérature, en termes du groupe de Brauer-Grothendieck, cf. §0 ci-après. Ce nouveau test pour l'existence (ou plutôt la non-existence) de points rationnels, qu'on peut voir aussi comme une obstruction au principe de Hasse - *l'obstruction de Manin au principe de Hasse* - n'est pas toujours facile à mettre en oeuvre. Dans son livre sur les *"formes cubiques"* paru en 1972, Manin [14] essaie en particulier d'interpréter l'exemple de Cassels et Guy au moyen de son obstruction, mais le calcul est peu convaincant. En outre, il s'appuie de façon essentielle sur les résultats de factorisation de Cassels et Guy, si bien qu'on ne voit pas comment généraliser les arguments à d'autres surfaces (1).

L'objet du présent article est de développer une méthode de calcul de l'obstruction de Manin dans le cas des surfaces cubiques diagonales définies sur $\mathbb{Q}$, méthode qui soit effective et relativement simple, au point d'aboutir à une exploitation systématique et à son calcul sur ordinateur.

Le calcul a été effectué par M. Vallino sur IBM 4341 pour toutes les équations (1) à coefficients entiers positifs <100 (ce qui couvre bien sûr très largement la liste originelle de Cassels et Guy). Ce calcul a donné, à équivalences élémentaires près, 245 contre-exemples au principe de Hasse, l'exemple de Bremner étant celui qui minimise le produit abcd. De plus, et c'est là sans doute le résultat le plus intéressant, on a trouvé une solution rationnelle non triviale pour toutes les équations ayant des solutions locales et sans obstruction de Manin. Ceci conforte, au moins pour les surfaces (1), la conjecture suivante: *pour les surfaces rationnelles définies sur un corps de nombres, l'obstruction de Manin est la seule obstruction au principe de Hasse.* Si tel était le cas, l'algorithme décrit dans notre article fournirait une méthode entièrement effective pour décider si une équation du type (1) a une solution non triviale.

Les grandes lignes de la méthode sont décrites au §0 ci-après. On y trouve aussi la description du contenu des autres paragraphes. Nous nous contentons de citer ici quelques résultats particulièrement simples (voir §5):

Proposition 2. *S'il existe un nombre premier* p *de mauvaise réduction pour la surface* V *d'équation* (1) *tel que* V *ne soit pas rationnelle sur* $\mathbb{Q}_p$, *alors il n'y a pas d'obstruction de Manin au principe de Hasse pour* V.

Dans un tel cas nous dirons plutôt: *l'obstruction de Manin est vide pour* V.

Corollaire. *S'il existe un nombre premier* p *qui divise un, et un seul, des coefficients de l'équation* (1), *supposés entiers sans facteur cubique, alors l'obstruction de Manin au principe de Hasse pour la surface* V *est vide.*

Conjecture. *Sous l'hypothèse du corollaire,* $\prod_{\ell} V(\mathbb{Q}_{\ell}) \neq \emptyset \Longrightarrow V(\mathbb{Q}) \neq \emptyset$.

Il s'agit évidemment d'un cas particulier de la conjecture générale ci-dessus, compte tenu du corollaire précédent. Cette conjecture particulière suffirait à entraîner la validité du *principe de Hasse pour les hypersurfaces cubiques diagonales dans* $\mathbb{P}^n_{\mathbb{Q}}$ *pour* $n \geq 4$, *et l'existence de points rationnels sur celles-ci pour* $n \geq 6$ (voir §9).

Voici enfin une série infinie de contre-exemples au principe de Hasse (voir §7):

Proposition 5. *Les surfaces cubiques* V *d'équation homogène*

(4) $$x^3 + p^2y^3 + pqz^3 + q^2t^3 = 0$$

où $p \equiv 2$ *et* $q \equiv 5 \bmod 9$ *sont premiers, ont un point dans chaque complété de* $\mathbb{Q}$, *mais n'en ont pas dans* $\mathbb{Q}$.

Le plan de l'article est le suivant:

Les résultats de cet article ont été annoncés dans un exposé au Séminaire de théorie des nombres de Paris [7], auquel on peut aussi se reporter pour la description - sans justification - de l'algorithme établi dans le présent travail.

§0. Plan de la méthode

Soit V une variété projective et lisse définie sur le corps de nombres k. On suppose $V(k_v)\neq\emptyset$ pour chaque complété k_v de k. L'obstruction de Manin au principe de Hasse (Manin [13], [14]) est la condition suivante:

(5) $$\text{pour tout } (x_v) \in \prod_v V(k_v)\,,\ \text{il existe } A \in \mathrm{Br}(V) \text{ telle que } \sum_v \mathrm{inv}_v A(x_v) \neq 0\,.$$

Dans cette expression, $A(x_v)\in \mathrm{Br}\, k_v$ désigne la fibre de A au point x_v et, inv_v désigne l'invariant local $\mathrm{Br}\, k_v \hookrightarrow \mathbb{Q}/\mathbb{Z}$. La formule de produit assure que cette condition (5) est une obstruction à l'existence d'un point rationnel pour V.

Si $\bar{k}$ désigne une clôture galoisienne de k et si $\bar{V}=V\times_k\bar{k}$, on dit que V est une variété rationnelle si $\bar{V}$ est birationnelle à l'espace projectif. C'est le cas des surfaces cubiques lisses. Pour une telle variété $\mathrm{Pic}\,\bar{V}$ est un module galoisien $\mathbb{Z}$-libre de rang fini et on a une suite exacte naturelle:

(6) $$0 \longrightarrow \mathrm{Br}(k) \longrightarrow \mathrm{Br}(V) \longrightarrow H^1(k,\mathrm{Pic}\,\bar{V}) \longrightarrow 0\,.$$

Le groupe de cohomologie galoisienne $H^1(k,\mathrm{Pic}\,\bar{V})$ est fini et on peut se limiter dans la condition (5) à la considération d'un nombre fini d'algèbres A_i qui engendrent $\mathrm{Br}(V)$ modulo $\mathrm{Br}(k)$.

L'objet du §1 est précisément le calcul de $\mathrm{Br}(V)/\mathrm{Br}(k)$ pour une surface cubique diagonale. On trouve 0, $\mathbb{Z}/3$ ou $(\mathbb{Z}/3)^2$, mais pour notre problème on s'intéresse uniquement à des cas où il vaut $\mathbb{Z}/3$, car dans les autres cas la surface V a de façon "évidente" un point dans k.

Soit alors $\mathcal{A}$ une algèbre dont la classe engendre $\mathrm{Br}(V)$ modulo $\mathrm{Br}(k)$. D'après la discussion précédente, la condition (5) s'écrit alors simplement:

(7) $$\text{pour tout } (x_v) \in \prod_v V(k_v) \text{ on a } \sum_v \mathrm{inv}_v \mathcal{A}(x_v) \neq 0\,.$$

Pour toute place v de bonne réduction pour V et $\mathcal{A}$, et pour tout $x_v\in V(k_v)$, on a (V est projective et $\mathrm{Br}\,\mathcal{O}_v=0$ pour l'anneau des entiers $\mathcal{O}_v$ de k_v):

(8) $$\mathrm{inv}_v \mathcal{A}(x_v) = 0\,.$$

Il suffit donc de considérer les places v, en nombre fini, de mauvaise réduction.

Pour continuer le calcul il faut disposer d'une représentation commode de $\mathcal{A}$ qui permette d'évaluer effectivement les invariants locaux $\mathrm{inv}_v \mathcal{A}(x_v)$. Un cas a priori très favorable est celui où la variété V devient rationnelle sur une extension cyclique k' de k, ce qui est le cas des surfaces cubiques diagonales. On dispose en effet alors du diagramme commutatif suivant dont les lignes sont exactes:

$$\begin{array}{ccccccccc}
 & 0 & \longrightarrow & \mathrm{Br}(k,k') & \longrightarrow & \mathrm{Br}(V,k') & \longrightarrow & H^1(G,\mathrm{Pic}\,V_{k'}) & \longrightarrow 0 \\
 & & & \| & & \downarrow & & \downarrow \partial & \\
(9) & 0 & \longrightarrow & H^2(G,k'^*) & \longrightarrow & H^2(G,k'(V)^*) & \longrightarrow & H^2(G,k'(V)^*/k'^*) & \longrightarrow 0 \\
 & & & & & \downarrow & & \downarrow & \\
 & & & & & H^2(G,\mathrm{Div}\,V_{k'}) & = & H^2(G,\mathrm{Div}\,V_{k'}) . &
\end{array}$$

On note Br(V,k') le noyau de $\mathrm{Br}(V) \longrightarrow \mathrm{Br}(V_{k'})$ et G le groupe de Galois de k'/k. On peut donc prendre $\mathcal{A}$ dans Br(V,k') et la représenter par un élément du noyau de

$$k(V)^*/Nk'(V)^* \longrightarrow \mathrm{Div}\,V/N(\mathrm{Div}\,V_{k'})$$

puisque $H^2(G,M)=M^G/N_G M$. Le calcul de $\mathcal{A}(x)$ est alors très simple: si $\mathcal{A}$ est représentée par la fonction f, l'élément $\mathcal{A}(x) \in H^2(G,k'^*) \simeq k^*/Nk'^*$ n'est autre que la classe de f(x) dans k^*/Nk'^* après modification éventuelle de f pour qu'elle soit inversible en x.

Dans le cas des surfaces de Châtelet cette méthode de calcul s'applique sans difficulté. Il n'en est pas de même pour les surfaces cubiques diagonales pour la simple raison suivante: dans le diagramme

$$\begin{array}{ccc}
 & & H^{-1}(G,\mathrm{Pic}\,V_{k'}) \\
 & & \downarrow \partial \\
k(V)^*/Nk'(V)^* & \longrightarrow & k(V)^*/k^*Nk'(V)^*
\end{array}$$

on sait trouver un générateur ζ du groupe $H^{-1}(G,\mathrm{Pic}\,V_{k'})$ mais on ne sait pas trouver simplement une fonction $\omega \in k(V)^*$ qui relève $\partial(\zeta)$. Il se trouve qu'on sait cependant le faire, de façon très simple, au niveau d'une extension bicyclique K/k de degré 9 contenant k'/k. Aussi a-t-on besoin de savoir calculer commodément la cohomologie des groupes bicycliques, ce qui est l'objet du §2.

Ayant ainsi représenté au §3 l'algèbre $\mathcal{A}$ qui engendre Br(V)/Br(k), il reste à calculer $\mathrm{inv}_v\mathcal{A}(x_v)$ pour les divers x_v relatifs aux places de mauvaise réduction. Ces invariants sont à valeurs dans $\mathbb{Z}/3$. On suppose désormais $k=\mathbb{Q}(\theta)$ où $\theta^2+\theta+1=0$. Il se produit alors une alternative qui simplifie les calculs:

1er cas: il existe une place v_0 de mauvaise réduction en laquelle $\mathrm{inv}_{v_0}\mathcal{A}(x_{v_0})$ prend toutes les valeurs possibles dans $\mathbb{Z}/3$;

2ème cas: pour toute place v de mauvaise réduction la surface V est rationnelle sur k_v.

Il est immédiat que dans le premier cas l'obstruction de Manin est vide. Dans le second cas, si v est de mauvaise réduction, $\mathcal{A}(x_v)$ ne dépend pas du point x_v et on peut poser:

(10) $$i_v = \mathrm{inv}_v\mathcal{A}(x_v) \quad \text{et} \quad i = \sum_v i_v \, .$$

L'obstruction de Manin s'écrit alors simplement:

(11) $$i \neq 0 \, .$$

Le §4 expose les détails du calcul de l'obstruction de Manin sans aboutir immédiatement à une procédure facile à utiliser dans tous les cas. Le §5 traite le 1er cas comme indiqué ci-dessus. Le §6 indique comment procéder à l'achèvement du calcul dans le cas où V est rationnelle aux places de mauvaise réduction. On montre au §7 sur des exemples, dont ceux de Cassels et Guy et de Bremner, comment la méthode s'applique. Au §8, on résume les étapes du calcul sous la forme d'une procédure proche du programme pour ordinateur, et on décrit les résultats des calculs de M.Vallino sur ordinateur. En particulier, on donne la table des contre-exemples au principe de Hasse dans le domaine étudié. Le §9 contient l'application aux hypersurfaces cubiques diagonales de la conjecture sur l'obstruction de Manin mentionnée dans l'introduction. Enfin, au §10, on applique aux surfaces cubiques diagonales la méthode de la descente [8], ce qui ramène dans ce cas la conjecture sur l'obstruction de Manin au problème de la validité du principe de Hasse pour certaines intersections très particulières de deux hypersurfaces cubiques (proposition 11).

§1. Calcul du groupe de Brauer

Dans ce paragraphe k est un corps de nombres ou un complété d'un tel corps. On suppose que k contient θ vérifiant $\theta^2+\theta+1=0$. On veut calculer Br(V) pour la surface cubique d'équation homogène

(1) $ax^3 + by^3 + cz^3 + dt^3 = 0$ $\qquad a, b, c, d \in k^*$.

Vu les hypothèses sur k, on a la suite exacte:

$$\mathrm{Br}(k) \longrightarrow \mathrm{Br}(V) \longrightarrow H^1(k, \mathrm{Pic}\,\overline{V}) \longrightarrow 0$$

avec même une injection au début si V a un point dans k ou dans chaque complété. L'objet de ce paragraphe est d'établir la proposition suivante:

Proposition 1. *Soit* V *la surface cubique diagonale d'équation* (1) *définie sur le corps* k. *On fait les hypothèses ci-dessus sur* k. *On trouve:*

$H^1(k,\mathrm{Pic}\,\overline{V}) = 0$ *si l'un des* ab/cd *est un cube,*
$(\mathbb{Z}/3)^2$ *pour* $x^3 + y^3 + z^3 + dt^3 = 0$ *si* d *n'est pas un cube,*
$\mathbb{Z}/3$ *sinon.*

Ce dernier cas est le cas générique. Dans le second cas il y a évidemment un point rationnel, dans le premier aussi lorsque k est un corps de nombres et que V a un point dans chaque complété de k. On a en effet le lemme suivant:

Lemme 1. *Soit* V *une surface cubique diagonale telle que* ad/bc *soit un cube. On a l'équivalence:*

$V(k) \neq \emptyset \iff$ V *est k-rationnelle.*

Si k *est un corps de nombres, une telle surface vérifie le principe de Hasse.*

En effet, on peut mettre l'équation de V sous la forme:

$$x^3 + \lambda y^3 + \mu(z^3 + \lambda t^3) = 0$$

ou encore, en considérant l'extension galoisienne $k'=k(\sqrt[3]{\lambda})$, sous la forme affine:

$$N_{k'/k}(\zeta) = \mu\,.$$

Ainsi, V contient un ouvert qui est un torseur sous le k-tore $R^1_{k'/k}\mathbb{G}_m$, et ce tore, de dimension 2, est une variété affine rationnelle sur k , cf. [25]. On a donc:

$$V(k) \neq \emptyset \iff \mu \in N_{k'/k}k'^* \iff V \text{ est k-rationnelle.}$$

D'autre part, on sait bien que le tore $R^1_{k'/k}\mathbb{G}_m$ vérifie le principe de Hasse pour k'/k de degré 3. C'est là le résultat de Selmer [17] évoqué dans l'introduction.

Avant de prouver la proposition 1, il nous faut introduire un certain nombre de <u>notations</u> qui serviront très souvent dans la suite.

On peut écrire l'équation de V sous la forme:

$$(1') \qquad \begin{aligned} &x^3 + \lambda y^3 + \mu z^3 + \lambda\mu\nu t^3 = 0 \\ &\lambda = b/a\,, \quad \mu = c/a\,, \quad \nu = ad/bc\,. \end{aligned}$$

On pose alors:

$$(12) \qquad \begin{aligned} &\alpha = \sqrt[3]{\lambda} && \gamma = \sqrt[3]{\nu} \\ &\beta = \alpha\gamma = \sqrt[3]{d/c} && \delta = \alpha/\gamma \sim \sqrt[3]{ac/bd}\,, \end{aligned}$$

où le signe $\sim$ veut dire "à un élément de k^* près". Dans le <u>cas générique</u>, le corps $K=k(\alpha,\gamma)$ est une extension de degré 9, de groupe de Galois $\simeq(\mathbb{Z}/3)^2$. On a alors le diagramme de corps suivant, où les notations sont claires:

(13)

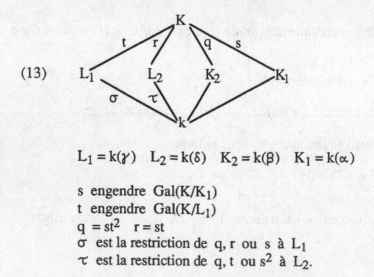

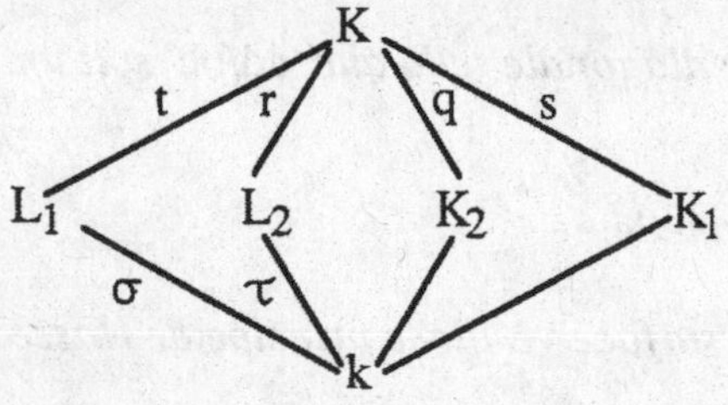

$L_1 = k(\gamma) \quad L_2 = k(\delta) \quad K_2 = k(\beta) \quad K_1 = k(\alpha)$

s engendre $Gal(K/K_1)$
t engendre $Gal(K/L_1)$
$q = st^2 \quad r = st$
σ est la restriction de q, r ou s à L_1
τ est la restriction de q, t ou s^2 à L_2.

La surface V étant donnée, il y a 3 choix possibles pour K, suivant la façon d'écrire l'équation de V sous la forme (1'). Pour le calcul du groupe de Brauer il nous faut encore considérer l'extension

$$K' = K(\gamma') \text{ où } \gamma' = \sqrt[3]{ab/cd}$$

non triviale dans le cas générique. On prolonge s,t,q,r à Gal(K'/k) en les faisant agir trivialement sur γ'. On introduit:

$$\alpha' = \sqrt[3]{\mu} = \sqrt[3]{c/a} \sim \gamma\gamma'$$

$$\beta' = \sqrt[3]{\mu\nu} = \sqrt[3]{d/b} \sim \gamma'/\gamma .$$

On choisit les générateurs s,t,w de telle sorte qu'ils agissent suivant le tableau ci-dessous:

(14)

$u \rightarrow$	α	β	γ	δ	α'	β'	γ'
${}^{s}u/u$	1	θ	θ	θ^2	θ	θ^2	1
${}^{t}u/u$	θ	θ	1	θ	1	1	1
${}^{w}u/u$	1	1	1	1	θ	θ	θ

On considère alors les droites suivantes sur $V_{K'}$:

$$L(i)\begin{cases} x+\theta^i\alpha y=0 \\ z+\theta^i\beta t=0 \end{cases} \qquad L'(i)\begin{cases} x+\theta^i\alpha y=0 \\ z+\theta^{i+1}\beta t=0 \end{cases} \qquad L''(i)\begin{cases} x+\theta^i\alpha y=0 \\ z+\theta^{i+2}\beta t=0 \end{cases}$$

$$M(i)\begin{cases} x+\theta^i\alpha' z=0 \\ y+\theta^{i+1}\beta' t=0 \end{cases} \qquad M'(i)\begin{cases} x+\theta^i\alpha' z=0 \\ y+\theta^{i+2}\beta' t=0 \end{cases} \qquad M''(i)\begin{cases} x+\theta^i\alpha' z=0 \\ y+\theta^{i}\beta' t=0 . \end{cases}$$

On pose:

$$L = L(0) + L(1) + L(2)$$

$$M = M(0) + M(1) + M(2)$$

et, si H est le diviseur d'une section hyperplane, on définit ℓ dans Pic $V_{K'}$ par:

(15) $\qquad H = 3\ell - L - M .$

On a:

(16) $$\mathrm{Pic}\,\overline{V} = \mathrm{Pic}\,V_{K'} = \mathbb{Z}\ell \oplus \bigoplus_{i=0}^{2}\mathbb{Z}.L(i) \oplus \bigoplus_{i=0}^{2}\mathbb{Z}.M(i)\,.$$

L'action du groupe de Galois de K'/k est donnée, toujours dans le cas générique, par:

(17)

$g \rightarrow$	w	t	s	r	q
$^{g}L(i)$	L(i)	L(i+1)	L'(i)	L'(i+1)	L'(i+2)
$^{g}M(i)$	M(i+1)	M(i)	M'(i+1)	M'(i+1)	M'(i+1)
$^{g}L'(i)$	L'(i)	L'(i+1)	L"(i)	L"(i+1)	L(i)
$^{g}M'(i)$	M'(i+1)	M'(i)	M"(i+1)	M"(i+1)	M"(i+1)
$^{g}L''(i)$	L"(i)	L"(i+1)	L(i)	L(i+1)	L(i+2)
$^{g}M''(i)$	M"(i+1)	M"(i)	M(i+1)	M(i+1)	M(i+1)

D'après le lemme 1, la variété V est K-rationnelle, d'où:

(18) $$\mathrm{Pic}\,V_K = (\mathrm{Pic}\,V_{K'})^{w} = \mathbb{Z}.\ell \oplus \mathbb{Z}.L(0) \oplus \mathbb{Z}.L(1) \oplus \mathbb{Z}.L(2) \oplus \mathbb{Z}.M\,,$$

puisque w agit trivialement sur L(i) et que $^{w}M(i)=M(i+1)$. D'après le tableau précédent, t permute les L(i) et agit trivialement sur L, M et donc sur ℓ d'après (15). Comme t-module, $\mathrm{Pic}\,V_K$ est donc de permutation:

$$\mathrm{Pic}\,V_K = \mathbb{Z}.\ell \oplus \mathbb{Z}[t].L(0) \oplus \mathbb{Z}.M \simeq \mathbb{Z}^2 \oplus \mathbb{Z}[t]$$

d'où:

$$H^1(k,\mathrm{Pic}\,\overline{V}) = H^1(\sigma, \mathrm{Pic}\,V_K)^t)$$

et:

$$(\mathrm{Pic}\,V_K)^t = \mathbb{Z}.\ell \oplus \mathbb{Z}.L \oplus \mathbb{Z}.M$$

comme groupe abélien. Pour obtenir les structures galoisiennes il est bon d'écrire un certain nombre de relations dans $\mathrm{Pic}\,\overline{V}$, relations obtenues par de simples calculs d'intersections:

(19)
$$\begin{aligned}
L'(i) &= 2\ell - L(i) - L(i+1) - M\\
L''(i) &= \ell - L(i) - L(i+2)\\
M'(i) &= \ell - M(i) - M(i+1)\\
M''(i) &= 2\ell - M(i) - M(i+2) - L\,.
\end{aligned}$$

On trouve ainsi que l'action de σ sur ℓ, L et M est donnée par:

$$
(20)\qquad \begin{aligned} {}^{\sigma}\ell &= 4\ell - L - 2M \\ {}^{\sigma}L &= 6\ell - 2L - 3M \\ {}^{\sigma}M &= 3\ell - 2M \end{aligned}
$$

et on note que:

$$
{}^{\sigma}(\ell - M) = \ell - L\,.
$$

Comme σ-module, on a donc la décomposition suivante:

$$
(21)\qquad (\mathrm{Pic}\ V_K)^t = \mathbb{Z}.H \oplus (\mathbb{Z}[\sigma]/N_\sigma).(\ell - L) \simeq \mathbb{Z} \oplus \mathbb{Z}[\sigma]/N_\sigma
$$

dont le second facteur est formé des $a\ell+bL+cM$ avec $a+b+c=0$.

Etant donné un groupe $\simeq \mathbb{Z}/3$ dont on a choisi un générateur σ, et un σ-module R, le groupe de cohomologie $H^1(\sigma,R)$ s'identifie au quotient ${}_{N_\sigma}R/\Delta_\sigma R$ du noyau de la norme

$$
N_\sigma = 1 + \sigma + \sigma^2
$$

par l'image de

$$
\Delta_\sigma = 1 - \sigma\,.
$$

D'où le lemme suivant qui résume cette discussion:

Lemme 2. *Dans le cas générique,*

$$
H^1(k,\mathrm{Pic}\ \overline{V}) = H^{-1}(\sigma,(\mathrm{Pic}\ V_K)^t) \simeq \mathbb{Z}/3
$$

avec pour générateur $\ell-L$ *ou* $\ell-M$.

Traitons ensuite le cas où K/k est de degré 9, mais $\sqrt[3]{\mu} \in K$. Autrement dit: K'=K. Il faut alors "oublier" w. En particulier: Pic V_K=Pic $\overline{V}$. Les actions de s,t,q,r sont encore données par (14) pour $\alpha,\beta,\gamma,\delta$ et par (17) pour L(i), L'(i) et L"(i). Si

$$
\alpha' = \sqrt[3]{\mu} \sim \alpha^m\gamma^n \qquad\qquad m\,,n \in \mathbb{Z}/3
$$

il faut modifier (14) et (17) comme suit:

$u \Rightarrow$	α'	β'	γ'
${}^{s}u/u$	θ^{n}	θ^{n+1}	θ^{n+2}
${}^{t}u/u$	θ^{m}	θ^{m}	θ^{m}

puis:

$g \Rightarrow$	t	s
${}^{g}M(i)$	M(i+m)	M'(i+n)
${}^{g}M'(i)$	M'(i+m)	M"(i+n)
${}^{g}M''(i)$	M"(i+m)	M(i+n)

Comme t-module:

$$\mathrm{Pic}\ V_K = \mathbb{Z}\ell \oplus \mathbb{Z}[t].L(0) \oplus \mathbb{Z}.M(0) \oplus \mathbb{Z}.M(1) \oplus \mathbb{Z}.M(2) \simeq \mathbb{Z}^4 \oplus \mathbb{Z}[t]$$

est encore de permutation, et comme groupe abélien:

$$(\mathrm{Pic}\ V_K)^t = \mathbb{Z}\ell \oplus \mathbb{Z}\cdot L \oplus \mathbb{Z}\cdot M \qquad \text{si } m \neq 0$$
$$(\mathrm{Pic}\ V_K)^t = \mathbb{Z}\ell \oplus \mathbb{Z}\cdot L \oplus \mathbb{Z}\cdot M(0) \oplus \mathbb{Z}\cdot M(1) \oplus \mathbb{Z}\cdot M(2) \qquad \text{si } m = 0\,.$$

Si $m \neq 0$, on a:

$${}^{s}L = L' = 6\ell - 2L - 3M \quad \text{et} \quad {}^{s}M = M' = 3\ell - 2M\,.$$

La situation est donc identique au cas générique, et la conclusion est la même. Si $m=0$, l'action de σ est donnée par les formules:

$${}^{\sigma}M(i) = M'(i+n) \quad {}^{\sigma}L = L' \quad {}^{\sigma}M = M' \quad {}^{\sigma}(\ell - M) = \ell - L\,.$$

Traitons d'abord le cas $n=0$, autrement dit, $\sqrt[3]{\mu} \in k$. La structure galoisienne de $(\mathrm{Pic}\ V_K)^t$ est alors donnée par:

(22) $$(\mathrm{Pic}\ V_K)^t = \mathbb{Z}[\sigma].M(0) \oplus (\mathbb{Z}[\sigma]/N_\sigma)\cdot(\ell - L) \simeq \mathbb{Z}[\sigma] \oplus \mathbb{Z}[\sigma]/N_\sigma\,.$$

Autrement dit, la structure galoisienne est différente du cas générique, mais rien ne change pour le H^1 et le lemme 2 vaut encore dans ce cas. Pour établir (22), on note que:

$$H = M(0) + M'(0) + M''(0) = 3\ell - L - M = \ell + (\ell - L) + (\ell - M) ,$$

ce qui prouve que ℓ appartient au sous-module engendré par $M(0)$ et $(\ell - L)$ dans $(\mathrm{Pic}\ V_K)^t$; on y trouve donc aussi L et M, et $M(1) = \ell - M(0) - M'(0)$, enfin aussi $M(2) = M - M(0) - M(1)$.

Le cas $n=2$ est analogue au précédent. La structure galoisienne de $(\mathrm{Pic}\ V_K)^t$ est encore donnée par (22), comme on peut voir en notant que:

$$H = M(0) + M'(2) + M''(1) = 3\ell - L - M = \ell + (\ell - L) + (\ell - M) ,$$

ce qui prouve que ℓ appartient au sous-module engendré dans $(\mathrm{Pic}\ V_K)^t$ par $M(0)$ et $(\ell - L)$; on y trouve donc aussi L et M, et $M(2) = \ell - M(0) - M'(2)$, enfin aussi $M(1) = M - M(0) - M(2)$.

Le cas $n=1$ est différent. Comme ${}^{\sigma}M(i) = M'(i+1) = \ell - M(i+1) - M(i+2)$, on a:

$$\Delta_{\sigma} M(i) = M - \ell .$$

En particulier, $M(1) - M(0)$ est invariant, et la structure galoisienne de $(\mathrm{Pic}\ V_K)^t$ est donnée par:

$$(23) \qquad (\mathrm{Pic}\ V_K)^t = \mathbb{Z}.H \oplus \mathbb{Z}(M(1) - M(0)) \oplus \mathbb{Z}[\sigma].M(0) \simeq \mathbb{Z}^2 \oplus \mathbb{Z}[\sigma] .$$

Il s'agit donc alors d'un module de permutation et $H^1(k, \mathrm{Pic}\ \overline{V}) = 0$ dans ce cas. Ce résultat n'est pas étonnant: ce cas correspond à l'hypothèse "ab/cd est un cube" et alors V est rationnelle sur k si $V(k) \neq \emptyset$, voir aussi [25], car dans ce cas-là V est un torseur sous un tore de dimension 2. Pour établir (23), on note d'abord que le sous-module de $(\mathrm{Pic}\ V_K)^t$ engendré par H, $M(1) - M(0)$ et $M(0)$, contient $M(1)$, donc $M'(1)$ et $M''(0)$; comme $H = M(0) + M'(0) + M''(0)$, il contient aussi $M'(0)$, et donc $M(2)$ et $\ell = M(1) + M(2) + M'(1)$ et $M = M(0) + M(1) + M(2)$, enfin $L = 3\ell - H - M$. En résumé:

Lemme 3. *Si l'extension* K/k *est de degré* 9:

$H^1(k, \mathrm{Pic}\ \overline{V}) = H^{-1}(\sigma, (\mathrm{Pic}\ V_K)^t) = 0$ *si* μ/ν *est un cube dans* k^*

$= \mathbb{Z}/3$ *sinon, et est alors engendré par la classe de* $\ell - L$ *ou encore par celle de* $\ell - M$.

Avec les notations de (1), la condition μ/ν est un cube se traduit par "ab/cd est un cube".

Si $\gamma = \sqrt[3]{ad/bc} \in k^*$, on trouve $H^1(k, \mathrm{Pic}\, \overline{V}) = 0$ comme lorsque ab/cd est un cube dans k^*. Le seul cas restant à traiter est donc celui où K/k est de degré 3 engendrée par γ. On peut même supposer $\sqrt[3]{\mu} \in K$, car sinon, par échange de λ et μ on retrouve une situation étudiée au lemme 3. On peut enfin supposer $\delta \sim \sqrt[3]{ac/bd} \notin k^*$, car sinon on trouve encore 0. Il s'agit donc d'étudier une équation du type:

$$x^3 + \nu^m y^3 + \nu^n z^3 + \nu^{m+n+1} t^3 = 0 \qquad m, n \in \mathbb{Z}/3 \qquad m \neq 1, n \neq 1$$

sachant que ν n'est pas un cube dans k^*. De fait, toutes ces équations se ramènent à l'équation

$$x^3 + y^3 + z^3 + \nu t^3 = 0$$

et on sait bien que dans ce cas (cf. [14] ou $[8]_{II}$)

$$H^1(k, \mathrm{Pic}\, \overline{V}) \simeq (\mathbb{Z}/3)^2 .$$

D'alleurs, ${}^{\sigma}L(i) = L'(i)$ et ${}^{\sigma}M(i) = M'(i)$, les formules (20) valent encore et $N_\sigma L(i) = N_\sigma M(i) = H$, ce qui donne aisément la structure galoisienne de $\mathrm{Pic}\, V_K$:

$$\begin{aligned} \mathrm{Pic}\, V_K &= \mathbb{Z}[\sigma]\cdot L(0) \oplus (\mathbb{Z}[\sigma]/N_\sigma)\cdot(\ell - L) \oplus (\mathbb{Z}[\sigma]/N_\sigma)\cdot(M(0) - L(0)) \\ &\simeq \mathbb{Z}[\sigma] \oplus ((\mathbb{Z}[\sigma]/N_\sigma)^2. \end{aligned}$$

Ceci achève la démonstration de la proposition 1.

Remarque 1. Pour notre sujet, il est raisonnable de faire l'hypothèse suivante sur l'équation (1):

(24) aucun des a/b ou ab/cd n'est un cube dans k^*.

En effet, si a/b est un cube, V(k) est non vide de manière évidente, et si ab/cd est un cube, V vérifie le principe de Hasse. Or, sous cette hypothèse (24), on vérifie aisément que l'extension K/k est toujours de degré 9 quelle que soit la manière d'écrire l'équation de V sous la forme (1'). En effet, si (24) est vérifiée et si K/k n'est pas de degré 9, nécessairement $K = k(\gamma)$ est de degré 3 et $\alpha \sim \gamma^m$; si m=0, alors a/b est un cube; si m=1, c'est ac/bd qui est un cube, et pour m=2, c'est c/d.

Remarque 2. Soit V d'équation (1) vérifiant l'hypothèse (24) et possédant un point dans chaque complété de k. D'après le lemme 1, elle devient rationnelle sur le corps $L_1 = k(\gamma)$, extension cyclique de k. Puisqu'on connaît un générateur du groupe $H^1(k, \mathrm{Pic}\, \overline{V}) = H^{-1}(\sigma, \mathrm{Pic}\, V_{L_1}) \simeq \mathbb{Z}/3$, à savoir la classe de $\ell - M \in \mathrm{Pic}\, V_{L_1}$, on doit pouvoir trouver $\mathcal{A}$ dans Br(V), et même dans $\mathrm{Br}(V, L_1)$, qui relève la classe de

ℓ–M dans $H^1(K,\mathrm{Pic}\,\overline{V})$. On a:

$$\begin{aligned}\mathrm{Br}(V,L_1) &= \ker(H^2(\sigma,L_1(V)^*) \longrightarrow H^2(\sigma,\mathrm{Div}\,V_{L_1}))\\ &\simeq \{h \in k(V)^* \mid \mathrm{div}(h) = N_\sigma(D')\}/N_\sigma(L_1(V)^*)\,.\end{aligned}$$

La procédure à suivre pour obtenir $\mathcal{A}=(h,L_1(V)/k(V),\sigma)$ est la suivante (cf. (9')):

(i) trouver $D' \in \mathrm{Div}\,V_{L_1}$ tel que: $\mathrm{cl}(D') = \ell - M$,

(ii) trouver $h \in k(V)^*$ telle que: $\mathrm{div}(h) = N_\sigma(D')$.

Ayant ainsi trouvé h, l'obstruction de Manin se calcule très simplement en calculant les $i_v(h(x_v))$ où $i_v\colon k_v^*/N_\sigma L_{1w}^* \hookrightarrow \mathbb{Z}/3$ est l'invariant local. Les calculs sont alors rendus possibles grâce aux traductions fournies par la cohomologie des groupes cycliques.

En fait, cette procédure achoppe ici sur le point suivant: trouver un diviseur D' explicite de façon à pouvoir ensuite calculer h. La difficulté tient au fait que ℓ est la classe de cubiques gauches L_1-rationnelles tracées sur V (si l'on contracte sur L_1 le S_6 constitué des L(i) et M(i) on obtient une variété de Severi-Brauer qui a des points dans chaque complété - c'est donc $\mathbb{P}^2_{L_1}$- et le système linéaire de cubiques sur V_{L_1} correspond à celui des droites dans le plan projectif). Il faudrait donc trouver explicitement une telle cubique pour appliquer directement la procédure esquissée ci-dessus.

En revanche, on sait répondre aisément à la question suivante:

(i') trouver $D \in \mathrm{Div}\,V_K$ tel que: $\mathrm{cl}(D) = \ell - M$.

On peut en effet prendre, d'après les relations (19):

(25) $D = L'(2) - L''(0)$.

On notera que, d'après le tableau (17):

(25') $D = \Delta_r L'(2)$.

Pour continuer, on peut chercher D' vérifiant (i) à partir de D. On peut aussi faire les calculs au niveau de l'extension bicyclique K/k suivant la procédure esquissée au niveau cyclique L_1/k. C'est la méthode que nous suivons. Elle suppose connue une interprétation de la cohomologie d'un groupe bicyclique analogue à celle d'un groupe cyclique.

§2. Cohomologie du groupe $(\mathbb{Z}/3)^2$

Nous rappelons simplement les résultats sur la cohomologie des groupes abéliens finis(Takahashi [23]) qui nous seront utiles pour la suite. Si G est un groupe abélien fini et si $g \in G$ est d'ordre n, on note:

(26) $$\Delta_g = 1 - g \quad \text{et} \quad N_g = 1 + g + \ldots + g^{n-1} \in \mathbb{Z}[G].$$

Si G est <u>cyclique</u>, le choix d'un générateur g détermine la résolution suivante du G-module trivial $\mathbb{Z}$:

(27) $$\begin{array}{ccccccccccc} \ldots & \mathbb{Z}[G] & \xrightarrow{\Delta} & \mathbb{Z}[G] & \xrightarrow{N} & \mathbb{Z}[G] & \xrightarrow{\Delta} & \mathbb{Z}[G] & \xrightarrow{\varepsilon} & \mathbb{Z}, \\ d^\circ: & 3 & & 2 & & 1 & & 0 & & \end{array}$$

où $\Delta = \Delta_g$ et $N = N_g$. Par définition, le complexe L. résolvant vérifie $L_i = \mathbb{Z}[G]$ et $d_{2i} = \Delta$, $d_{2i+1} = N$. Ceci détermine, pour tout G-module M, une présentation des groupes de cohomologie; ce sont les groupes d'homologie du complexe:

$$\begin{array}{cccccccccccc} & M & \xrightarrow{\Delta} & M & \xrightarrow{N} & M & \xrightarrow{\Delta} & M & \xrightarrow{N} & M & \xrightarrow{\Delta} & \ldots \\ d^\circ: & 0 & & 1 & & 2 & & 3 & & 4 & & \end{array}$$

On a donc la présentation suivante:

(28) $$\begin{aligned} H^{2i}(G,M) &= M^g/N_g M \\ H^{2i+1}(G,M) &= {}_N M/\Delta M, \end{aligned}$$

où ${}_N M$ désigne le noyau de la norme.

Si G est <u>bicyclique</u>, le choix de deux générateurs s, d'ordre m, et t, d'ordre n, détermine la résolution suivante du G-module trivial $\mathbb{Z}$:

(29) $$\ldots \mathbb{Z}[G]^5 \xrightarrow{d_3} \mathbb{Z}[G]^4 \xrightarrow{d_2} \mathbb{Z}[G]^3 \xrightarrow{d_1} \mathbb{Z}[G]^2 \xrightarrow{d_0} \mathbb{Z}[G] \xrightarrow{\varepsilon} \mathbb{Z},$$

où, par définition, le complexe résolvant L. vérifie: $L_i = \mathbb{Z}[G]^{i+1}$ et, avec des notations évidentes:

$$\begin{aligned} d_{2i}e_{2j} &= N_s e_{2j-1} + \Delta_t e_{2j} \\ d_{2i}e_{2j+1} &= \Delta_s e_{2j} - N_t e_{2j+1} \\ d_{2i+1}e_{2j} &= N_s e_{2j-1} + N_t e_{2j} \\ d_{2i+1}e_{2j+1} &= \Delta_s e_{2j} - \Delta_t e_{2j+1} \end{aligned}$$

où l'on convient que $e_{-1}=0$ et $(e_0,\dots,e_i)$ est la base canonique de $L_i=\mathbb{Z}[G]^{i+1}$. Si M est un G-module, les groupes $H^i(G,M)$ sont les groupes d'homologie du complexe:

$$\begin{array}{ccccccccccc} M & \xrightarrow{d_0^*} & M^2 & \xrightarrow{d_1^*} & M^3 & \xrightarrow{d_2^*} & M^4 & \xrightarrow{d_3^*} & M^5 & \longrightarrow & \dots \\ d°: \; 0 & & 1 & & 2 & & 3 & & 4 & & \end{array}$$

où les applications d_i^* sont obtenues par transposition à partir des d_i. Explicitons les premières applications sous forme matricielle:

$$d_0^* = \begin{bmatrix}\Delta_t\\ \Delta_s\end{bmatrix} \quad d_1^* = \begin{bmatrix}N_t & \\ \Delta_s & -\Delta_t \\ & N_s\end{bmatrix} \quad d_2^* = \begin{bmatrix}\Delta_t & & \\ \Delta_s & -N_t & \\ & N_s & \Delta_t \\ & & \Delta_s\end{bmatrix} \quad d_3^* = \begin{bmatrix}N_t & & & \\ \Delta_s & -\Delta_t & & \\ & N_s & N_t & \\ & & \Delta_s & -\Delta_t \\ & & & N_s\end{bmatrix}.$$

Pour la suite, on est surtout intéressé par d_1^*:

$$\begin{array}{ccc} C^1(G,M) & \longrightarrow & C^2(G,M) \\ (a,b) & \longrightarrow & (N_ta,\Delta_sa-\Delta_tb,N_sb) \end{array}$$

et d_2^*:

$$\begin{array}{ccc} C^2(G,M) & \longrightarrow & C^3(G,M) \\ (a,b,c) & \longrightarrow & (\Delta_ta,\Delta_sa-N_tb,N_sb-\Delta_tc,\Delta_sc)\ . \end{array}$$

Les i-cocycles sont donc donnés pour $i=1,2$ par:

$$\begin{aligned} &Z^1(G,M) = \{(a,b)\in M^2 \mid N_ta = N_sb = 1 \text{ et } \Delta_sa = \Delta_tb\}\ , \\ (30)\qquad & \\ &Z^2(G,M) = \{(a,b,c)\in M^3 \mid a\in M^t,\ c\in M^s,\ N_tb = \Delta_sa \text{ et } N_sb = -\Delta_tc\}\ . \end{aligned}$$

On aura également besoin dans la suite de la description explicite de morphismes d'inflation et de restriction. Ceux relatifs aux sous-groupes <s> et <t> sont très faciles à décrire. Le morphisme $(27)_t \longrightarrow (29)_G$ défini par l'injection de $\mathbb{Z}[t]$ dans le premier facteur $\mathbb{Z}[G]$ de $\mathbb{Z}[G]^{i+1}$ induit la restriction

$$(31)\qquad \begin{array}{ccc} H^i(G,M) & \longrightarrow & H^i(t,M) \\ (a_0,\dots,a_i) & \longrightarrow & a_0\ . \end{array}$$

De même, le morphisme $(27)_s \longrightarrow (29)_G$ défini par l'injection $\mathbb{Z}[s] \longrightarrow \mathbb{Z}[G]^{i+1}$ dans le dernier facteur $\mathbb{Z}[G]$, induit la restriction

$$(32) \qquad \begin{array}{ccc} H^i(G,M) & \longrightarrow & H^i(s,M) \\ (a_0,\ldots,a_i) & \longrightarrow & a_i\,. \end{array}$$

Le morphisme $(29)_G \longrightarrow (27)_{G/t}$ défini par la projection $\mathbb{Z}[G]^{i+1} \longrightarrow \mathbb{Z}[G/t]$ du dernier facteur $\mathbb{Z}[G]$ sur $\mathbb{Z}[G/t]$, définit l'inflation

$$(33) \qquad \begin{array}{ccc} H^i(\sigma,M^t) & \longrightarrow & H^i(G,M) \\ a & \longrightarrow & (0,\ldots,0,a) \end{array}$$

où σ désigne la classe de s mod t. Pour la suite, on a également besoin de décrire les morphismes de restriction aux sous-groupes $\langle st\rangle$ et $\langle st^2\rangle$. On construit un morphisme de complexes $(27)_{st} \longrightarrow (29)_G$ en commençant comme suit:

$$\begin{array}{ccccccc} \mathbb{Z}[st] & \xrightarrow{N_{st}} & \mathbb{Z}[st] & \xrightarrow{\Delta_{st}} & \mathbb{Z}[st] & \xrightarrow{\varepsilon} & \mathbb{Z} \\ \omega_2\downarrow & & \omega_1\downarrow & & \omega_0\downarrow & & \| \\ \mathbb{Z}[G]^3 & \xrightarrow[d_1]{} & \mathbb{Z}[G]^2 & \xrightarrow[d_0]{} & \mathbb{Z}[G] & \xrightarrow{\varepsilon} & \mathbb{Z} \end{array}$$

$$\begin{aligned} \omega_0(1) &= 1 \\ \omega_1(1) &= (1,t) \\ \omega_2(1) &= (st,1-st^2,1)\,. \end{aligned}$$

En particulier, la restriction

$$H^2(G,M) \longrightarrow H^2(st,M) = M^{st}/N_{st}M$$

est donnée par:

$$(34) \qquad (a,b,c) \longrightarrow {}^{st}a + {}^{1-st^2}b + c\,.$$

Dans ce dernier cas, on a supposé implicitement $m=n=3$. On posera alors:

$$(35) \qquad \begin{aligned} q &= st^2 \\ r &= st\,. \end{aligned}$$

On construit enfin un morphisme de complexes $(27)_q \longrightarrow (29)_G$ en commençant comme suit:

$$\begin{array}{ccccccc}
\mathbb{Z}[st^2] & \xrightarrow{N_q} & \mathbb{Z}[st^2] & \xrightarrow{\Delta_q} & \mathbb{Z}[st^2] & \xrightarrow{\varepsilon} & \mathbb{Z} \\
\downarrow \omega'_2 & & \downarrow \omega'_1 & & \downarrow \omega'_0 & & \| \\
\mathbb{Z}[G]^3 & \xrightarrow[d_1]{} & \mathbb{Z}[G]^2 & \xrightarrow[d_0]{} & \mathbb{Z}[G] & \xrightarrow{\varepsilon} & \mathbb{Z}
\end{array}$$

$$\omega'_0(1) = 1$$
$$\omega'_1(1) = (-q,1)$$
$$\omega'_2(1) = (-s,-1+r,t) .$$

La restriction

$$H^2(G,M) \longrightarrow H^2(q,M) = M^q/N_qM$$

est donc donnée par

(36) $\quad (a,b,c) \longrightarrow {}^{-s}a + {}^{r-1}b + {}^{t}c .$

§3. L'algèbre d'Azumaya $\mathcal{A}$

On considère désormais une surface cubique V d'équation homogène

(1) $\quad ax^3 + by^3 + cz^3 + dt^3 = 0 \qquad a, b, c, d \in \mathbb{Q}^* .$

On met cette équation sous la forme:

(1)' $\quad x^3 + \lambda y^3 + \mu z^3 + \lambda\mu\nu t^3 = 0$

$$\lambda = b/a , \quad \mu = c/a , \quad \nu = ad/bc .$$

On considère le corps $k=\mathbb{Q}(\theta)$ et on définit α, β, γ et δ comme en (12) au §1. On fait l'hypothèse

(24) aucun des a/b ou ab/cd n'est un cube dans $\mathbb{Q}^*$

et on suppose $\prod V(\mathbb{Q}_p)\neq\emptyset$. Dès lors, le corps $K=k(\alpha,\gamma)$ est une extension de degré 9 de k et on introduit les notations du diagramme (13). De plus, V est rationnelle sur L_1, et comme L_1/k est cyclique, on a une suite exacte naturelle

$$0 \longrightarrow Br(k,L_1) \longrightarrow Br(V_k,L_1) \longrightarrow H^1(\sigma,\text{Pic } V_{L_1}) \longrightarrow 0$$

où

$$H^1(\sigma,\mathrm{Pic}\,V_{L_1}) = {}_{N_\sigma}\mathrm{Pic}\,V_{L_1}/\Delta_\sigma \mathrm{Pic}\,V_{L_1} \simeq \mathbb{Z}/3$$

est engendré par la classe de $\ell-M$ (cf. §2, remarques 1 et 2). De fait, comme déjà indiqué, on doit opérer au niveau de l'extension K/k plutôt que L_1/k. On a le diagramme naturel suivant:

$$\begin{array}{ccccccc}
H^1(\sigma,\mathrm{Pic}\,V_{L_1}) & & & & & & \\
\kappa \big\uparrow & \searrow^{\mathrm{Inf}}_{\simeq} & & & & & \\
 & & H^1(G,\mathrm{Pic}\,V_K) & \xrightarrow{\partial} & H^2(G,K(V)^*/K^*) & \xrightarrow{j} & H^2(G,\mathrm{Div}\,V_K) \\
\mathrm{Br}(V_k,L_1) & & \kappa_K \big\uparrow & & \pi \big\uparrow & & \big\| \\
 & \searrow & & & & & \\
(37) & & \mathrm{Br}(V_k,K) & \xrightarrow{\iota} & H^2(G,K(V)^*) & \longrightarrow & H^2(G,\mathrm{Div}\,V_K)
\end{array}$$

dont les lignes sont exactes et où κ, et par suite κ_K aussi, est surjective. Le groupe

$$H^1(\sigma,\mathrm{Pic}\,V_{L_1}) = {}_{N_\sigma}\mathrm{Pic}\,V_{L_1}/\Delta_\sigma \mathrm{Pic}\,V_{L_1} \simeq \mathbb{Z}/3$$

est engendré, d'après le §1, par la classe de $\ell-M$, ou de $\ell-L={}^\sigma(\ell-M)$, ou encore par celle de $L+M-2\ell={}^\sigma(\ell-L)$. Soit ξ le générateur défini par l'une de ces trois classes. D'après (33):

$$\mathrm{Inf}(\xi) = (0,\xi)\ .$$

Prenons pour ξ la classe de $L+M-2\ell \in \mathrm{Pic}\,V_K$. D'après (19), si

(38) $$D := L(1) - L'(2) \in \mathrm{Div}\,V_K\,,$$

on a:

$$\mathrm{cl}(D) = \mathrm{cl}(L + M - 2\ell) = \xi \in \mathrm{Pic}\,V_K\,.$$

On a: $N_s(D)=L(1)+L'(1)+L''(1)-L(2)-L'(2)-L''(2)$, soit:

(39) $$N_s(D) = \mathrm{div}(f)$$

où:

(40) $$f = \frac{x+\theta\alpha y}{x+\theta^2\alpha y} \in K_1(V)^*\,.$$

Le morphisme ∂ est le bord déduit de la suite exacte de G-modules:

$$1 \longrightarrow K(V)^*/K^* \longrightarrow \mathrm{Div}\, V_K \longrightarrow \mathrm{Pic}\, V_K \longrightarrow 0 ,$$

$$\partial(0,\xi) = d_1^*(0,D) = (0,-\Delta_t D, N_s D) .$$

Or, $-\Delta_t D = L(2)+L'(2)-L(1)-L'(0) = L(2)+L'(2)+L''(2)-L(1)-L'(0)-L''(2)$, soit:

(41) $\quad -\Delta_t D = \mathrm{div}(g)$

où:

(42) $\quad g = \dfrac{x+\theta^2\alpha y}{z+\theta\beta t} \in K(V)^* .$

Ainsi:

$$\partial(\mathrm{Inf}(\xi)) = (0,\mathrm{div}(g),\mathrm{div}(f)) \in Z^2(G,K(V)^*/K^*) .$$

Ce 2-cocycle admet comme relèvement la 2-cochaîne

$$(1,g,f) \in C^2(G,K(V)^*) .$$

Un relèvement quelconque s'écrit $(1,g/\varepsilon,f/\eta)$ avec ε et $\eta \in K^*$. Son bord vaut:

$$d_2^*(1,g/\varepsilon,f/\eta) = (1,N_t(\varepsilon)/N_t(g),N_s(g)\cdot\Delta_t(f)/N_s(\varepsilon)\cdot\Delta_t(\eta),\Delta_s(f)/\Delta_s(\eta)) .$$

Or:

(43') $\quad N_s(g)\cdot\Delta_t(f) = -\mu$

(43) $\quad N_t(g) = -\mu ,$

comme on le déduit aisément de l'équation (1)' et des expressions (40) et (42) de f et g. On en déduit:

$$d_2^*(1,g/\varepsilon,f/\eta) = (1,-N_t(\varepsilon)/\mu,-\mu/(N_s(\varepsilon)\cdot\Delta_t(\eta)),1/\Delta_s(\eta)) .$$

Finalement,

$$(1,g/\varepsilon,f/\eta) \in Z^2(G,K(V)^*)$$

équivaut aux conditions:

(44) $\quad N_t(\varepsilon) = -\mu$

(45) $\quad \eta \in K_1^*$ et $\Delta_t(\eta) = -\mu/N_s(\varepsilon) .$

On est ainsi ramené au problème suivant:

Problème: Trouver $\varepsilon \in K^*$ et $\eta \in K_1^*$ vérifiant les conditions (44) et (45).

Il est clair que ce problème a toujours des solutions (ε,η). En effet, V est rationnelle sur L_1 et son équation (1)' s'écrit sur L_1, en posant $t'=\gamma t$:

$$x^3 + \lambda y^3 + \mu(z^3 + \lambda t'^3) = 0\,,$$

ce qui montre que $-\mu$ est une norme de l'extension $K=L_1(\alpha)/L_1$. Ceci assure l'existence de $\varepsilon \in K^*$ vérifiant (44). Dès lors $-\mu/N_s(\varepsilon) \in K_1{}^*$ et $N_t(-\mu/N_s(\varepsilon))=1$ ce qui assure l'existence de η vérifiant (45) par application du théorème 90 de Hilbert à l'extension cyclique K_1/k. On notera que deux relèvements de $(1,g,f)$ diffèrent par un élément du type $(1,\varepsilon,\eta)$ avec $\varepsilon \in K^*, \eta \in K_1^*, N_t\varepsilon=1$ et enfin $\Delta_t\eta=1/N_s\varepsilon$, autrement dit par un élément de $Br(k,K)$.

Définition. Soient $\varepsilon \in K^*$ et $\eta \in K_1^*$ vérifiant les conditions (44) et (45). On note $\mathcal{A} \in Br(V_k,K)$ la classe d'algèbres d'Azumaya sur V telle que:

(46) $\qquad \iota(\mathcal{A}) = cl(1,g/\varepsilon,f/\eta) \in H^2(G,K(V)^*.$

Cette algèbre $\mathcal{A}$ est définie à un élément de $Br(k,K)$ près. D'après (31), sa restriction à L_1 est triviale. On a donc:

$$\mathcal{A} \in Br(V_k,L_1)$$

et $\kappa(\mathcal{A})$ engendre $H^1(\sigma,\mathrm{Pic}\,V_{L_1})$.

Remarque 3. On peut se demander s'il est possible de trouver $\mathcal{A}$, autrement dit ε et η, qui ait bonne réduction aux places où V a elle-même bonne réduction, i.e. aux places v qui ne divisent pas 3abcd. En pratique, ce sera toujours le cas pour tous les exemples numériques traités dans le cadre de la procédure algorithmique décrite au §8, mais on ne voit pas de raison a priori pour qu'il en soit ainsi!

En effet, pour ce qui est tout d'abord de ε, l'équation

(44) $\qquad N_{K/L_1}(\varepsilon) = -\mu$

a une solution ε dans K^*, mais elle n'en a pas a priori qui soit une unité aux places où K/L_1 est étale et où μ est une unité: en termes de cohomologie étale, si U'/U est un revêtement étale qui prolonge l'extension générique K/L_1, l'application naturelle

$$H^1(U,R^1_{U'/U}\mathbb{G}_m) \longrightarrow H^1(L_1,R^1_{K/L_1}\mathbb{G}_m),$$

où $R^1_{U'/U}\mathbb{G}_m$ désigne le tore noyau de la norme $R_{U'/U}\mathbb{G}_{mU} \longrightarrow \mathbb{G}_{mU}$, n'est pas nécessairement injective.

Pour ce qui est de η, la question de savoir si l'équation

(45) $\quad \Delta_t(\eta) = -\mu/N_s(\varepsilon)$

a une solution η dans K_1^* qui soit inversible aux places où μ et ε le sont et où K_1/k est étale revient à la validité du théorème 90 pour un certain revêtement étale U_1/U_0 prolongeant l'extension K_1/k, et cela non plus n'est pas automatique a priori!

§4. Le calcul de l'obstruction

Dans ce paragraphe, nous indiquons simplement les détails du calcul de l'obstruction sans tenir compte d'un certain nombre de phénomènes particuliers qui seront établis dans les paragraphes suivants et faciliteront les calculs dans tous les cas.

Avant de détailler le calcul de l'obstruction, pas à pas, il est bon de faire quelques remarques à propos de l'obstruction de Brauer-Manin dont la définition a été donnée au début du §0 pour V projective et lisse sur le corps de nombres k. Soit

(46) $\quad \mathcal{V} = \prod_v V(k_v)/\mathrm{Br}$

où $V(k_v)/\mathrm{Br}$ désigne l'ensemble des classes pour l'équivalence de Brauer, celle-ci étant définie par l'accouplement naturel $\mathrm{Br}\, V_v \times V(k_v) \longrightarrow \mathrm{Br}\, k_v$. Toute algèbre $A \in \mathrm{Br}(V)$ définit une application

$$i_A : \mathcal{V} \longrightarrow \mathbb{Q}/\mathbb{Z}$$

par:

(47) $\quad i_A(x_v) = \sum_v \mathrm{inv}_v A(x_v)$.

Celle-ci ne dépend que de la classe de A mod $\mathrm{Br}(k)$. Si B est une partie de $\mathrm{Br}(V)$, on note i_B l'application $(i_A)_{A\in B}$ et i l'application $i_{Br(V)}$. On a évidemment:

$$V(k) \subset \ker(i)$$

et l'obstruction de Manin n'est autre que la condition

(5') $\qquad \ker(i) = \emptyset$.

Si au contraire $\ker(i) \neq \emptyset$, on dit que l'obstruction de Manin est "vide".

Lemme 4. *On conserve les hypothèses et notations ci-dessus.*

(i) *Si* B *engendre* Br(V)/Br(k), *alors* $\ker i_B = \ker i$.
(ii) *Soit* k'/k *une extension finie, et soit* $V' = V_{k'}$. *Si l'obstruction de Manin est vide pour* V, *elle est vide pour* V'.
(iii) *On suppose en outre que* V *est une surface cubique et* k'/k *quadratique. Alors l'obstruction de Manin est vide pour* V *si et seulement si elle est vide pour* V'.

L'assertion (i) est immédiate d'après la formule (47). Ceci justifie le fait que dans le cas qui nous intéresse on considère la condition (5) pour une seule algèbre d'Azumaya $\mathcal{A}$.

Pour (ii), soit $j: \mathcal{V} \longrightarrow \mathcal{V}'$ l'application canonique. Soient $A' \in Br(V')$ et $A := cor_{V'/V}(A') \in Br(V)$. Si $(x_v) \in \mathcal{V}$, on a la formule

(48) $\qquad i_A(x_v) = i_{A'}(j(x_v))$

qui prouve aussitôt l'assertion (ii): si $i_A(x_v)=0$ pour un certain $(x_v)\in\mathcal{V}$ et pour tout $A \in Br(V)$, alors $i_{A'}(j(x_v))=0$ pour tout $A' \in Br(V')$. La formule (48) résulte de formules locales: soit $k'_v = k_v \otimes_k k'$; on a une "formule de projection"

$$(\mathrm{cor}\ A')(x_v) = \mathrm{cor}(A'(x_v)) \qquad\qquad \text{dans } Br(k_v)$$

où $A'(x_v) \in Br(k'_v)$ et, suivant que v se prolonge à k' en une place w ou deux places w et w', la corestriction d'une algèbre $\mathfrak{a} \in Br(k'_v)$ a pour invariant

$$\mathrm{inv}_v(\mathrm{cor}\ \mathfrak{a}) = \mathrm{inv}_w(\mathfrak{a}) \quad \text{ou} \quad \mathrm{inv}_w(\mathfrak{a}) + \mathrm{inv}_{w'}(\mathfrak{a}) \qquad\qquad \text{dans } \mathbb{Q}/\mathbb{Z} .$$

La démonstration de (iii) utilise le sous-lemme suivant que nous admettons pour l'instant:

Sous-lemme. *Soit* $G=G_1(\mathbb{P}^3)$ *la grassmannienne des droites de* $\mathbb{P}^3$. *Soit* $A\in Br(V)$. *Il existe une algèbre constante* $\mathfrak{a}\in Br(k)$ *telle que, pour tout corps* K/k *et toute droite* $D\in G(K)$ *non tracée sur* V_K, *on ait:*

(49) $$A(D\cdot V) = j_K(\mathfrak{a}) \qquad \textit{dans } Br(K),$$

où $j_K: Br(k) \longrightarrow Br(K)$ *est l'application canonique et* $D\cdot V$ *désigne le 0-cycle intersection.*

Ajoutons que si $\sum_i n_i x_i$ est un 0-cycle de V_K, on a, par définition:

(50) $$A(\sum_i n_i x_i) = \sum_i n_i \mathrm{cor}_{K(x_i)/K}(A(x_i)) \in Br(K) .$$

Démontrons (iii). Par hypothèse, il existe $(x'_w)\in\mathcal{V}'$ tel que, pour toute algèbre $A'\in Br(V')$, on ait:

(51) $$i_{A'}(x'_w) = 0 .$$

Les classes pour l'équivalence de Brauer dans $V(k'_w)$ étant ouvertes, on peut supposer qu'aucun des x'_w n'appartient à une droite tracée sur V.

Soient $A\in Br(V)$ quelconque et $\mathfrak{a}\in Br(k)$ l'algèbre constante associée à A par le sous-lemme ci-dessus.

On va associer à (x'_w) un point $(x_v)\in\mathcal{V}$ (indépendamment de l'algèbre A précédente), de telle sorte que, pour toute algèbre $A\in Br(V)$:

(52) $$i_A(x_v) = 0 .$$

Pour définir x_v on distingue plusieurs cas.

Supposons d'abord que v admet un seul prolongement w à k'. Notons alors $\bar{x}'_w$ le conjugué de x'_w (k'_w/k_v est alors quadratique). Si $x'_w\neq\bar{x}'_w$, la droite D_v qui joint ces deux points est définie sur k_v et $D_v\cdot V=x'_w+\bar{x}'_w+x_v$ pour un certain $x_v\in V(k_v)$; on en déduit par le sous-lemme:

(53) $$\mathrm{cor}(A(x'_w)) + A(x_v) = j_v(\mathfrak{a}) \qquad \text{dans } Br(k_v) .$$

Si $x'_w\in V(k_v)$, on choisit une tangente D_v à V en x'_w, qui soit définie sur k_v; on a alors $D_v\cdot V=2x'_w+x_v$ pour un certain $x_v\in V(k_v)$ et, par application du sous-lemme:

$$2A(x'_w) + A(x_v) = j_v(\mathfrak{a}) \qquad \text{dans } Br(k_v)$$

ce qui s'écrit encore sous la forme précédente (53) si au lieu de considérer $A(x'_w)$ dans $Br(k_v)$ on la considère dans $Br(k'_w)$, compte tenu de l'égalité $cor\circ res=2\times$.

Supposons ensuite que v admet deux prolongements w et w' à k'. Soit D_v la droite joignant x'_w à $x'_{w'}\in V(k_v)$ si ces deux points sont distincts, et sinon une tangente à V en x'_w qui soit k_v-rationnelle. Alors: $D_v\cdot V=x'_w+x'_{w'}+x_v$ pour un certain $x_v\in V(k_v)$, et par application du sous-lemme:

$$(54) \qquad A(x'_w) + A(x'_{w'}) + A(x_v) = j_v(\mathfrak{a}) \qquad \text{dans } Br(k_v).$$

On déduit alors successivement de (53) et (54), suivant que v a un ou deux prolongements à k':

$$\mathrm{inv}_w A(x'_w) + \mathrm{inv}_v A(x_v) = \mathrm{inv}_v \mathfrak{a}$$

$$\mathrm{inv}_w A(x'_w) + \mathrm{inv}_{w'} A(x'_{w'}) + \mathrm{inv}_v A(x_v) = \mathrm{inv}_v \mathfrak{a},$$

ce qui, par sommation sur v, et compte tenu de $\sum_v \mathrm{inv}_v\mathfrak{a}=0$, donne finalement:

$$i_A(x'_w) + i_A(x_v) = 0 \qquad \text{dans } \mathbb{Q}/\mathbb{Z}.$$

L'hypothèse (51) donne $i_A(x'_w)=0$, d'où $i_A(x_v)=0$, ce qui achève la preuve de l'assertion (iii), modulo celle du sous-lemme.

Voici enfin la *démonstration du sous-lemme*, qui vaut naturellement dans un cadre plus large. Soit G_0 l'ouvert de $G=Gr_1(\mathbb{P}^3)$ formé des droites non tracées sur V: le fermé complémentaire est de dimension 0, et par suite

$$(55) \qquad Br(G_0) = Br(G) = Br(k).$$

Soit $I\subset G\times V$ la variété d'incidence:

$$\begin{array}{ccc} & I\subset G\times V & \\ {}^{\pi}\swarrow & & \searrow{}^{q} \\ G & & V. \end{array}$$

Considérons alors le diagramme induit au-dessus de l'ouvert G_0:

$$\begin{array}{ccc} & I_0 \subset G_0 \times V & \\ {}^{\pi_0}\swarrow & & \searrow^{q_0} \\ G_0 & & V\,. \end{array}$$

La variété d'incidence I étant définie par une seule équation dans la variété régulière $G \times V$ est de Cohen-Macaulay. Le morphisme π est propre et surjectif, de même π_0 par changement de base. Le choix de G_0 fait que π_0 est quasi-fini, donc fini. Comme G_0 est un schéma régulier et que I_0 est de Cohen-Macaulay, un critère connu (cf. [9] EGA IV 15.4.2 et 6.1.5) assure que π_0 est plat, il est donc fini localement libre: il est de rang constant 3.

Si $D \in G(k)$, on établit alors aisément la formule:

$$A(D \cdot V) = (\pi_{0*} q_0^* A)(D) \qquad \text{dans } Br(k),$$

où $D \cdot V$ désigne le 0-cycle intersection de V avec la droite D, tandis que dans le membre de droite D est vu comme un point rationnel de G. D'après (55),

$$(56) \qquad \pi_{0*} q_0^* A = \mathfrak{a} \in Br(k) = Br(G)\,.$$

Ceci prouve la formule (49) pour $K=k$. Le cas général résulte du fait que toute la situation commute par image réciproque au changement de base Spec K $\longrightarrow$ Spec k. Ceci achève la démonstration du sous-lemme, et donc celle du lemme 4.

Dans le cas qui nous intéresse, le lemme 4 (iii) assure qu'il revient au même de calculer l'obstruction de Brauer-Manin sur $\mathbb{Q}$ ou sur $\mathbb{Q}(\theta)$.

Après ces considérations préliminaires sur l'obstruction de Manin, nous revenons à l'étude d'une surface cubique V d'équation homogène

$$(1) \qquad ax^3 + by^3 + cz^3 + dt^3 = 0 \qquad a, b, c, d \in \mathbb{Q}^*$$

et reprenons les notations des §§3 et 1. En particulier, $k=\mathbb{Q}(\theta)$, et nous allons calculer l'obstruction de Manin sur k. Voici les diverses étapes de ce calcul.

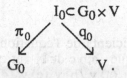

Pas n°0: *Conditions locales.*

Par multiplication des coordonnées et des coefficients de l'équation par des nombres rationnels convenables, on peut supposer que $a,b,c,d \in \mathbb{N}$ et que, pour chaque nombre premier p, l'ensemble des valuations $(v_p(a),v_p(b),v_p(c),v_p(d))$ soit, à permutation près, de l'un des types suivants:

(57) $(0,0,0,0)$ $(0,0,0,1)$ $(0,0,0,2)$ $(0,0,1,1)$ $(0,0,1,2)$.

Il existe d'ailleurs une unique <u>équation réduite</u> de ce type pour V: elle minimise $abcd$ parmi toutes les équations (1) à coefficients entiers équivalentes.

Notons alors $\mathcal{S}$ l'ensemble des places (de $\mathbb{Q}$, ou de $\mathbb{Q}(\theta)$, suivant le contexte) de mauvaise réduction pour V, i.e. l'ensemble des places divisant $3abcd$.

On doit vérifier si les conditions locales $V(\mathbb{Q}_p) \neq \emptyset$ sont satisfaites. C'est automatique par bonne réduction si $p \notin \mathcal{S}$. C'est également toujours vérifié pour $p \equiv 2 \bmod 3$. Pour $p \equiv 1 \bmod 3$, les cas où $V(\mathbb{Q}_p) = \emptyset$ sont les suivants:

(a) à permutation près, le type de l'équation est $(0,0,1,1)$ en p, i.e. les coefficients sont (a',b',pc',pd') avec a',b',c',d' unités p-adiques, et, ni a'/b', ni c'/d' n'est un cube dans $\mathbb{F}_p$;

(b) à permutation près, le type en p de l'équation est $(0,0,1,2)$, i.e. les coefficients sont (a',b',pc',p^2d') avec a',b',c',d' unités p-adiques, et a'/b' n'est pas un cube dans $\mathbb{F}_p$.

Pour $p=3$ enfin, $V(\mathbb{Q}_3) = \emptyset$ dans le seul cas suivant:

(c) à permutation près, le type en 3 de l'équation est $(0,0,0,2)$, i.e. les coefficients sont $(a',b',c',9d')$ avec a',b',c',d' unités 3-adiques, et, à permutation près, $(a',b',c') \equiv (1,2,4) \bmod 9$.

On voit donc que les conditions locales se vérifient de façon parfaitement effective. On suppose désormais qu'elles sont vérifiées pour l'équation (1). On peut également supposer que celle-ci vérifie les conditions:

(24) aucun des a/b ou ab/cd n'est un cube dans $\mathbb{Q}^*$,

car, sinon, le principe de Hasse vaut et l'obstruction de Manin est donc vide. On est alors dans la situation du §3, on met l'équation réduite (1), dont on a éventuellement permuté les coefficients a,b,c,d, sous la forme

(1)' $$x^3 + \lambda y^3 + \mu z^3 + \lambda\mu\nu t^3 = 0$$

et on introduit les notations de (12) et (13) et celles du §3.

Pas n°1: *Calcul de* ε.

Il s'agit de trouver $\varepsilon \in K^*$ solution de l'équation

(44) $$N_t(\varepsilon) = N_{K/L_1}(\varepsilon) = -\mu\ .$$

D'après Siegel [19] qui a traité le cas général d'une extension galoisienne de corps de nombres, la recherche d'une telle solution est parfaitement effective et se ramène à un nombre fini de vérifications élémentaires. En pratique, il est plus simple de remplacer K/L_1 par l'extension non galoisienne $\mathbb{Q}(\alpha,\gamma)/\mathbb{Q}(\gamma)$, ce qui est loisible, car l'équation (44) a aussi une solution dans $\mathbb{Q}(\alpha,\gamma)$ et, d'autre part, la recherche d'une solution de l'équation $N_{\mathbb{Q}(\alpha,\gamma)/\mathbb{Q}(\gamma)}(\varepsilon) = -\mu$ est encore une question effective d'après Bartels [2] qui a étendu au cas non galoisien les résultats de Siegel.

En pratique, on opère comme suit. Parmi les éléments de $\mathbb{Q}^*$ qui appartiennent à $N_t\mathbb{Q}(\alpha,\gamma)^*$, il y a successivement $N_{\mathbb{Q}(\alpha)/\mathbb{Q}}\mathbb{Q}(\alpha)^*$, $N_{\mathbb{Q}(\beta)/\mathbb{Q}}\mathbb{Q}(\beta)^*$ et $N_{\mathbb{Q}(\delta)/\mathbb{Q}}\mathbb{Q}(\delta)^*$, autrement dit les éléments de la forme:

$$\frac{x^3+\lambda y^3}{x'^3+\lambda y'^3} \qquad \frac{x^3+\lambda\nu y^3}{x'^3+\lambda\nu y'^3} \qquad \frac{x^3+\lambda\nu^2 y^3}{x'^3+\lambda\nu^2 y'^3} \qquad\qquad x\,,\,y\,,\,x'\,,\,y' \in \mathbb{Q}$$

et il y a aussi ν. On fait alors des tables d'entiers de la forme

(58) $$x^3+\lambda y^3 \qquad x^3+\lambda\nu y^3 \qquad x^3+\lambda\nu^2 y^3$$

puis des tables de produits et quotients de tels nombres, parmi lesquels il y a évidemment λ et ν, et on y cherche μ à des cubes près.

On peut en fait montrer que l'équation (44) admet une solution

(59) $$\varepsilon = \varepsilon_\alpha \varepsilon_\beta$$

avec $\varepsilon_\alpha \in \mathbb{Q}(\alpha)$ et $\varepsilon_\beta \in \mathbb{Q}(\beta)$, mais ce résultat ne nous sera pas utile. Il justifie néanmoins la méthode pratique indiquée ci-dessus.

Notons enfin que tout point $P=(x,y,z,t) \in V(L_1)$ situé hors du diviseur de g donne une solution

(60) $$\varepsilon = g(P)$$

de (44). De fait, on peut aussi bien prendre $P \in V(\mathbb{Q}(\gamma))$.

<u>Pas n°2:</u> *Calcul de* η.

Il s'agit de trouver $\eta \in K_1^*$ solution de l'équation

(45) $\eta/{}^t\eta = -\mu/N_s(\varepsilon)$.

Cela revient simplement à résoudre un système d'équations linéaires homogènes 3×3. Une telle solution est également donnée par la résolvante de Lagrange-Hilbert définie par un élément $a_0 \in K_1^*$, pourvu qu'elle soit non nulle:

$$\eta = a_0 - {}^t a_0 \mu/N_s(\varepsilon) - {}^{t^2} a_0 \mu^2/{}^{1+t}N_s(\varepsilon) .$$

Pour la suite, il est important de faire la remarque suivant. Si ε est de la forme

(61) $\varepsilon = \varepsilon_\beta \varepsilon_\delta$

avec $\varepsilon_\beta \in K_2^*$ et $\varepsilon_\delta \in L_2^*$ on a: $N_s(\varepsilon) = N_t(\varepsilon)$ car il en est ainsi pour ε_β et ε_δ. Comme $N_t(\varepsilon) = -\mu$ d'après l'équation (44), on en déduit alors $-\mu/N_s(\varepsilon) = 1$, et

$$\eta = 1$$

est dans ce cas solution de (45).

<u>Pas n°3:</u> *Calcul de* $V(k_v)/Br$.

Ayant ainsi déterminé ε et η, on obtient $\mathcal{A} = \mathcal{A}_{\varepsilon,\eta} \in Br(V_k)$ caractérisée par:

(46) $\iota(\mathcal{A}) = cl(1, g/\varepsilon, f/\eta) \in H^2(G, K(V)^*)$,

et calculer l'obstruction de Brauer consiste à étudier le noyau de l'application

$$i = i_{\mathcal{A}} : \mathcal{V}_k = \prod_v V(k_v)/Br \longrightarrow \mathbb{Q}/\mathbb{Z}$$

définie par la formule (47). Il faut donc calculer les $V(k_v)/Br$.

Lemme 5. *Soit* $P_v \in V(k_v)$ *un point quelconque.*

(i) *Si* V *est* k_v*-rationnelle, i.e. si l'un des* ab/cd, ac/bd, ad/bc *est un cube dans* k_v*, alors:*

$$V(k_v)/Br = \{P_v\} .$$

(ii) *Si* v *est une place de bonne réduction pour* V, *i.e.* $v \notin \mathcal{S}$, *alors:*

$$V(k_v)/Br = \{P_v\} .$$

En effet, on a a priori le diagramme suivant:

$$\begin{array}{ccc} V(k_v)/Z & \hookrightarrow & A_0(V_v) \\ \Big\downarrow\!\!\!\downarrow & & \\ V(k_v)/Br & & \end{array}$$

où $A_0(V_v)$ désigne le groupe des classes pour l'équivalence rationnelle de 0-cycles de degré 0 de V_v et où Z désigne l'équivalence induite sur $V(k_v)$ par l'application $V(k_v) \longrightarrow A_0(V_v)$ définie par $P \longrightarrow cl(P-P_v)$. Or, sous chacune des hypothèses (i) ou (ii):

$$A_0(V_v) = 0 .$$

Pour (i) c'est bien connu, et pour (ii) c'est le théorème A (iii) de [6].

Dans les cas où $v \in \mathcal{S}$ et où V n'est pas k_v-rationnelle, on est amené en principe à faire une étude directe dans chaque cas. De fait, ces cas-là seront réglés par les résultats du §5.

<u>Pas n°4:</u> *Calcul des invariants locaux* $inv_v \mathcal{A}(P_v)$.

Rappelons d'abord que, si v est une place de bonne réduction pour V et $\mathcal{A}$, alors:

(8) $\qquad inv_v \mathcal{A}(P_v) = 0 .$

On commence donc par déterminer l'ensemble $\mathcal{S}_{\mathcal{A}}$ des places de *mauvaise réduction* pour V et $\mathcal{A}$. La mauvaise réduction $\mathcal{S}$ de V étant connue, $\mathcal{S}_{\mathcal{A}}$ s'obtient en lui ajoutant les places de mauvaise réduction de ε et de η hors de $\mathcal{S}$. Par exemple, si $\varepsilon=\varepsilon'/\varepsilon''$ où ε' et ε'' sont des produits d'entiers de $\mathbb{Q}(\alpha)$, $\mathbb{Q}(\beta)$ et $\mathbb{Q}(\delta)$, les places de mauvaise réduction pour ε sont parmi les diviseurs premiers de $N_t(\varepsilon')\cdot N_t(\varepsilon'')$. En particulier, <u>si</u> $\varepsilon=\varepsilon'$ <u>ou</u> $1/\varepsilon''$, <u>et si</u> $\eta=1$, <u>alors</u> $\mathcal{S}_{\mathcal{A}}=\mathcal{S}$.

Soient v une place de k - qu'on peut supposer finie - et w un prolongement quelconque de v à K. On a le diagramme naturel, et commutatif, suivant:

$$\begin{array}{ccc}
\mathrm{Br}(V_k, K) & \longrightarrow & \mathrm{Br}(V_v, K_w^*) \\
\downarrow & H^2(G, K(V) \otimes_k k_v)^*) = H^2(G, \prod_{w|v} K_w(V)^*) & \downarrow \\
 & \nearrow \qquad\qquad\qquad \searrow \simeq & \\
H^2(G, K(V)^*) & \xrightarrow{\quad \rho \quad} & H^2(G^v, K_w(V)^*)
\end{array}$$

où $G^v = \mathrm{Gal}(K_w/k_v)$ est le groupe de décomposition de K/k en v et où ρ est la restriction de G au sous-groupe G^v. L'isomorphisme $\simeq$ provient du lemme de Shapiro (voir par exemple [24] p. 176 pour un argument analogue). Si $G^v = \langle\gamma\rangle$ est cyclique, le choix d'un générateur γ détermine un isomorphisme (28):

$$H^2(G^v, K_w(V)^*) \simeq k_v(V)^*/N_\gamma K_w(V)^* .$$

Lemme 6. *On conserve les hypothèses et notations ci-dessus. Si* $G^v = \langle\gamma\rangle$, *et si l'on fait l'identification* (28) *ci-dessus, on a les formules suivantes, à valeurs dans* $k_v(V)^*/N_\gamma K_w(V)^*$:

$$\begin{aligned}
\rho(\mathcal{A}_{\varepsilon,\eta}) &= 1 && \textit{si } \gamma = t \\
&= f/\eta && \textit{si } \gamma = s \\
&= {}^q\varepsilon/\varepsilon\eta && \textit{si } \gamma = r \\
&= h\cdot\varepsilon/{}^r\varepsilon\,{}^t\eta && \textit{si } \gamma = q ,
\end{aligned}$$

où $h = \dfrac{z+\theta\beta t}{z+\beta t}$.

Dans le mode de représentation (30) associé au choix du couple (s,t) de générateurs de G, l'algèbre $\mathcal{A}_{\varepsilon,\eta}$ est représentée par le 2-cocycle $(1, g/\varepsilon, f/\eta)$. Les formules (31) et (32) donnent aussitôt les deux premières formules, pour $\gamma=t$ et $\gamma=s$. Pour $\gamma=r$, la formule (34) donne $(gf\,{}^q\varepsilon)/(\varepsilon\eta\,{}^q g)$, mais les expressions (40) et (42) de f et g et l'action de $q=st^2$ déduite de (14) donnent $gf={}^q g$. Pour $\gamma=q$ enfin, la formule (36) donne $(\varepsilon\,{}^r g\,{}^t f)/({}^r\varepsilon\,{}^t\eta g)$, et le calcul donne ${}^t f\cdot{}^r g/g = h$.

Soit $\psi \in k_v(V)^*$ une fonction telle que $\rho(\mathcal{A}_{\varepsilon,\eta}) = \psi$ dans $k_v(V)^*/N_\gamma K_w(V)^*$. Si $P_v \in V(k_v)$ n'appartient pas au diviseur de ψ, alors:

$$\mathcal{A}_{\varepsilon,\eta}(P_v) = \psi(P_v) \qquad\qquad \text{dans } k_v^*/N_\gamma K_w^* .$$

Avant de poursuivre le calcul, il nous faut disposer d'un <u>formulaire</u> précis et explicite pour calculer les invariants locaux, qui sont donnés par des <u>symboles de restes normiques</u> qu'on écrira additivement:

$$[a,b]_v \in \mathbb{Z}/3 .$$

De fait, ce symbole dépend du choix d'une racine cubique θ de l'unité. On <u>fixe</u> une telle racine θ et considère $k=\mathbb{Q}(\theta)$, ce qui fixe, pour chaque place v, une certaine racine cubique de l'unité dans k_v. On adopte les conventions d'Artin-Tate [1], §12, p. 150, et d'Iwasawa, identiques à celles de Serre ([18], XIV, §2, voir remarque p. 215) et Tate ([24] p. 351) pour le symbole d'Artin ψ_v, mais opposées pour le symbole de restes normiques. On a le diagramme commutatif:

$$\begin{array}{ccc} H^2(G^v,K_w^*) & \xrightarrow{\mathrm{inv}_v} & \mathbb{Z}/3 \\ \simeq\downarrow & & \downarrow\simeq \\ k_v^*/N_\gamma K_w^* & \xrightarrow{\psi_v} & G^v=\langle\gamma\rangle \end{array}$$

où les flèches verticales sont définies par le choix du générateur γ. Soit $b\in k_v^*$ tel que:

(62) $$K_w = k_v(\sqrt[3]{b}) \quad \text{et} \quad {}^{\gamma}\sqrt[3]{b} = \theta\,\sqrt[3]{b} .$$

Par définition ([1], §12, p. 150), le symbole de restes normiques $(a,b)_v=\theta^{[a,b]_v}$ vérifie:

$${}^{\psi_v(a)}\sqrt[3]{b} = (a,b)_v\sqrt[3]{b} ,$$

où $a\in k_v^*/N_\gamma K_w^*$. On en déduit:

$$\psi_v(a) = \gamma^{[a,b]_v} ,$$

et par suite:

$$\mathrm{inv}_v(A_\theta(a,\gamma)) = [a,b]_v \in \mathbb{Z}/3 ,$$

où $A_\theta(a,\gamma)$ désigne l'algèbre cyclique définie par a et γ moyennant θ. En particulier:

(63) $$\mathrm{inv}_v\mathcal{A}_{\varepsilon,\eta}(P_v) = [\psi(P_v),b]_v .$$

Le symbole $[\ ,\ ]_v$: $(k_v^*/N_\gamma K_w^*)\times(k_v^*/N_\gamma K_w^*) \longrightarrow \mathbb{Z}/3$ est biadditif et anticommutatif.

Le symbole $[\ ,\ ]_v$: $(k_v^*/N_\gamma K_w^*)\times(k_v^*/N_\gamma K_w^*) \longrightarrow \mathbb{Z}/3$ est biadditif et anticommutatif. On peut ainsi le calculer grâce aux formules qui suivent (pour **F1** et **F2**, voir [24], p. 352, ex. 2.8, et p. 349, ex. 1.4, sans oublier de changer les signes; pour **F3**, voir [10]):

F1. Soient $p\equiv 1 \bmod 3$ et v_p l'un des prolongements, qu'on notera aussi p_1 et p_2, de $\mathbb{Q}$ à k de la valuation p-adique; si u est une unité de $\mathbb{Q}_p=k_v$, on a la formule:

(64) $\quad [u,p]_{v_p} = -i$ tel que $u^{(p-1)/3}\equiv\theta^i \bmod p$

où $\theta=\theta_{v_p}$ est l'image de la racine θ initiale par $k \xrightarrow{v_p} \mathbb{Q}_p$. En particulier:

(65) $\quad [\theta,p]_{v_p} \equiv -(p-1)/3 \bmod p$.

F2. Soient $p\equiv 2 \bmod 3$ et v_p le prolongement, qu'on notera parfois p, de la valuation p-adique de $\mathbb{Q}$ à k; si u est une unité de $\mathbb{Q}_p(\theta)=k_{v_p}$, on a la formule:

(66) $\quad [u,p]_{v_p} = -i$ tel que $u^{(p^2-1)/3} \equiv \theta^i \bmod p$.

En particulier:

(67) $\quad [\theta,p]_{v_p} \equiv -(p^2-1)/3 \bmod p$.

F3. Pour $p=3$ on a les formules suivantes, où $\lambda=1-\theta$ et $a,b,c\in\mathbb{Z}$:

(68) $\quad [\theta,1+a\lambda+b\lambda^2]_3 \equiv a+a^2+b \bmod 3$

(69) $\quad [1+a\lambda+\dots,1+b\lambda^2]_3 \equiv -ab \bmod 3$

(70) $\quad [\lambda,1+a\lambda+b\lambda^2+c\lambda^3]_3 \equiv ((a-a^3)/3)+ab-c \bmod 3$

(71) $\quad [1+a\lambda+b\lambda^2+c\lambda^3,3]_3 \equiv ((a-a^3)/3)+a+a^2+ab+b-c \bmod 3$.

Voici quelques relations utiles pour les calculs en $p=3$:

(72)
$$\begin{aligned} \theta &= 1-\lambda \\ 3 &= -\theta^2\lambda^2 \\ 3 &\equiv -\lambda^2-\lambda^3 \bmod \lambda^4 \\ -2 &\equiv 1+\lambda^2+\lambda^3 \bmod \lambda^4 . \end{aligned}$$

Les formules ci-dessus dérivent aisément de [10] ou d'Artin-Tate ([1], thm. 10, p. 164, où il faut néanmoins corriger ζ en ζ/λ dans la formule 3, cf. p. 168 et p. 173): si u est une unité de $\mathbb{Q}_3(\theta)$,

$$[u,\lambda]_3 \equiv (1/3)\cdot S(\frac{\theta}{\lambda}\log u) \quad \text{et} \quad [\theta,u]_3 \equiv (1/3)\cdot S(\log u)\,.$$

Voici encore des formules qui se déduisent des précédentes toujours pour $p=3$:

(73) $$[1 + a\lambda + b\lambda^2 + \ldots ,1 + c\lambda + d\lambda^2 + \ldots]_3 \equiv ac(a-c) - ad + bc \mod 3$$

(74) $$[1+a\lambda+b\lambda^2+c\lambda^3,3^m(1+3d)]_3 \equiv m(((a-a^3)/3)+a+a^2+ab+b-c)+ad \mod 3$$

(75) $$[\theta,1 + 3a]_3 \equiv 2a \mod 3\,.$$

Lemme 7. *On note* $\overline{}$ *la conjugaison dans* $\mathbb{Q}(\theta)$. *Soient* p *un nombre premier et* $\boldsymbol{p}$ *un facteur premier dans* $\mathbb{Q}(\theta)$. *Si* $u,v\in\mathbb{Q}(\theta)^*$, *on a:*

(76) $$[u,v]_{\boldsymbol{p}} = -[\bar{u},\bar{v}]_{\bar{\boldsymbol{p}}}\,.$$

Autrement dit, avec les notations intoduites en **F1**, **F2**, **F3**:

$$[u,v]_p = -[\bar{u},\bar{v}]_p \qquad \text{si } p \not\equiv 1 \mod 3$$
$$[u,v]_{p_1} = -[\bar{u},\bar{v}]_{p_2} \qquad \text{si } p \equiv 1 \mod 3\,.$$

Etendons la <u>notation</u> $[u,v]_p$, qui a un sens pour $p\not\equiv 1 \mod 3$, en posant:

$$[u,v]_p = [u,v]_{p_1} + [u,v]_{p_2} \qquad \text{si } p \equiv 1 \mod 3\,.$$

Corollaire. *Si* $u,v\in\mathbb{Q}^*$, *on a, pour tout* p*:*

(77) $$[u,v]_p = 0\,.$$

Avant de démontrer le lemme, il est bon de préciser clairement la définition du symbole $[\ ,\ ]_{\boldsymbol{p}}$ où $\boldsymbol{p}$ désigne un facteur premier de p dans $k=\mathbb{Q}(\theta)$. Il dépend du choix d'une racine cubique $\theta_{\boldsymbol{p}}$ non triviale de 1 dans $k_{\boldsymbol{p}}$ et c'est l'application

$$[\ ,\ ]_{\boldsymbol{p}}^{\theta_{\boldsymbol{p}}} : k_{\boldsymbol{p}}^* \times k_{\boldsymbol{p}}^* \longrightarrow \mathbb{Z}/3$$

obtenue en considérant la suite d'applications canoniques:

$$k_{\mathfrak{p}}^*/k_{\mathfrak{p}}^{*3} \times k_{\mathfrak{p}}^*/k_{\mathfrak{p}}^{*3} \xrightarrow{\simeq} H^1(k_{\mathfrak{p}},\mu_3) \times H^1(k_{\mathfrak{p}},\mu_3) \xrightarrow{\cup} H^2(k_{\mathfrak{p}},\mu_3^{\otimes 2})$$

$$\xleftarrow{\simeq} H^2(k_{\mathfrak{p}},\mu_3)\otimes\mu_3 \xrightarrow[\simeq]{\mathrm{inv}_{\mathfrak{p}}\otimes\,\mathrm{id}} \mathbb{Z}/3\otimes\mu_3 \xleftarrow[\simeq]{1\otimes\theta_{\mathfrak{p}}} \mathbb{Z}/3$$

et définie, avec des notations abrégées, par

(78) $\mathrm{inv}_{\mathfrak{p}}(u \cup v) = [u,v]_{\mathfrak{p}}^{\theta_{\mathfrak{p}}} \otimes \theta_{\mathfrak{p}}$,

ce qui a été noté plus haut $(u,v)_{\mathfrak{p}}$. La valeur de ce symbole sur $k^*\times k^*$ s'obtient par composition avec le plongement canonique $k^* \longrightarrow k_{\mathfrak{p}}^*$:

$$[\ ,\ ]_{\mathfrak{p}}^{\theta} : k^* \times k^* \xrightarrow{v_{\mathfrak{p}}} k_{\mathfrak{p}}^* \times k_{\mathfrak{p}}^* \longrightarrow \mathbb{Z}/3 ,$$

avec pour choix de $\theta_{\mathfrak{p}}$ l'image par ce plongement de la racine $\theta\in k^*$ fixée initialement. Voir [18], chap. XIV.

Pour $u,v\in k^*$, on a donc:

(79) $[u,v]_{\mathfrak{p}}^{\theta^2} = -[u,v]_{\mathfrak{p}}^{\theta}$,

puisque, dans $(\mathbb{Z}/3)\otimes\mu_3$, on a: $c\otimes\theta^2=(-c)\otimes\theta$. La conjugaison de $k=\mathbb{Q}(\theta)$ induit le diagramme commutatif suivant:

$$\begin{array}{ccccc}
k^*/k^{*3}\times k^*/k^{*3} & = & H^1(k,\mu_3)\times H^1(k,\mu_3) & \xrightarrow{\cup} & H^2(k,\mu_3^{\otimes 2}) \\
\downarrow - & & \downarrow - & & \downarrow - \\
k^*/k^{*3}\times k^*/k^{*3} & = & H^1(k,\mu_3)\times H^1(k,\mu_3) & \xrightarrow{\cup} & H^2(k,\mu_3^{\otimes 2})
\end{array}$$

$$\begin{array}{ccc}
\longrightarrow H^2(k_{\mathfrak{p}},\mu_3)\otimes\mu_3 & \xrightarrow{\mathrm{inv}_{\mathfrak{p}}} & \mathbb{Z}/3\otimes\mu_3 \\
\downarrow - & & \downarrow - \\
\longrightarrow H^2(k_{\overline{\mathfrak{p}}},\mu_3)\otimes\mu_3 & \xrightarrow{\mathrm{inv}_{\overline{\mathfrak{p}}}} & \mathbb{Z}/3\otimes\mu_3 .
\end{array}$$

Autrement dit, on a l'égalité:

$$[u,v]_{\mathfrak{p}}^{\theta} \otimes \theta^2 = [\bar{u},\bar{v}]_{\overline{\mathfrak{p}}}^{\theta^2} \otimes \theta^2$$

dans $(\mathbb{Z}/3)\otimes\mu_3$, pour $u,v\in k^*$. Compte tenu de la relation (79) ci-dessus, ceci donne aussitôt:

(76) $[u,v]^{\theta}_{\mathfrak{p}} = -[\bar{u},\bar{v}]^{\theta}_{\bar{\mathfrak{p}}}$.

Après cette digression sur les symboles de restes normiques, nous allons commencer l'analyse des valeurs de $\mathrm{inv}_v\mathcal{A}(P_v)$ lorsque G^v est cyclique.

e *Le cas* $G^v=\langle e\rangle$. Ceci équivaut aux conditions suivantes:

a/b et c/d sont des cubes dans $\mathbb{Q}_p$,
ou: ad/bc et ac/bd sont des cubes dans $\mathbb{Q}_p$,
ou: λ et ν sont des cubes dans $\mathbb{Q}_p$.

On a alors nécessairement:

(80) $\mathrm{inv}_v\mathcal{A}(P_v) = 0$.

On suppose *désormais* qu'on n'est pas dans ce cas-là.

t *Le cas* $G^v=\langle t\rangle$. Ceci équivaut aux conditions suivantes:

ad/bc est un cube dans $\mathbb{Q}_p$,
ou: ν est un cube dans $\mathbb{Q}_p$.

La variété V est alors $\mathbb{Q}_p$-rationnelle, et, de toute manière, d'après le lemme 6:

(80) $\mathrm{inv}_v\mathcal{A}(P_v) = 0$,

puisque $\rho(\mathcal{A})=1$.

r *Le cas* $G^v=\langle r\rangle$. Ceci équivaut aux conditions suivantes:

ac/bd est un cube dans $\mathbb{Q}_p$,
ou: λ/ν est un cube dans $\mathbb{Q}_p$.

D'après les lemmes 1 et 5, la variété V est $\mathbb{Q}_p$-rationnelle et $\mathrm{inv}_v\mathcal{A}(P_v)$ prend toujours la même valeur, quel que soit $P_v\in V(k_v)$. Du reste, d'après le lemme 6, la restriction $\rho(\mathcal{A})$ est l'algèbre constante définie par ${}^q\varepsilon/\varepsilon\eta\in k_v^*/NK_w^*$ et par l'extension K_w/k_v moyennant les choix de θ comme racine cubique de 1 et de r comme générateur de G^v. Puisque $K_w=k_v(\alpha)=k_v(\gamma)$ et que ${}^r\alpha/\alpha={}^r\gamma/\gamma=\theta$, on obtient finalement la valeur prise par $\mathrm{inv}_v\mathcal{A}(P_v)$ quel que soit P_v:

(81) $$\mathrm{inv}_v\mathcal{A}(P_v) = [{}^q\varepsilon/\varepsilon\eta,\lambda]_v = [{}^q\varepsilon/\varepsilon\eta,\nu]_v\,.$$

s *Le cas* $G^v=\langle s\rangle$. Ceci équivaut aux conditions suivantes:

b/a est un cube dans $\mathbb{Q}_p$,
ou: λ est un cube dans $\mathbb{Q}_p$.

En ce cas, V n'est plus nécessairement $\mathbb{Q}_p$-rationnelle et $\mathrm{inv}_v\mathcal{A}(P_v)$ peut, comme on le verra plus loin, prendre plusieurs valeurs. D'après le lemme 6, la restriction $\rho(\mathcal{A})$ est l'algèbre définie par $f/\eta \in k_v(V)^*/NK_w(V)^*$ et par l'extension $K_w(V)/k_v(V)$ moyennant les choix de θ et de s comme générateur de G^v. Comme $K_w=k_v(\gamma)$ et que ${}^s\gamma/\gamma = \theta$, on a donc:

(82) $$\mathrm{inv}_v\mathcal{A}(P_v) = [f(P_v)/\eta,\nu]_v\,,$$

si du moins $f(P_v)$ est bien définie (sinon, il faut bouger P_v dans la même classe pour l'équivalence de Brauer, ou bien bouger f par une norme de $K_w(V)^*$). Par un changement de variables, l'équation de V peut s'écrire sur $\mathbb{Q}_p$:

$$x^3 + y^3 + \mu z^3 + \mu\nu t^3 = 0\,,$$

et alors:
$$f = \frac{x+\theta y}{x+\theta^2 y}.$$

Pour $P_v=(1,-1,0,0)\in V(k_v)$, on trouve donc $f(P_v)=-\theta$. Dans ce cas-là, on obtient donc <u>au moins</u> la valeur

(83) $$\mathrm{inv}_v\mathcal{A}(P_v) = [\theta/\eta,\nu]_v\,.$$

Le cas **e** étant supposé exclu, ν n'est pas un cube dans $\mathbb{Q}_p$, et V est $\mathbb{Q}_p$-rationnelle si, et seulement si, μ/ν est un cube dans $\mathbb{Q}_p$. S'il en est ainsi, la valeur (83) est la seule valeur prise par $\mathrm{inv}_v\mathcal{A}(P_v)$. Sinon, il faut chercher d'autres valeurs éventuelles, comme on le verra au §5.

q *Le cas* $G^v=\langle q\rangle$. Ceci équivaut aux conditions suivantes:

d/c est un cube dans $\mathbb{Q}_p$,
ou: $\lambda\nu$ est un cube dans $\mathbb{Q}_p$.

Ce cas est semblable au précédent. D'après le lemme 6, la restriction $\rho(\mathcal{A})$ est l'algèbre définie par $h\varepsilon/{}^r\varepsilon^t\eta \in k_v(V)^*/NK_w(V)^*$ et par l'extension $K_w(V)/k_v(V)$, moyennant les choix de θ et de q comme générateur de G^v. Comme $K_w=k_v(\gamma)=k_v(\alpha)$ et que ${}^q\gamma/\gamma=\theta$ et ${}^q\alpha/\alpha=\theta^2$, on a donc:

$$\text{(84)} \qquad \mathrm{inv}_v\mathcal{A}(P_v) = [h(P_v)\cdot\varepsilon/{}^r\varepsilon^t\eta,\nu]_v = [{}^r\varepsilon^t\eta/\varepsilon h(P_v),\lambda]_v\,,$$

du moins si $h(P_v)$ est bien défini. Par un changement de variables, l'équation de V peut s'écrire sur $\mathbb{Q}_p$:

$$x^3 + \lambda y^3 + \mu z^3 + \mu t^3 = 0\,,$$

et alors:

$$h = \frac{z+\theta t}{z+t}\,.$$

Pour $P_v=(0,0,1,-\theta)$, on trouve donc $h(P_v)=-\theta^2$. Dans ce cas-là, on obtient donc toujours <u>au moins</u> la valeur

$$\text{(85)} \qquad \mathrm{inv}_v\mathcal{A}(P_v) = [\theta^2\cdot\varepsilon/{}^r\varepsilon^t\eta,\nu]_v = [\theta\cdot{}^r\varepsilon^t\eta/\varepsilon,\lambda]_v\,.$$

Le cas e étant supposé exclu, λ n'est pas un cube dans $\mathbb{Q}_p$, et V est $\mathbb{Q}_p$-rationnelle si, et seulement si, μ/ν est un cube dans $\mathbb{Q}_p$. S'il en est ainsi, la valeur (85) est la seule valeur prise par $\mathrm{inv}_v\mathcal{A}(P_v)$. Sinon, il faut chercher d'autres valeurs éventuelles, voir §5.

G *Le cas* $G^v=G$. Indiquons tout de suite que ce cas pourra toujours être évité dans tous les calculs pratiques effectués après le §5. Voici néanmoins une méthode de calcul. Quitte à bouger $P_v \in V(k_v)$ dans la même classe pour l'équivalence de Brauer, on peut faire en sorte que le 2-cocycle

$$(1,g(P_v)/\varepsilon,f(P_v)/\eta) \in Z^2(G,K_w^*)$$

soit bien défini, et il s'agit de calculer son invariant en v. D'après (30), ou (43) et (44), on a: $N_t(g(P_v)/\varepsilon)=1$. Le théorème 90 de Hilbert appliqué à l'extension cyclique K_w/L_{1v} affirme alors l'existence de $\omega(P_v) \in K_w^*$ tel que

$$\text{(86)} \qquad \Delta_t(\omega(P_v)) = g(P_v)/\varepsilon\,.$$

D'après les définitions du §2, on a: $d_1^*(1,\omega(P_v))=(1,1/\Delta_t(\omega(P_v)),N_s(\omega(P_v)))$ et, par suite, le 2-cocycle initial est homologue à

$$(1,1,N_s(\omega(P_v))\cdot f(P_v)/\eta) = \mathrm{Inf}_{G/t,G}(N_s(\omega(P_v))\cdot f(P_v)/\eta)$$

où

$$N_s(\omega(P_v))\cdot f(P_v)/\eta \in k_v^*/N_\sigma L_{1v}^*$$

définit un élément de $H^2(<\sigma>,L_{1v}^*)$ via le choix du générateur σ. Comme ${}^\sigma\gamma/\gamma=\theta$, on en déduit finalement:

$$\text{(87)} \qquad \mathrm{inv}_v \mathcal{A}_{\varepsilon,\eta}(P_v) = [N_s(\omega(P_v))\cdot f(P_v)/\eta, \nu]_v .$$

Remarque 4. Le calcul ci-dessus peut s'appliquer, de façon générique, au 2-cocycle

$$(1,g/\varepsilon,f/\eta) \in Z^2(G,K(V)^*) .$$

Pour les même raisons que ci-dessus, $N_t(g/\varepsilon)=1$, et le théorème 90 appliqué à l'extension cyclique $K(V)/L_1(V)$ affirme l'existence de $\omega \in K(V)^*$ telle que:

$$\text{(86')} \qquad \Delta_t(\omega) = g/\varepsilon .$$

De même que ci-dessus, on trouve alors que le 2-cocycle $(1,g/\varepsilon,f/\eta)$ est homologue au cocycle $\mathrm{Inf}_{G/t,G}(N_s(\omega)\cdot f/\eta)$, où

$$\text{(88)} \qquad h = N_s(\omega)\cdot f/\eta \in k(V)^*$$

définit, moyennant le choix de σ, un élément de $H^2(<\sigma>,L_1(V)^*)$. Cette méthode de calcul permet de trouver $h \in k(V)^*$ vérifiant la condition (ii) de la remarque 2 avec

$$D' = D + \mathrm{div}(\omega) \in \mathrm{Div}(V_{L_1})$$

où $D=L'(2)-L''(0)\in \mathrm{Div}\ V_K$ a été considéré dans la même remarque 2. On obtient ainsi finalement:

$$\mathcal{A}_{\varepsilon,\eta} = (h,L_1(V)/k(V),\sigma) \in \mathrm{Br}(V,L_1) ,$$

et les invariants locaux $\mathrm{inv}_v\mathcal{A}(P_v)$ sont alors faciles à calculer sous la forme (87). Cependant, cette méthode de calcul, en apparence simple et naturelle, achoppe sur le calcul de ω, qui se révèle fort pénible en pratique et conduit ensuite à des calculs très complexes.

§5. Le premier cas

C'est le cas, a priori peu favorable pour les calculs, où il existe un p de mauvaise réduction sur V, tel que V ne soit pas $\mathbb{Q}_p$-rationnelle. L'objet de ce § est d'établir la proposition suivante qui donne une réponse uniforme dans ce cas.

Proposition 2. *Si la surface cubique diagonale* V *a mauvaise réduction en* p*, et si* V *n'est pas* $\mathbb{Q}_p$*-rationnelle, l'obstruction de Manin au principe de Hasse est vide pour* V.

On en déduit aussitôt le corollaire cité dans l'introduction:

Corollaire. *Soit* V *une surface cubique diagonale sur* $\mathbb{Q}$ *d'équation homogène:*

(1) $$ax^3 + by^3 + cz^3 + dt^3 = 0 \qquad a, b, c, d \in \mathbb{Q}^* .$$

S'il existe un nombre premier p *qui divise un, et un seul, coefficient de l'équation, les coefficients étant supposés sans cube, alors l'obstruction de Manin au principe de Hasse est vide pour* V.

Le démonstration du corollaire utilise le lemme suivant:

Lemme 8. *Soit* V *une surface cubique diagonale d'équation* (1) *avec* $a,b,c,d \in F^*$, *où* F *est un corps de caractéristique* $\neq 3$. *On a l'équivalence:*

$$V \textit{ est } F\textit{-rationnelle} \iff V(F) \neq \emptyset \textit{ et l'un des } ab/cd \textit{ est un cube dans } F^*.$$

L'implication $\Leftarrow$ utilise le même argument que dans la démonstration du lemme 1: si $F' = F(\sqrt[3]{\lambda})$, le tore $R^1_{F'/F}\mathbb{G}_m$ est une variété F-rationnelle, même si F'/F n'est pas galoisienne. Inversement, si V est F-rationnelle, la proposition 1 montre que l'un des ab/cd est un cube dans $F(\theta)$: c'est donc aussi un cube dans F.

Sous l'hypothèse du corollaire, $v_p(ab/cd) = \pm v_p(ac/bd) = \pm v_p(ad/bc) = \pm 1$, aucun des ab/cd n'est un cube dans $\mathbb{Q}_p$ et, d'après le lemme, V n'est pas $\mathbb{Q}_p$-rationnelle.

Il reste à établir la proposition, et c'est l'objet du reste de ce paragraphe.

Principe de la démonstration.

(a) Soient $k=\mathbb{Q}(\theta)$ et $v_0=v_p$ une place de k prolongeant la valuation p-adique de $\mathbb{Q}$. On va établir l'assertion suivante: l'application

$$(89)\qquad \begin{aligned} V(k_{v_0}) &\longrightarrow \mathbb{Z}/3 \\ P_{v_0} &\longrightarrow \mathrm{inv}_{v_0}\mathcal{A}(P_{v_0}) \end{aligned}$$

est surjective. Cette assertion implique évidemment la proposition, car elle entraîne a fortiori que l'application $i_{\mathcal{A}}: \mathcal{V} \longrightarrow \mathbb{Z}/3$ définie par la somme (47) des invariants locaux des $\mathcal{A}(P_v)$ est surjective, et atteint donc la valeur 0.

(b) Pour établir l'assertion (a), on va chercher une courbe E de genre 1 tracée sur $V_{k_{v_0}}$, telle que $E(k_{v_0})\neq\emptyset$ et que la restriction de $\mathcal{A}$ à E ne soit pas constante .

En effet, soit E une telle courbe. Fixons $O\in E(k_{v_0})$. L'application

$$\begin{aligned} E(k_{v_0}) &\xrightarrow{\ \simeq\ } \mathrm{Pic}^0(E) \xrightarrow{\ \rho^*(\mathcal{A})\ } \mathbb{Z}/3 \\ P &\longrightarrow P-O \longrightarrow \mathrm{inv}_{v_0}(\mathcal{A}(P)-\mathcal{A}(O)) \end{aligned}$$

n'est pas l'application 0, car, d'après le théorème de dualité locale de Lichtenbaum ([12], corollaire 1 et théorème 4), l'application $\rho^*: \mathrm{Br}\,E/\mathrm{Br}\,k_{v_0} \longrightarrow (\mathrm{Pic}^0E)^*$ est un isomorphisme. Comme $\rho^*(\mathcal{A})$ est un homomorphisme, c'est donc une surjection. L'existence de E comme ci-dessus implique donc l'assertion (a).

(c) Pour établir que la restriction de $\mathcal{A}$ à E n'est pas constante, il suffit par exemple de voir que la restriction de $\mathcal{A}$ à $E_{K_{1v'_0}}$ n'est pas constante, ce qui est a priori plus simple à vérifier. En effet,

$$\mathcal{A}_{K_{1v'_0}} = A_\theta(f/\eta, K_{w_0}(V)/K_{1v'_0}(V), s) \in \mathrm{Br}(V_{K_{1v'_0}}, K_{w_0}))$$

(lemme 6), et la restriction de $\mathcal{A}$ à $E_{K_{1v'_0}}$ n'est pas constante si, et seulement si,

$$(90)\qquad f: E(K_{1v'_0}) \longrightarrow K^*_{1v'_0}/N_s K^*_{w_0}$$

n'est pas constante.

Lemme 9. *Soient* F *un corps local de caractéristique résiduelle* $\neq 3$ *et* $\mathcal{O}$ *son anneau d'entiers. On suppose* $\theta \in F$. *Soit* E *une courbe elliptique sur* F *d'équation homogène*

$$x^3 + y^3 + dz^3 = 0 \qquad\qquad d \in \mathcal{O}^* .$$

Alors, la fonction

$$f = (x + \theta y)/(x + \theta^2 y)$$

définit une application

$$E(F) \xrightarrow{\ f\ } \mathcal{O}^*/\mathcal{O}^{*3} \simeq \mathbb{Z}/3$$

qui est surjective.

L'application définie par f prend a priori ses valeurs dans F^*/F^{*3}, mais en fait aboutit dans $\mathcal{O}^*/\mathcal{O}^{*3} \hookrightarrow F^*/F^{*3}$.

Soit $\mathfrak{E}$ une courbe elliptique lisse sur $\mathcal{O}$ dont la fibre générique soit E. La relation:

(91) $\qquad \operatorname{div}(f) = 3(P'-P'')$

où $P'=(\theta,-1,0)$ et $P''=(\theta^2,-1,0) \in E(F)=\mathfrak{E}(\mathcal{O})$, montre que le diviseur $P'-P''$ définit sur $\mathcal{O}$ une isogénie

$$(\mathbb{Z}/3)_{\mathcal{O}} \hookrightarrow \underline{\mathrm{Pic}}^0_{\mathfrak{E}/\mathcal{O}} \twoheadrightarrow \mathfrak{E}' .$$

L'isogénie duale définit alors un torseur sur $\mathfrak{E}$ sous μ_3:

$$\mu_3 \hookrightarrow \hat{\mathfrak{E}}' \twoheadrightarrow \mathfrak{E} .$$

Par lissité de $\mathfrak{E}'$ et par le théorème de Lang, $H^1(\mathcal{O},\mathfrak{E}')=0$, et le torseur ci-dessus définit donc une surjection

$$E(F) = \mathfrak{E}(\mathcal{O}) \twoheadrightarrow \mathcal{O}^*/\mathcal{O}^{*3}$$

dont on voit, sur F par exemple, qu'en raison de la relation (91) elle est donnée par la fonction f.

Lemme 10. *Soit* F'/F *une extension cyclique de degré* 3 *de corps locaux. On suppose* $\theta \in F$ *et* F'/F *modérément ramifiée. On note* $\mathcal{O}$ *et* $\mathcal{O}'$ *les anneaux d'entiers de* F *et* F' *respectivement et* $N=N_{F'/F}$. *On a alors:*

$$\mathcal{O}^*/\mathcal{O}^{*3} = \mathcal{O}^*/N\mathcal{O}'^* = F^*/NF'^* .$$

On a: $\mathcal{O}^*/N\mathcal{O}'^* = F^*/NF'^*$ parce que F'/F est ramifiée et $\mathcal{O}^*/\mathcal{O}^{*3} \simeq \mathbb{Z}/3$ parce que $\theta \in F$ et que la caractéristique résiduelle est $\neq 3$. D'où $\mathcal{O}^*/\mathcal{O}^{*3} = \mathcal{O}^*/N\mathcal{O}'^*$.

Lemme 11. *Soit* V *la surface cubique diagonale d'équation homogène:*

$$x^3 + y^3 + \mu z^3 + \mu\nu t^3 = 0 \qquad \mu, \nu \in F^* ,$$

définie sur le corps F *de caractéristique* $\neq 3$. *On suppose* $\theta \in F$ *et on note* $F'=F(\sqrt[3]{\nu})$. *La fonction*

$$f = (x + \theta y)/(x + \theta^2 y)$$

définit une application

$$\text{(92)} \qquad V(F) \xrightarrow{\ f\ } F^*/NF'^*$$

qui prend au moins les valeurs θ, $\mu^2\theta$, $\mu\theta$.

La fonction f définit une application $V(F) \longrightarrow F^*/NF'^*$ - où $N=N_{F'/F}$ - parce que son diviseur appartient à $N_{F'/F}\mathrm{Div}(V_{F'})$, comme on le déduit aussitôt de l'équation ci-dessus. Celle-ci s'écrit:

$$z^3 + \nu t^3 = -(x^3 + y^3)/\mu$$

et la fonction $(x^3+y^3)/\mu$ appartient donc à $N_{F'/F}F'(V)^*$. On en déduit aisément les expressions suivantes de f modulo $N_{F'/F}F'(V)^*$:

$$f = (x + \theta y)/(x + \theta^2 y) = \mu(x + \theta^2 y)/(x + y) = \mu^2(x + y)/(x + \theta y)$$

dans $F(V)^*/N_{F'/F}F'(V)^*$. En utilisant ces diverses expressions, on en déduit:

$$f(1,-1,0,0) = \theta$$
$$f(1,-\theta,0,0) = \mu^2\theta$$
$$f(1,-\theta^2,0,0) = \mu\theta$$

dans F^*/NF'^*.

(d) Pour démontrer l'assertion (a), et donc la proposition, <u>on peut permuter</u> a,b,c,d à son gré dans l'équation (1), car ceci modifie seulement $\mathcal{A}$ par une algèbre constante ce qui est sans importance pour la surjectivité de l'application (89).

<u>Démonstration de la proposition pour</u> $p \neq 3$.

Par hypothèse, V a mauvaise réduction en p. Il résulte donc de l'étude des conditions locales du §4 qu'à permutation près des coefficients (a,b,c,d), ceux-ci sont nécessairement de l'un des types suivants, où a',b',c',d' désignent toujours des unités p-adiques:

(i) (a',b',c',pd')
(ii) (a',b',c',p^2d')
(iii) (a',b',pc',pd') avec, si $p \equiv 1 \bmod 3$, a'/b' ou c'/d' cube mod p
(iv) (a',b',pc',p^2d') avec, si $p \equiv 1 \bmod 3$, a'/b' cube mod p.

Si l'on considère l'équation de V sous la forme:

(1') $$x^3 + \lambda y^3 + \mu z^3 + \lambda\mu\nu t^3 = 0$$

l'hypothèse d'irrationalité de V sur $\mathbb{Q}_p$ se traduit d'après le lemme 8 par la condition:

(93) aucun des nombres ν , λ/ν , μ/ν n'est un cube dans $\mathbb{Q}_p$.

On en déduit, quitte à faire encore des permutations, qu'on peut se limiter pour l'équation (1') aux trois cas suivants, où unité=unité p-adique, où i=1,2, où cube=cube dans $\mathbb{Q}_p$ et où $\|$ = divise exactement:

(1) λ et μ sont des unités et $p^i \| \nu$
(2) λ est un cube et une unité, $p^i \| \mu$ et ν est une unité mais pas un cube;
(3) λ est un cube et une unité, $p \| \mu$, $p \| \nu$, mais μ/ν n'est pas un cube.

On notera que les types (i) et (ii) se ramènent au cas (1), le type (iii) au cas (2) et le type (iv) au cas (3).

<u>Le cas</u> (1)

On considère la courbe E d'équation t=0 sur V. C'est la courbe plane d'équation:

$$x^3 + \lambda y^3 + \mu z^3 = 0 .$$

Comme λ et μ sont des unités p-adiques, la courbe E a bonne réduction en p, et par suite: $E(\mathbb{Q}_p)\neq\emptyset$. Quitte alors à passer à $K_1=k(\sqrt[3]{\lambda})$ et à changer de variable, on peut supposer que E a pour équation:

$$x^3+y^3+\mu z^3=0\,.$$

Puisque μ est une unité, on peut alors appliquer le lemme 9 à E sur le corps $F:=K_{1v_0'}$. Soit $F'=K_{w_0}=F(\sqrt[3]{\nu})$. Les hypothèses sur p et ν montrent que l'extension F'/F est modérément ramifiée. On peut donc appliquer le lemme 10, ce qui assure finalement que l'application composée:

$$(90)\qquad E(F)\xrightarrow{\ f\ }\mathcal{O}^*/\mathcal{O}^{*3}\xrightarrow{\ \simeq\ }F^*/NF'^*$$

est encore surjective, ce qui prouve l'assertion (c), et donc la proposition dans ce cas.

<u>Le cas</u> (2)

Quitte à faire un changement de variable sur $\mathbb{Q}_p$, on peut supposer V d'équation:

$$x^3+y^3+\mu z^3+\mu\nu t^3=0.$$

Soient $F=k_{v_0}$ et $F'=F(\sqrt[3]{\nu})$. Les hypothèses du lemme 11 sont vérifiées et, dans ce cas, d'après le lemme 6, l'assertion (a) revient à établir que l'application

$$(92)\qquad V(F)\xrightarrow{\ f\ }F^*/NF'^*.$$

n'est pas constante. L'hypothèse sur ν implique $F^*/NF'^*\simeq\mathbb{Z}/3$. D'après le lemme 11, il suffit donc d'établir:

$$\mu\notin NF'^*.$$

Comme ν est une unité, l'extension F'/F est non ramifiée, et comme $v_p(\mu)=1$ ou 2, μ n'est pas une norme de F'^*.

<u>Le cas</u> (3)

Le début du raisonnement est identique à celui relatif au cas (2). L'hypothèse sur ν, quoique différente, implique encore: $F^*/NF'^*\simeq\mathbb{Z}/3$. Mais ici F'/F est modérément ramifiée. D'après le lemme 10, on a donc:

$$\mathcal{O}^*\cap NF'^*=\mathcal{O}^{*3}.$$

Comme, par hypothèse, μ/ν est une unité qui n'est pas un cube, $\mu/\nu \notin NF'^*$. En revanche, $\nu \in NF'^*$ par définition même de F'. On en déduit: $\mu \notin NF'^*$, ce qui achève la démonstration dans ce cas par application du lemme 11 comme dans le cas (2).

Démonstration de la proposition pour p=3.

La démonstration consiste en une longue analyse, cas par cas, des divers types d'équation possibles. On cherche ici encore à démontrer l'assertion (a) et, pour ce faire, on peut considérer comme équivalentes deux équations

$$a_1x^3 + b_1y^3 + c_1z^3 + d_1t^3 = 0 \quad \text{et} \quad a_2x^3 + b_2y^3 + c_2z^3 + d_2t^3 = 0$$

dont les coefficients, à permutation près, sont tels que a_1/a_2, b_1/b_2, c_1/c_2, d_1/d_2, soient tous des cubes dans $\mathbb{Q}_3$. A cet égard, on fera un usage constant et implicite du fait suivant:

$$u \in \mathbb{Z}_3^* \text{ est un cube} \iff u \equiv \pm 1 \bmod 9 .$$

Il résulte de l'étude des conditions locales du §4 qu'à permutation près des coefficients (a,b,c,d) de l'équation (1), ceux-ci sont nécessairement de l'un des types suivants, où a',b',c',d' désignent toujours des unités 3-adiques:

(I) (a',b',c',d')
(II) (a',b',c',3d')
(III) (a',b',c',9d') avec $(a',b',c') \not\equiv (1,2,4)$ et de tous les permutés mod 9
(IV) (a',b',3c',3d')
(V) (a',b',3c',9d') .

De fait, compte tenu de l'hypothèse d'irrationalité de V sur $\mathbb{Q}_3$, on peut se limiter aux équations dont les coefficients $(a,b,c,d)=(1,\lambda,\mu,\lambda\mu\nu)$ sont de l'un des types suivants:

(1) $(1,1,1,2^i)$	avec $i \neq 0$	$\nu = 2^i$	
(2) $(1,1,2^i,3\cdot 2^j)$		$\nu = 3\cdot 2^{j-i}$	
(3) $(1,2,4,3\cdot 2^j)$		$\nu = 3\cdot 2^j$	
(4) $(1,1,2^i,9\cdot 2^j)$		$\nu = 9\cdot 2^{j-i}$	
(5) $(1,1,3\cdot 2^i,3\cdot 2^j)$	avec $i \neq j$	$\nu = 2^{j-i}$	
(6) $(1,1,9\cdot 2^i,9\cdot 2^j)$	avec $i \neq j$	$\nu = 2^{j-i}$	
(7) $(1,1,3\cdot 2^i,9\cdot 2^j)$	avec $i+j \neq 0$	$\nu = 3\cdot 2^{j-i}$	$\mu/\nu = 2^{2i-j}$
(8) $(1,2^h,3\cdot 2^i,9\cdot 2^j)$	avec $h \neq 0$ et $i+j \neq h$	$\nu = 3\cdot 2^{j-h-i}$	$\mu/\nu = 2^{2i-j+h}$.

Dans la réduction expliquée ci-après des types (I) à (V) aux cas (1) à (8), on s'autorise évidemment des permutations des coefficients et des modifications par des cubes de $\mathbb{Q}_3^*$ et on utilise implicitement les faits suivants:

$$\mathbb{Z}_3^*/\mathbb{Z}_3^{*3} = \{1,2,4\} ,$$

$$\text{pour } u \in \mathbb{Z}_3^* , \qquad u \text{ non cube} \Longleftrightarrow u \equiv \pm 2 \text{ ou } \pm 4 \bmod 9 .$$

Type (I). Deux des coefficients a,b,c,d coïncident nécessairement dans $\mathbb{Z}_3^*/\mathbb{Z}_3^{*3}$, par exemple a et b. On peut donc supposer les coefficients de la forme $(1,1,2^i,2^j)$, mais la condition (93) impose que ν et μ/ν ne soient pas des cubes, autrement dit $i \pm j \neq 0$ dans $\mathbb{Z}/3$. Ceci implique que i ou j soit nul et l'autre non nul. On est ainsi réduit au cas (1).

Type (II). Ou bien deux des coefficients a,b,c coïncident dans $\mathbb{Z}_3^*/\mathbb{Z}_3^{*3}$, par exemple a et b, ce qui donne le cas (2), ou bien, à permutation près, (a,b,c)=(1,2,4) dans $\mathbb{Z}_3^*/\mathbb{Z}_3^{*3}$, ce qui donne le cas (3).

Type (III). Le cas (a,b,c)=(1,2,4) ou l'un des permutés, dans $\mathbb{Z}_3^*/\mathbb{Z}_3^{*3}$, étant exclu a priori, deux des coefficients a,b,c coïncident dans $\mathbb{Z}_3^*/\mathbb{Z}_3^{*3}$, ce qui donne le cas (4).

Type (IV). Deux des nombres a',b',c',d' coïncident nécessairement dans $\mathbb{Z}_3^*/\mathbb{Z}_3^{*3}$. Si ce sont a' et b', on obtient le cas (5). Si ce sont c' et d', on obtient le cas (6). Sinon, on peut supposer, à permutation permise près, que ce sont a' et c'. On est ainsi ramené aux coefficients $(1,2^i,3,3\cdot 2^j)$ avec i et $j \neq 0$. Mais la condition (93) impose alors $i \pm j \neq 0$, ce qui est impossible.

Type (V). On a simplement distingué le cas h=0 du cas $h \neq 0$ pour la suite de la discussion, ce qui donne les cas (7) et (8).

Les conditions indiquées par ailleurs sur i, j et éventuellement h, traduisent simplement la condition (93).

La démonstration de la proposition se fait cas par cas, en commençant par les cas où $\lambda=1$. Soient $F=\mathbb{Q}_3(\theta)$ et $F'=F(\sqrt[3]{\nu})$. Il s'agit d'établir l'assertion (a), à savoir que l'application (89): $V(F) \longrightarrow \mathbb{Z}/3$ est surjective. Or, si $\lambda=1$, cette application n'est autre, d'après le lemme 6 et la formule (82), que l'application composée:

$$V(F) \xrightarrow{f/\eta} F^*/NF'^* \xrightarrow{\sim} \mathbb{Z}/3$$
$$u \longrightarrow [u,\nu]_3 .$$

Pour établir que cette application est surjective, on peut oublier la constante η et il suffit même de voir que l'application composée $\tilde{f}$:

$$(92) \qquad V(F) \xrightarrow{f} F^*/NF'^* \simeq \mathbb{Z}/3$$

n'est pas constante (on peut appliquer le même argument qu'en (b) pour une courbe elliptique auxiliaire E section plane de V par deux points en lesquels f ne prend pas la même valeur). D'après le lemme 7, si x et $y \in \mathbb{Z}$ sont deux entiers non tous deux nuls:

$$\tilde{f}(x,y) := [f(x,y),\nu]_3 = -[x + \theta y,\nu]_3 .$$

Le formulaire **F3** du §4 permet alors de calculer les valeurs suivantes de $\tilde{f}(x,y)$, valeurs qui seront utilisées ci-après dans l'étude cas par cas:

$(x,y) \longrightarrow$	(1,0)	(1,–1)	(2,1)	(1,3)	(1,–4)	(1,5)	
$\nu = 2 \longrightarrow$	0	1	2	0	2	2	$\longleftarrow \tilde{f}(x,y)$
$\nu = 3 \longrightarrow$	0	0	0	1	1	2	$\longleftarrow \tilde{f}(x,y)$

Rappelons que:

$$[\theta,3]_3 = 0 \quad \text{et} \quad [\theta,2]_3 = 1 .$$

Voici simplement deux exemples de calcul de $\tilde{f}(x,y)$. Avec les notations du formulaire **F3**, en particulier $\lambda = 1 - \theta$, on a:

$$1 + 3\theta = 1 - \lambda^2 \quad \text{et} \quad 1 - 4\theta = \lambda + \lambda^2 ,$$

d'où:

$$\tilde{f}(1,3) = -[1 - \lambda^2,\nu]_3 = -[\theta,\nu]_3 - [1 + \lambda,\nu]_3 ,$$

soit:

pour $\nu = 2$, $\qquad -1 - [1 + \lambda, 1 + \lambda^2 + \lambda^3]_3 = -1 + 1 = 0$

pour $\nu = 3$, $\qquad -[1 + \lambda,3]_3 = 1$

et:

$$\tilde{f}(1,-4) = -[\lambda,\nu]_3 - [1+\lambda,\nu]_3$$

soit:

pour $\nu = 2$, $\quad -[\lambda, 1+\lambda^2+\lambda^3]_3 - [1+\lambda, 1+\lambda^2+\lambda^3]_3 = 1+1 = 2$
pour $\nu = 3$, $\quad -[\lambda,3]_3 - [1+\lambda,3]_3 = 0-2 = 1$

Le cas (1)

$$x^3+y^3+z^3+2^it^3 = 0 \qquad \nu = 2^i.$$

On trouve deux valeurs différentes:

$$\tilde{f}(1,-1,0,0) = i \neq 0 \quad \text{et} \quad \tilde{f}(1,0,-1,0) = 0.$$

Le cas (2)

$$x^3+y^3+2^iz^3+3\cdot 2^jt^3 = 0 \qquad \nu = 3\cdot 2^{j-i}.$$

On vérifie sans peine qu'étant donné un entier $u \equiv \pm 1 \bmod 9$, et i et j étant fixés, l'équation $2^iz^3+2^jt^3=u$ a toujours une solution (z_0,t_0) dans $\mathbb{Z}_3$ (prendre t_0 égal à 0, 1 ou -1 suivant les cas). Il existe donc des points $(1,0,z_0,t_0)$ et $(1,3,z_1,t_1)$ dans $V(\mathbb{Q}_3)$, et on trouve ainsi deux valeurs distinctes:

$$\tilde{f}(1,0,-,-) = 0 \quad \text{et} \quad \tilde{f}(1,3,-,-) = 1.$$

Le cas (4)

$$x^3+y^3+2^iz^3+9\cdot 2^jt^3 = 0 \qquad \nu = 9\cdot 2^{j-i}.$$

Etant donné $u \in \mathbb{Z}_3^*$ et i et j étant fixés, l'équation $3\cdot 2^iz^3+2^jt^3=u$ a toujours une solution (z_0,t_0) dans $\mathbb{Z}_3$; il suffit d'ailleurs de considérer le cas $i=j=0$. On en déduit qu'il en est de même pour l'équation $9u+2^iz^3+9\cdot 2^jt^3=0$. Il existe donc des points $(2,1,z_0,t_0)$ et $(1,5,z_1,t_1)$ dans $V(\mathbb{Q}_3)$, et on trouve ainsi deux valeurs distinctes:

$$\tilde{f}(2,1,-,-) = i-j \quad \text{et} \quad \tilde{f}(1,5,-,-) = 1+i-j.$$

Le cas (5)

$$x^3+y^3+3\cdot 2^iz^3+3\cdot 2^jt^3 = 0 \qquad \nu = 2^{j-i}.$$

Dans ce cas, $i \neq j$, et pour $u \in \mathbb{Z}_3^*$ donné, l'équation $2^iz^3+2^jt^3=3u$ a toujours une solution (z_0,t_0) dans $\mathbb{Z}_3$. Il existe donc un point $(2,1,z_0,t_0)$ dans $V(\mathbb{Q}_3)$ et on obtient ainsi deux valeurs distinctes:

$$\tilde{f}(1,-1,0,0) = j - i \quad \text{et} \quad \tilde{f}(2,1,-,-) = i - j$$

puisque $i \neq j$.

Le cas (6)

$$x^3 + y^3 + 9\cdot 2^i z^3 + 9\cdot 2^j t^3 = 0 \qquad \nu = 2^{j-i}.$$

On a encore $i \neq j$ et l'équation $2^i z^3 + 2^j t^3 = u$ qu'on est conduit à considérer a une solution (z_0,t_0) pour $u \equiv \pm 1 \bmod 9$ si i ou j est nul, sinon elle a une solution (z_1,t_1) par exemple pour $u \equiv \pm 2 \bmod 9$. Suivant les valeurs de i et j, il existe donc un point $(2,1,z_0,t_0)$ ou un point $(1,-4,z_1,t_1)$ dans $V(\mathbb{Q}_3)$. Comme $i \neq j$, on trouve ainsi toujours deux valeurs distinctes:

$$\tilde{f}(1,-1,0,0) = j - i \quad \text{et} \quad \tilde{f}(2,1,-,-) = \tilde{f}(1,-4,-,-) = i - j\,.$$

Le cas (7)

$$x^3 + y^3 + 3\cdot 2^i z^3 + 9\cdot 2^j t^3 = 0 \qquad \nu = 3\cdot 2^{j-i}.$$

On cherche une solution avec $z=0$. L'équation $2^j t^3 = u$ a une solution pour $u \equiv \pm 1$ mod 9 si $j=0$, pour $u \equiv \pm 2$ mod 9 si $j=1$, et pour $u \equiv \pm 4$ mod 9 si $j=2$. On en déduit l'existence dans $V(\mathbb{Q}_3)$ de points $(2,1,0,t_0)$ si $j=0$, puis $(1,-4,0,t_1)$ si $j=1$, enfin $(1,5,0,t_2)$ si $j=2$. On trouve ainsi les valeurs suivantes:

$$\begin{array}{ll} & \tilde{f}(1,-1,0,0) = j - i \\ \text{si } j = 0, & \tilde{f}(2,1,0,-) \;\; = i - j = i \\ \text{si } j = 1, & \tilde{f}(1,-4,0,-) = 1 + i - j = i \\ \text{si } j = 2, & \tilde{f}(1,5,0,-) \;\; = 2 + i - j = i, \end{array}$$

ce qui donne dans tous les cas les valeurs $j-i$ et i, distinctes puisque $i+j \neq 0$.

<u>Il reste</u> à traiter <u>les cas où</u> λ <u>n'est pas un cube</u>, à savoir les cas (3) et (8). L'extension K/k reste alors de rang 9 en 3, ce qui rend pénibles les calculs d'invariants locaux $\mathrm{inv}_3 \mathcal{A}(P)$, soit que l'on procède directement pour établir l'assertion (a) selon la méthode indiquée en **G** à la fin du §4, soit que l'on applique la méthode (b) en considérant une courbe elliptique E convenable tracée sur V_{k_3} et en passant à l'extension $k_3(\alpha)$. Aussi allons-nous utiliser une méthode différente et indirecte pour éviter des calculs trop fastidieux. Si $P \in V(k)$ et si v est une place de k, l'algèbre $\mathcal{A}$ ayant été fixée au préalable, on note

(94) $\qquad i_v(P) := \mathrm{inv}_v \mathcal{A}(P)$

et, si p est un nombre premier, on pose:

$$
(95) \qquad \begin{aligned} i_p(P) &= i_v(P) \\ i_p(P) &= i_{v_1}(P) + i_{v_2}(P) \end{aligned}
$$

selon que la valuation p-adique de $\mathbb{Q}$ a un seul prolongement v de $\mathbb{Q}$ à k, ou deux prolongements v_1 et v_2. On a évidemment la relation:

$$
(96) \qquad i(P) := \sum_p i_p(P) = 0 .
$$

La méthode que nous allons utiliser consiste à trouver, pour chaque sorte d'équation sur $\mathbb{Q}_3$ un <u>modèle</u> convenable V sur $\mathbb{Q}$, de telle sorte que l'application

$$
(97) \qquad i_3 : V(\mathbb{Q}) \longrightarrow \mathbb{Z}/3
$$

<u>ne soit pas constante</u>, et à calculer $i_3(P)$ pour $P \in V(\mathbb{Q})$ en commençant par le calcul des $i_p(P)$ pour $p \neq 3$ de mauvaise réduction et en utilisant ensuite la relation (96).

<u>Le cas</u> (3)

$$
x^3 + 2y^3 + 4z^3 + 3 \cdot 2^j t^3 = 0 .
$$

De fait, ces diverses équations sont équivalentes, et nous prenons comme modèle sur $\mathbb{Q}$ la variété V d'équation

$$
x^3 + 4y^3 + 2z^3 + 15t^3 = 0
$$

qui correspond directement au cas j=2. A des cubes près, on trouve:

$$
\begin{aligned} \lambda &= 4 & \nu &= 15 \\ \varepsilon &= \alpha^2 & \eta &= 1 . \end{aligned}
$$

Les premiers de mauvaise réduction sont 2,3,5. Comme ν est un cube en 2, le groupe de décomposition de K/k en 2 est engendré par t. En 5, il est engendré par s car λ y est un cube. Les formules (80) et (82) du §4 donnent alors:

$$
\begin{aligned} i_2(P) &= 0 \\ i_5(P) &= [f(P), \nu]_5 . \end{aligned}
$$

Soient $P_0=(1,0,-2,1)$ et $P_1=(3,1,-2,-1) \in V(\mathbb{Q})$. En utilisant les règles de calcul du symbole de restes normiques, en particulier les formules (66) et (67), et le lemme 7 pour $f(x,y)=(x+\theta\alpha y)/(x+\theta^2\alpha y)$, où l'on prend $\alpha \in \mathbb{Q}_p^*$, ici pour p=5, on trouve, en utilisant les congruences $\alpha = \sqrt[3]{4} \equiv -1 \bmod 5$ et $(2+\theta)^8 \equiv \theta^2 \bmod 5$:

$$i_5(P_0) = [1,\nu]_5 = 0$$
$$i_5(P_1) = -[3+\theta\alpha,\nu]_5 = -[-2-\theta,5]_5 = 2 .$$

Finalement:

$$i_3(P_0) = 0 \neq i_3(P_1) = 1 ,$$

et l'application i_3 n'est donc pas constante, ce qui prouve la proposition dans ce cas.
Le dernier cas:

$$x^3 + 2^h y^3 + 3\cdot 2^i z^3 + 9\cdot 2^j t^3 = 0 \qquad h \neq 0 \text{ et } i+j \neq h$$

se réduit à 6 équations inéquivalentes, par exemple celles où $h=1$.

Le cas $(8)_1$

$$x^3 + 2y^3 + 3z^3 + 9t^3 = 0 .$$

Prenons comme modèle sur $\mathbb{Q}$ la variété V d'équation

$$x^3 + 3y^3 + 9z^3 + 2t^3 = 0 .$$

A des cubes près: $\lambda = 3$ $\quad \nu = 2$
et on peut prendre: $\varepsilon = \alpha^2$ $\quad \eta = 1 .$

Les premiers de mauvaise réduction sont 2 et 3. En 2 où λ est un cube le groupe de décomposition de K/k est engendré par s et

$$i_2(P) = [f(P),\nu]_2 .$$

Soient $P_0=(1,-1,0,1)$ et $P_1=(2,1,-1,-1)\in V(\mathbb{Q})$. Comme $\alpha \equiv 1 \bmod 2$, on trouve:

$$i_2(P_0) = -[1-\theta,2]_2 = -[1+\theta,2]_2 = [\theta,2]_2 = 2$$
$$i_2(P_1) = -[2-\theta,2]_2 = -[\theta,2]_2 = 1 .$$

Finalement:

$$i_3(P_0) = 1 \neq i_3(P_1) = 2$$

et l'application i_3 n'est donc pas constante.

<u>*Le cas*</u> $(8)_2$

$$x^3 + 2y^3 + 3z^3 + 36t^3 = 0 .$$

Prenons comme modèle sur $\mathbb{Q}$ la variété V d'équation

$$x^3 + 2y^3 + 30z^3 + 36t^3 = 0 .$$

On a: $\lambda = 2$ $\quad \eta = \frac{3}{5}$

et on peut prendre: $\varepsilon = 5\beta(1-\beta)$ $\quad \eta = 1$.

Les premiers de mauvaise réduction sont 2,3,5. En 2, le groupe de décomposition de K/k est engendré par t car ν y est un cube, et en 5 il est engendré par s, d'où:

$$i_2(P) = 0 \qquad i_5(P) = [f(P),\nu]_5 .$$

Soient $P_0=(2,-1,1,-1)$ et $P_1=(4,1,-1,-1) \in V(\mathbb{Q})$. Comme $\alpha = \sqrt[3]{2} \equiv 3 \bmod 5$ et que $2-3\theta \equiv 2(1+\theta) \equiv -2\theta^2 \bmod 5$ et $(1+2\theta)^8 \equiv 1 \bmod 5$, on trouve, en utilisant en particulier les formules (66), (67), et (77):

$$i_5(P_0) = -[2-3\theta,\tfrac{3}{5}]_5 = [-2\theta^2,5]_5 = -[\theta,5]_5 = 2$$

$$i_5(P_1) = -[4+3\theta,\tfrac{3}{5}]_5 = [1+2\theta,5]_5 = 0 .$$

Ainsi:

$$i_3(P_0) = 1 \neq i_3(P_1) = 0 .$$

<u>*Le cas*</u> $(8)_3$

$$x^3 + 2y^3 + 6z^3 + 18t^3 = 0 .$$

Prenons comme modèle sur $\mathbb{Q}$ la variété V d'équation

$$x^3 + 3y^3 + 9z^3 + 4t^3 = 0 .$$

A des cubes près: $\lambda = 3$ $\quad \nu = 4$

et on peut prendre: $\varepsilon = \alpha^2$ $\quad \eta = 1$.

En dehors de 3, le seul premier de mauvaise réduction est 2, où le groupe de décomposition est engendré par s:

$$i_2(P) = [f(P),\nu]_2 .$$

Soient $P_0=(1,1,0,-1)$ et $P_1=(2,-1,-1,1)\in V(\mathbb{Q})$. Comme $\alpha\equiv 1 \bmod 2$, on a:

$$i_2(P_0) = -[1+\theta,4]_2 = [\theta^2,2]_2 = 1$$
$$i_2(P_1) = -[2-\theta,4]_2 = [\theta,2]_2 = 2\,.$$

Ainsi, i_2 n'est pas constante, et i_3 ne l'est donc pas non plus.

Le cas $(8)_4$

$$x^3+2y^3+6z^3+36t^3=0\,.$$

Choisissons comme modèle sur $\mathbb{Q}$ la variété V d'équation

$$x^3+3y^3+18z^3+5t^3=0\,.$$

A des cubes près: $\lambda = 3$ $\qquad \nu = \dfrac{5}{2}$

et on peut prendre: $\varepsilon = \dfrac{\alpha\delta}{\delta-1}$ $\qquad \eta = 1\,.$

En dehors de 3, les premiers de mauvaise réduction sont 2 et 5, et les groupes de décomposition de K/k y sont engendrés par s, d'où:

$$i_2(P) = [f(P),\nu]_2 \qquad i_5 = [f(P),\nu]_5\,.$$

Soient $P_0=(2,-1,0,-1)$ et $P_1=(1,-2,1,1)\in V(\mathbb{Q})$. On a $\alpha\equiv 1 \bmod 2$ et $\alpha\equiv 2 \bmod 5$. Comme en outre $(1-\theta)^8\equiv\theta \bmod 5$, on trouve:

$$i_2(P_0) = -[2-\theta,\tfrac{5}{2}]_2 = [\theta,2]_2 = 2$$
$$i_2(P_1) = -[1-2\theta,\tfrac{5}{2}]_2 = [1,2]_2 = 0$$
$$i_5(P_0) = -[2-2\theta,\tfrac{5}{2}]_5 = -[1-\theta,5]_5 = 1$$
$$i_5(P_1) = -[1+\theta,\tfrac{5}{2}]_5 = -[\theta^2,5]_5 = [\theta,5]_5 = 1\,.$$

D'où:

$$i_3(P_0) = 0 \neq i_3(P_1) = 2\,.$$

Le cas $(8)_5$

$$x^3+2y^3+12z^3+9t^3=0\,.$$

Prenons comme modèle sur $\mathbb{Q}$ la variété V d'équation

$$x^3 + 9y^3 + 12z^3 + 20t^3 = 0\,.$$

A des cubes près: $\lambda = 9$ $\quad \nu = 5$

et on peut prendre: $\varepsilon = \dfrac{6}{3-\beta}$ $\quad \eta = 1\,.$

Hors 3, les premiers de mauvaise réduction sont 2 et 5. En 2, le groupe de décomposition de K/k est engendré par t, et en 5 par s, d'où:

$$i_2(P) = 0 \qquad i_5(P) = [f(P),\nu]_5\,.$$

Soient $P_0=(2,0,1,-1)$ et $P_1=(1,-1,-1,1)\in V(\mathbb{Q})$. Comme $\alpha\equiv-1$ mod 5, on trouve:

$$i_5(P_0) = -[2,5]_5 = 0 \quad \text{et} \quad i_5(P_1) = -[1+\theta,5]_5 = [\theta,5]_5 = 1\,.$$

L'application i_5 n'est donc pas constante et $i_3=-i_5$ non plus.

Le cas $(8)_6$

$$x^3 + 2y^3 + 12z^3 + 18t^3 = 0\,.$$

Prenons comme modèle sur $\mathbb{Q}$ la variété V d'équation

$$x^3 + 9y^3 + 60z^3 + 4t^3 = 0\,.$$

A des cubes près: $\lambda = 9$ $\quad \nu = \dfrac{1}{5}$

et on peut prendre: $\varepsilon = \dfrac{3(1-\beta)}{\beta\gamma}$ $\quad \eta = 1\,.$

Le mauvaises places sont 2,3,5. Comme ν est un cube en 2 et λ en 5, on a:

$$i_2(P) = 0 \qquad i_5(P) = [f(P),\nu]_5\,.$$

Soient alors $P_0=(4,0,-1,-1)$ et $P_1=(2,-2,1,1)\in V(\mathbb{Q})$. Comme $\alpha\equiv-1$ mod 5, on a:

$$i_5(P_0) = -[4,\tfrac{1}{5}]_5 = 0 \quad \text{et} \quad i_5(P_1) = -[2+2\theta,\tfrac{1}{5}]_5 = [\theta^2,5]_5 = 2\,.$$

L'application i_5 n'est donc pas constante et $i_3=-i_5$ non plus.

§6. Le second cas.

C'est le cas où, pour tout premier p de mauvaise réduction pour V, la variété V est $\mathbb{Q}_p$-rationnelle. Autrement dit, sous l'hypothèse $V(\mathbb{Q}_p)\neq\emptyset$ pour tout premier p, on a, selon l'équation (1) ou (1') considérée - où a,b,c,d,λ,μ,ν sont supposés sans cube - la propriété (98) ou (98') suivante:

(98) pour tout diviseur premier p de 3abcd, l'un des ab/cd est un cube dans $\mathbb{Q}_p$;

(98') pour tout $p|3\lambda\mu\nu$, l'un des nombres ν, λ/ν ou μ/ν est un cube dans $\mathbb{Q}_p$.

L'objet de ce § est de montrer comment la méthode de calcul de l'obstruction exposée au §4 se simplifie dans ce second cas.

Proposition 3. *Soit* V *une surface cubique diagonale sur* $\mathbb{Q}$, *d'équation homogène:*

(1') $x^3 + \lambda y^3 + \mu z^3 + \lambda\mu\nu t^3 = 0$ $\lambda, \mu, \nu \in \mathbb{Q}^*$.

Soient $k=\mathbb{Q}(\theta)$ *et* $K=k(\sqrt[3]{\lambda}, \sqrt[3]{\nu})$. *Si* $p\neq 3$ *est un premier de mauvaise réduction pour* V *et si* V *est* $\mathbb{Q}_p$*-rationnelle, l'extension locale* K_w/k_v *en* p *est de degré* 3.

L'équation $x^3+6y^3+2z^3+3t^3=0$ montre que la conclusion est fausse pour p=3.

Corollaire. *Soient* V, k *et* K *comme au début de la proposition. Si on est dans le* second cas *pour* V, *l'extension locale* K_w/k_v *est cyclique pour toute place* $v\neq v_3$ *de* k.

En effet, pour v de bonne réduction, l'extension K_w/k_v est non ramifiée, donc cyclique.

Démontrons la proposition. Soit $v=v_p$. Supposons K_w/k_v de degré 9. Ni ν, ni λ/ν n'est donc un cube dans $\mathbb{Q}_p$. Puisque, par hypothèse, la propriété (98') est satisfaite, μ/ν est nécessairement un cube dans $\mathbb{Q}_p$. Quitte à changer de variables, on peut donc écrire sur $\mathbb{Q}_p$ l'équation de V sous la forme:

$$x^3 + \lambda y^3 + \nu z^3 + \lambda\nu^2 t^3 = 0 .$$

Quitte à changer encore de variables sur $\mathbb{Q}_p$, on peut supposer $v_p(\lambda)$ et $v_p(\nu)=0$, 1 ou -1. L'un des quatre nombres $\lambda,\nu,\lambda\nu$ ou λ/ν est alors nécessairement une unité p-adique. C'est donc un cube si $p\equiv 2 \bmod 3$, ce qui contredit dans ce cas l'hypothèse sur le degré de K_w/k_v. Si $p\equiv 1 \bmod 3$, on est dans l'un des cas suivants:

(i) $v_p(\lambda)=\pm 1$ et $v_p(\nu)=0$; une norme non nulle de l'extension non ramifiée $k_v(\sqrt[3]{\nu})/k_v$ est de valuation multiple de 3 et λ est de valuation ± 1; comme $V(\mathbb{Q}_p)\neq\emptyset$, l'équation $x^3+\nu z^3=-\lambda(y^3+\nu^2t^3)$ impose $x^3+\nu z^3=y^3+\nu^2t^3=0$, et ν doit donc être un cube dans $\mathbb{Q}_p$, ce qui contredit l'hypothèse sur K_w/k_v;

(ii) $v_p(\lambda)=0$ et $v_p(\nu)=\pm 1$; l'extension $k_v(\sqrt[3]{\lambda})/k_v$ étant non ramifiée, $x^3+\lambda y^3$ est nul ou de valuation multiple de 3, et, s'ils sont non nuls, νz^3 et $\lambda\nu^2t^3$ sont de valuation congrue à 1 et 2 mod 3, ou à -1 et -2; on en déduit, comme $V(\mathbb{Q}_p)\neq\emptyset$, que l'équation $x^3+\lambda y^3=-\nu z^3-\lambda\nu^2t^3$ impose $x^3+\lambda y^3=z=t=0$, et λ doit donc être un cube dans $\mathbb{Q}_p$, ce qui contredit l'hypothèse sur K_w/k_v;

(iii) $v_p(\lambda)=v_p(\nu)=\pm 1$; l'extension $k_v(\sqrt[3]{\lambda/\nu})/k_v$ est non ramifiée et, quitte à changer de variables, l'équation s'écrit: $x^3+\lambda\nu^2t^3=-\nu(z^3+\lambda\nu^2y^3)$; par un argument semblable à celui utilisé en (i), on en déduit que $(x,y,z,t)\in V(\mathbb{Q}_p)$ implique: $x^3+\lambda\nu^2t^3=z^3+\lambda\nu^2y^3=0$; l'hypothèse $V(\mathbb{Q}_p)\neq\emptyset$ impose donc que λ/ν soit un cube dans $\mathbb{Q}_p$, ce qui contredit l'hypothèse sur K_w/k_v;

(iv) $v_p(\lambda)=-v_p(\nu)=\pm 1$; l'extension $k_v(\sqrt[3]{\lambda\nu})/k_v$ est non ramifiée et, quitte à changer de variables, l'équation s'écrit: $z^3+\lambda\nu t^3=-\lambda\nu^2y^3-\nu^2x^3$; par un argument semblable à celui utilisé en (ii), on en déduit que $(x,y,z,t)\in V(\mathbb{Q}_p)$ implique: $z^3+\lambda\nu t^3=y=x=0$; l'hypothèse $V(\mathbb{Q}_p)\neq\emptyset$ impose donc que $\lambda\nu$ soit un cube dans $\mathbb{Q}_p$, ce qui contredit encore l'hypothèse sur K_w/k_v, et achève la démonstration.

Proposition 4. *Soit* V *une surface cubique diagonale sur* $\mathbb{Q}$, *d'équation homogène:*

(1') $$x^3+\lambda y^3+\mu z^3+\lambda\mu\nu t^3=0 \qquad \lambda, \mu, \nu\in\mathbb{Q}^*.$$

Soient $\beta=\sqrt[3]{\lambda\nu}$ *et* $\delta=\sqrt[3]{\lambda/\nu}$. *Si on est dans le* second cas *pour* V, *alors:*

$$\mu\in N_{\mathbb{Q}(\beta)/\mathbb{Q}}\mathbb{Q}(\beta)^* \quad et \quad \mu\in N_{\mathbb{Q}(\delta)/\mathbb{Q}}\mathbb{Q}(\delta)^*.$$

Reprenons les notations du diagramme (13) du §1: soient $K_2=k(\beta)$ et $L_2=k(\delta)$. Un argument élémentaire montre qu'il suffit d'établir

$$\mu \in N_{K_2/k}K_2^* \quad \text{et} \quad \mu \in N_{L_2/k}L_2^* .$$

Puisque le principe de Hasse vaut pour la norme d'une extension cyclique, il est équivalent d'établir que, pour toute place v de k, le nombre μ est une norme de l'extension $K_{2w'}/k_v$ et aussi de l'extension $L_{2w''}/k_v$. Il suffit même de le vérifier pour $v \neq v_3$ en raison de la loi de réciprocité. Par hypothèse, la propriété (98') est satisfaite. On est donc nécessairement dans l'un des cas suivants:

(i) v est une place de bonne réduction pour V; les extensions $K_{2w'}/k_v$ et $L_{2w''}/k_v$ sont alors non ramifiées et μ est une unité v-adique: c'est donc une norme de chacune de ces deux extensions;

(ii) ν est un cube dans k_v^*; on peut alors écrire $K_{2w'}=L_{2w''}=k_v(\sqrt[3]{\lambda})$ et, quitte à changer de variables, l'équation devient: $x^3+\lambda y^3=-\mu(z^3+\lambda t^3)$; si λ n'est pas un cube dans k_v, cette équation assure que μ est une norme de l'extension $k_v(\sqrt[3]{\lambda})/k_v$ puisque $V(k_v)\neq\emptyset$; le cas où λ est un cube dans k_v est évidemment trivial;

(iii) λ/ν est un cube dans k_v^*; on peut alors écrire $K_{2w'}=k_v(\sqrt[3]{\nu})$ et $L_{2w''}=k_v$; le résultat est donc immédiat à propos de l'extension $L_{2w''}/k_v$; quitte à changer de variables, l'équation s'écrit: $x^3+\nu y^3=-\mu(z^3+\nu^2 t^3)$; si ν n'est pas un cube dans k_v, cette équation assure que μ est une norme de l'extension $k_v(\sqrt[3]{\nu})/k_v$ puisque $V(k_v)\neq\emptyset$; le cas où ν est un cube dans k_v a déjà été considéré en (ii), et en fait ici c'est un cas trivial;

(iv) μ/ν est un cube dans k_v, mais ν n'est pas un cube dans k_v et $v\neq v_3$; d'après le corollaire de la proposition 3, on a donc $K_w=k_v(\sqrt[3]{\nu})$, et on peut écrire $K_{2w'}=k_v$ ou $k_v(\sqrt[3]{\nu})$ et de même $L_{2w''}=k_v$ ou $k_v(\sqrt[3]{\nu})$; comme ν est évidemment une norme à la fois pour l'extension k_v/k_v et l'extension $k_v(\sqrt[3]{\nu})/k_v$, il en est de même de μ.

Le calcul de l'obstruction dans le second cas.

Nous indiquons les modifications et simplifications à apporter au calcul exposé au §4 en fonction des résultats ci-dessus.

Pas n°0: Par hypothèse, l'un des ab/cd est un cube dans $\mathbb{Q}_3$. A la fin du pas n°0 on ordonne donc, ce qui est possible, les coefficients a,b,c,d de telle sorte que

(99) ad/bc soit un cube dans $\mathbb{Q}_3$.

Pas n°1: On cherche $\varepsilon \in K^*$ sous la forme

(61) $\varepsilon = \varepsilon_\beta \varepsilon_\delta \qquad \varepsilon_\beta \in \mathbb{Q}(\beta)^* \quad \text{et} \quad \varepsilon_\delta \in \mathbb{Q}(\delta)^*$,

ce qui est possible d'après la proposition 4 qui affirme qu'on peut même trouver ε sous la forme $\varepsilon = \varepsilon_\beta$, ou encore sous la forme $\varepsilon = \varepsilon_\delta$.

Pas n°2: On prend

$$\eta = 1$$

comme il est possible d'après le choix de ε, voir le pas n°2 au §4.

Pas n°3: Pour chaque place v de k,

$$V(k_v)/Br = \{P_v\}$$

où $P_v \in V(k_v)$ est un point arbitraire, comme l'assure le lemme 5 lorsqu'on est dans le second cas. On pose donc, pour chaque place v de k:

$$i_v = inv_v \mathcal{A}(P_v)$$

qui ne dépend pas du point P_v choisi, puis, pour tout premier p:

(100) $i_p = i_v \quad \text{ou} \quad i_p = i_{v_1} + i_{v_2}$,

suivant que la valuation p-adique de $\mathbb{Q}$ a un seul prolongement $v=v_p$ à k, ou deux prolongements v_1 et v_2, i.e. suivant que $p \equiv 2$ ou 3 mod 3, ou que $p \equiv 1$ mod 3. On pose enfin:

(101) $i = \sum_p i_p$.

Pas n°4: Comme on est dans le second cas, la proposition 3 s'applique et, pour toute place v de k, même v_3 en raison de l'hypothèse (99), le groupe de décomposition $G^v=Gal(K_w/k_v)$ est cyclique. Faisons alors l'inventaire des divers cas. Dans ce qui suit, p désigne un nombre premier et v un prolongement à k de la valuation p-adique de $\mathbb{Q}$.

e t *Le cas* $G^v=\langle e\rangle$ *ou* $G^v=\langle t\rangle$. Ceci équivaut à la condition

ν est un cube dans $\mathbb{Q}_p$,

et en ce cas:

(102) $i_p = 0$.

En particulier, la condition (99) implique:

$i_3 = 0$.

On suppose *désormais*, pour l'analyse des autres cas, que ν n'est pas un cube dans $\mathbb{Q}_p$. On peut aussi supposer $p\neq 3$.

s *Le cas* $G^v=\langle s\rangle$. Puisque ν n'est pas un cube dans $\mathbb{Q}_p$, ceci équivaut à:

λ est un cube dans $\mathbb{Q}_p$.

On trouve alors, d'après la formule (83):

(103) $i_v = [\theta,\nu]_v$.

r *Le cas* $G^v=\langle r\rangle$. Compte tenu de l'hypothèse sur ν, ceci équivaut à:

λ/ν est un cube dans $\mathbb{Q}_p$.

D'après le calcul (81), on trouve alors:

(104) $i_v = [{}^q\varepsilon/\varepsilon,\nu]_v = [{}^q\varepsilon_\delta/\varepsilon_\delta,\nu]_v$.

q *Le cas* $G^v=\langle q\rangle$. Vu l'hypothèse sur ν, ceci équivaut à la condition:

$\lambda\nu$ est un cube dans $\mathbb{Q}_p$.

D'après le calcul (84), on trouve alors:

(105) $\quad i_v = -[\theta\cdot{}^r\varepsilon/\varepsilon,\nu]_v = -[\theta\cdot{}^r\varepsilon_\beta/\varepsilon_\beta,\nu]_v$.

Lemme 12. *Avec les notations, hypothèses et choix ci-dessus, les expressions*

$$[\theta,\nu]_v\,,\quad [{}^q\varepsilon_\delta/\varepsilon_\delta,\nu]_v\,,\quad [\theta\cdot{}^r\varepsilon_\beta/\varepsilon_\beta,\nu]_v\,,$$

figurant dans les formules ci-dessus, sont indépendantes des plongements dans $\mathbb{Q}_p(\theta)$ *ou* $\mathbb{Q}_p$ *qui servent à les calculer. De façon précise, si* $p\equiv 2 \bmod 3$, *elles sont indépendantes des* k-plongements *de* $k(\delta)$ *et* $k(\beta)$ *dans* $\mathbb{Q}_p(\theta)$, *et si* $p\equiv 1 \bmod 3$ *des plongements de* k, $k(\delta)$ *et* $k(\beta)$ *dans* $\mathbb{Q}_p$. *En particulier, si* $p\equiv 1 \bmod 3$:

(106) $\quad i_p = -i_{v_1} = -i_{v_2}$.

Ce résultat est bien naturel pour ce qui est des k-plongements de $k(\delta)$ et $k(\beta)$ dans k_v, puisque c'est alors une conséquence implicite des formules (104) et (105). On va néanmoins traiter directement tous les cas.

Dans le cas **s**, l'énoncé n'a de contenu réel que pour $p\equiv 1 \bmod 3$. Il ne dépend d'ailleurs pas des hypothèses liées à ce cas et résulte aussitôt du lemme 7:

$$[\theta,\nu]_v = -[\theta^2,\nu]_{v_1} = [\theta,\nu]_{v_2}\,.$$

Dans le cas **r**, notons $^-$ l'automorphisme non trivial de $k(\delta)/\mathbb{Q}(\delta)$. Le groupe de Galois de $k(\delta)/\mathbb{Q}$ est $\simeq\mathfrak{S}_3$, il est engendré par q et $^-$, avec $^-\circ q = q^2\circ{}^-$. Suivant que $p\equiv 2$ ou $1 \bmod 3$, on considère les 3 k-plongements de $k(\delta)$ dans $\mathbb{Q}_p(\theta)$, ou les 6 plongements de $k(\delta)$ dans $\mathbb{Q}_p$. Fixons l'un de ces plongements ι. Soit v la valuation de k associée et soit encore $\iota: K \hookrightarrow k_v(\beta)=k_v(\sqrt[3]{\nu}\,)$ un prolongement du plongement précédent. Posons, par définition, pour $u\in k(\delta)$:

$$[u,\nu]_v := [u,\nu]_v^\iota = [\iota(u),\nu]_v \quad\text{et}\quad [u,\nu]_{\bar v} := [u,\nu]_{\bar v}^{\bar\iota} = [\iota(\bar u),\nu]_{\bar v}$$

où $\bar\iota = \iota\circ{}^-$. Les divers plongements considérés sont les $\iota\gamma$ où $\gamma\in\langle q\rangle$ ou $\langle q,{}^-\rangle$ selon les cas. Il s'agit donc de prouver les égalités:

(107) $\quad [\omega,\nu]_v = [{}^q\omega,\nu]_v = [{}^{q^2}\omega,\nu]_v = [\omega,\nu]_{\bar v} = [{}^q\omega,\nu]_{\bar v} = [{}^{q^2}\omega,\nu]_{\bar v}$,

pour $\omega = {}^q\varepsilon_\delta/\varepsilon_\delta$. Par hypothèse:

(108) $N_t(\varepsilon) = N_r(\varepsilon_\beta)\cdot N_q(\varepsilon_\delta) = -\mu$.

Comme $k_v(\beta)=k_v(\sqrt[3]{\nu})$, on a: $[N_r(\varepsilon_\beta),\nu]_v=[N_r(\iota(\varepsilon_\beta)),\nu]_v=0$. Après changement de variables, l'équation (1') s'écrit sur $\mathbb{Q}_p$: $x^3+\nu y^3=-\mu(z^3+\nu^2t^3)$. Ainsi: $[\mu,\nu]_v=0$ et, finalement:

$$[N_q(\varepsilon_\delta),\nu]_v=0.$$

Or: $\omega/{}^q\omega=({}^q\varepsilon_\delta)^3/N_q(\varepsilon_\delta)$. Comme $\iota({}^q\varepsilon_\delta)\in k_v$, on a: $[({}^q\varepsilon_\delta)^3,\nu]_v=0$. D'où la première égalité:

$$[\omega,\nu]_v = [{}^q\omega,\nu]_v .$$

La seconde égalité et les deux dernières s'obtiennent de même. Pour finir, notons que $\overline{\varepsilon_\delta}=\varepsilon_\delta$ puisque $\varepsilon_\delta\in\mathbb{Q}(\delta)$. Ainsi:

$$\overline{\omega} = \overline{{}^q\varepsilon_\delta/\varepsilon_\delta} = {}^{q^2}\overline{\varepsilon_\delta}/\overline{\varepsilon_\delta} = {}^{q^2}\varepsilon_\delta/\varepsilon_\delta = {}^{q^2}\omega^{-1} .$$

D'où finalement, en utilisant le lemme 7 et l'une des égalités déjà obtenues:

$$[\omega,\nu]_v = [-\overline{\omega},\nu]_{\overline{v}} = -[{}^{q^2}\omega^{-1},\nu]_{\overline{v}} = [{}^{q^2}\omega,\nu]_{\overline{v}} = [\omega,\nu]_{\overline{v}} .$$

Dans le cas **q**, on adopte les mêmes notations que dans le cas **r**, en remplaçant partout δ par β, et q par r. Comme l'invariance de $[\theta,\nu]_v$ a déjà été établie, il s'agit donc de prouver les égalités:

(107') $[\omega,\nu]_v = [{}^r\omega,\nu]_v = [{}^{r^2}\omega,\nu]_v = [\omega,\nu]_{\overline{v}} = [{}^r\omega,\nu]_{\overline{v}} = [{}^{r^2}\omega,\nu]_{\overline{v}}$,

où $\omega={}^r\varepsilon_\beta/\varepsilon_\beta$. Comme $k_v(\delta)=k_v(\sqrt[3]{\nu})$, on trouve: $[N_q(\varepsilon_\delta),\nu]_v=0$. D'autre part, étant à la fois dans le *second cas* et dans le cas **q**, ni ν, ni λ/ν n'est un cube dans $\mathbb{Q}_p$, et μ/ν est donc un cube dans $\mathbb{Q}_p$ d'après (98'). Ainsi, $[\mu,\nu]_v=0$, et:

$$[N_r(\varepsilon_\beta),\nu]_v = 0 .$$

La fin est identique au cas précédent, quitte à changer δ en β, et q en r.

<u>Remarque 5.</u> Il faut prendre garde au fait qu'une expression du type

$$[u,\nu]_v$$

où, suivant les cas, $u\in k^*$ ou $k(\beta)^*$ ou $k(\delta)^*$, et où v prolonge à k la valuation p-adique de $\mathbb{Q}$, *dépend* en général du choix du plongement de k, $k(\beta)$ ou $k(\delta)$, dans $\mathbb{Q}_p(\theta)$, ou $\mathbb{Q}_p$, selon que $p\equiv 2$ ou 3, ou $p\equiv 1 \bmod 3$. Par exemple, si $p\equiv 1 \bmod 3$, et si v_1 et v_2 sont les deux prolongements à k de la valuation p-adique de $\mathbb{Q}$, et θ_1, $\theta_2\in\mathbb{Q}_p$ les images de θ par les deux plongements $k \hookrightarrow \mathbb{Q}_p$ respectivement associés, si enfin u est une unité p-adique, la formule (64) donne précisément:

$$(109)\quad \begin{aligned} [u,p]_{v_1} &= -m_1 \qquad \text{tel que} \quad u^{(p-1)/3} \equiv \theta_1^{m_1} \bmod p \\ [u,p]_{v_2} &= -m_2 \qquad \text{tel que} \quad u^{(p-1)/3} \equiv \theta_2^{m_2} \bmod p\,. \end{aligned}$$

Si en outre $u\in\mathbb{Q}^*$, on a donc $[u,p]_{v_1}=-[u,p]_{v_2}$ comme il résulte d'ailleurs du lemme 7. Ainsi, si $\theta_1\equiv 2 \bmod 7$:

$$[2,7]_{7_1} = 1 \neq [2,7]_{7_2} = 2\,.$$

Voici un autre exemple. Soient $\delta=\sqrt[3]{3}$ et δ_1 son image par l'unique plongement de $\mathbb{Q}(\delta)$ dans $\mathbb{Q}_2$. Il y a en revanche trois k-plongements de $k(\delta)$ dans $\mathbb{Q}_2(\theta)$, et l'expression $[\delta,2]_2$ prend successivement, suivant le plongement, les valeurs suivantes:

$$[\delta_1,2]_2 = 0 \qquad [\theta\delta_1,2]_2 = 2 \qquad [\theta^2\delta_1,2]_2 = 1\,.$$

r q *Complément pour les cas* $G^v=\langle r\rangle$ *et* $G^v=\langle q\rangle$. Le lemme et la remarque ci-dessus permettent de préciser comment poursuivre le calcul de i_p dans les cas **r** et **q**. De fait, nous nous limitons ici au cas **r**, l'autre cas étant tout à fait analogue. Tout d'abord, si $p\equiv 1 \bmod 3$, on choisit l'une des racines θ_1 de l'équation $\theta^2+\theta+1=0$ dans $\mathbb{Q}_p$. On peut même fixer un tel choix une fois pour toutes pour chaque $p\equiv 1 \bmod 3$. Ce choix détermine l'une des valuations v, soit v_1, qui prolonge à k la valuation p-adique de $\mathbb{Q}$. Ensuite, quel que soit le reste de $p \bmod 3$, on choisit une racine cubique δ_0 de λ/ν dans $\mathbb{Q}_p$, ou $\mathbb{Q}_p(\theta)$, autrement dit, on choisit un plongement de $k(\delta)$ dans $\mathbb{Q}_p$, ou $\mathbb{Q}_p(\theta)$, expédiant θ sur θ_1, ou θ, ce qui fixe un sens pour l'expression $[{}^q\varepsilon_\delta/\varepsilon_\delta,\nu]_v$ et permet de calculer, en prenant garde de ne pas changer les choix de θ_1 et de δ_0 au cours de ce calcul. De façon plus précise, voici comment opérer si $\varepsilon=u+v\delta$ avec $u,v\in\mathbb{Z}$ et $(u,v)=1$. On supposera, ce qui est loisible, que λ/ν est une unité p-adique et que μ est sans cube.

Si $p\equiv 2$ ou 3 mod 3, on note δ_1 l'unique racine cubique de λ/ν dans $\mathbb{Q}_p$. Alors:

$$(110) \qquad [{}^q\varepsilon/\varepsilon,\nu]_p = [u + v\theta\delta_1,\nu]_p$$

car par exemple, par le corollaire du lemme 7 et par continuité: $[u+v\delta_1,\nu]_p=0$. On notera que, compte tenu des hypothèses, $u+v\theta\delta_1$ est une unité p-adique.

Si $p\equiv 1$ mod 3, soit δ_1 l'une des racines cubiques de λ/ν dans $\mathbb{Q}_p$. Alors:

$$(111) \qquad [{}^q\varepsilon/\varepsilon,\nu]_p = [u + v\delta_1,\nu]_{v_1} - [u + v\delta_2,\nu]_{v_1},$$

où $\delta_2=\theta_1\delta_1$. De plus, compte tenu des hypothèses, on peut choisir δ_1 de sorte que $u+v\delta_1$ et $u+v\delta_2$ soient des unités p-adiques, ce qui peut faciliter les calculs.

§7. Exemples

Nous indiquons ici sur des exemples simples comment s'appliquc la méthode de calcul de l'obstruction de Manin exposée aux §§4,5,6 sans nécessairement tenir compte des simplifications suggérées au §6. L'application de cette méthode à une variété V donnée suppose plusieurs choix: d'abord le choix de l'ordre des coefficients a,b,c,d, ensuite le choix de ε et celui de η, et le calcul de l'obstruction se révèle plus ou moins simple dans chaque cas en fonction de ces choix. Alors qu'ici nous traitons quelques exemples suivant diverses variantes, la procédure qui sera indiquée au §8 sera plus contraignante et limitera les choix à opérer.

Voici à titre indicatif le début de la liste d'exemples numériques établie par Cassels et Guy en 1966, et formée de contre-exemples potentiels au principe de Hasse:

$$\begin{aligned}
x^3 &+ 4y^3 + 10z^3 + 25t^3 = 0\\
5x^3 &+ 9y^3 + 10z^3 + 12t^3 = 0\\
3x^3 &+ 4y^3 + 10z^3 + 15t^3 = 0\\
3x^3 &+ 6y^3 + 10z^3 + 25t^3 = 0\\
5x^3 &+ 6y^3 + 12z^3 + 25t^3 = 0\\
&\ldots\ldots\ldots\ldots\ldots\ldots
\end{aligned}$$

Cassels et Guy [4] traitèrent effectivement le second exemple de cette liste en montrant que dans ce cas $V(\mathbb{Q})=\emptyset$, puis Bremner [3] traita de même le premier dix ans plus tard. Nous allons d'abord reprendre ces deux exemples-là de diverses manières.

Il est commode de faire au préalable quelques rappels et de fixer quelques notations et conventions. Pour $u,v \in \mathbb{Q}^*$, la notation

$$u \sim v$$

signifiera $\frac{u}{v} = w^3$ avec $w \in \mathbb{Q}$. Ayant fixé pour V une équation de la forme:

$$(1') \qquad x^3 + \lambda y^3 + \mu z^3 + \lambda\mu\nu t^3 = 0\,,$$

on cherchera $\varepsilon \in K^*$ tel que:

$$(44') \qquad N_t(\varepsilon) = \mu$$

au lieu de $N_t(\varepsilon) = -\mu$, ceci afin d'alléger l'écriture. Cette modification est tout à fait licite, puisqu'il eût suffi de changer g en $-g$ dans la définition (42) i.e. $g = -(x+\theta^2\alpha y)/(z+\theta\beta t)$, pour obtenir ensuite la condition (44'), sans rien changer ultérieurement ni pour η, ni pour $\mathcal{A}$. Voici enfin deux tableaux fort utiles pour les calculs. Le premier s'obtient à partir des formules (65), (67) et (71):

(112)

p mod 9	1	2	3	4	5	7	8
$[\theta,p]_p$	0	2	0	1	1	2	0

Le second dérive du tableau (14) compte tenu des définitions de q et de r:

(113)

$u \longrightarrow$	α	β	γ	δ
${}^s u/u$	1	θ	θ	θ^2
${}^t u/u$	θ	θ	1	θ
${}^q u/u$	θ^2	1	θ	θ
${}^r u/u$	θ	θ^2	θ	1

(a) *L'exemple de Cassels et Guy.*

$$(2) \qquad 5x^3 + 9y^3 + 10z^3 + 12t^3 = 0\,.$$

Ecrivons l'équation sous la forme:

$$x^3 + 15y^3 + 30z^3 + 36t^3 = 0\,.$$

On obtient:

$$\nu = 2/25 \sim 10 \qquad \lambda\nu = 6/5 \qquad \mu = 30$$
$$\lambda = 15 \qquad \lambda/\nu = 375/2 \sim 3/2$$

et on peut prendre:

$$\varepsilon = \beta/(\beta-1)^2 \qquad \eta = 1.$$

Comme $N(\beta)=6/5$ et que $N(\beta-1)=1/5$, les premiers de mauvaise réduction sont 2,3,5: dans ce cas $\mathcal{S}_{\mathfrak{A}}=\mathcal{S}$. On voit aussitôt que la condition (98') est vérifiée pour chacun de ces nombres premiers: ν, ou λ/ν, ou μ/ν est un cube dans $\mathbb{Q}_p$ pour $p\in\mathcal{S}$. On est donc dans le *second cas*. On a alors le tableau suivant:

en 2	λ	cube	$G^v = \langle s\rangle$	$i_2 = [\theta,\nu]_2 = [\theta,2]_2 = 2$
en 3	ν	cube	$G^v = \langle t\rangle$	$i_3 = 0$
en 5	λ/ν	cube	$G^v = \langle r\rangle$	$i_5 = [{}^q\varepsilon_\delta/\varepsilon_\delta,\nu]_5 = 0$ car $\varepsilon = \varepsilon_\beta$.

Finalement:

$$i = i_2 = 2 \neq 0\ ,$$

et l'obstruction de Manin est "non-vide" - ou "non nulle" - pour la surface V de Cassels et Guy, ce qui prouve $V(\mathbb{Q})=\emptyset$.

(b) *L'exemple de Bremner.*

(3) $$x^3 + 4y^3 + 10z^3 + 25t^3 = 0\ .$$

De fait, c'est le premier exemple - c'est le cas p=2 et q=5 - d'une série infinie (4) de contre-exemples au principe de Hasse parmi les surfaces cubiques diagonales:

Proposition 5. *Les surfaces cubiques* V *d'équation homogène*

(4) $$x^3 + p^2y^3 + pqz^3 + q^2t^3 = 0$$

où $p\equiv 2$ *et* $q\equiv 5 \bmod 9$ *sont deux entiers premiers, ont un point dans chaque complété* $\mathbb{Q}_\ell$*, mais n'en ont pas dans* $\mathbb{Q}$.

Ecrivons l'équation (4) sous la forme:

$$x^3 + py^3 + p^2qz^3 + p^4q^2t^3 = 0\ .$$

On obtient:

$$\nu = pq \qquad \lambda\nu = p^2q \qquad \mu = p^2q$$
$$\lambda = p \qquad \lambda/\nu = 1/q$$

et on peut prendre

$$\varepsilon = \beta \qquad \eta = 1 .$$

Il est immédiat que $\mathcal{S}_{\mathcal{A}}=\mathcal{S}=\{3,p,q\}$. Comme $pq\equiv 1 \bmod 9$, le nombre ν est un cube dans $\mathbb{Q}_3$. Comme $p\equiv 2 \bmod 3$, le nombre λ/ν, qui est une unité p-adique, est un cube dans $\mathbb{Q}_p$. Comme $q\equiv 2 \bmod 3$, le nombre $\mu/\nu=p$, qui est une unité q-adique, est un cube dans $\mathbb{Q}_q$. Les conditions (98') sont donc vérifiées et on est donc dans le *second cas*. En particulier, pour chaque nombre premier ℓ, on a $V(\mathbb{Q}_\ell)\neq\emptyset$. On a le tableau suivant:

en 3	ν	cube	$G^v = \langle t\rangle$	$i_3 = 0$
en p	λ	cube	$G^v = \langle s\rangle$	$i_p = [\theta,\nu]_p = [\theta,p]_p = 2$ car $p \equiv 2 \bmod 9$
en q	λ/ν	cube	$G^v = \langle r\rangle$	$i_q = [{}^q\varepsilon_\delta/\varepsilon_\delta,\nu]_q = 0$ car $\varepsilon = \varepsilon_\beta$.

Finalement:

$$i = i_p = 2 \neq 0$$

et l'obstruction de Manin est donc "non nulle". Par conséquent: $V(\mathbb{Q})=\emptyset$.

(c) *Variante pour l'exemple de Bremner généralisé.*

Il s'agit d'une variante de la démonstration de la proposition ci-dessus. On écrit l'équation (4) sous la forme:

$$x^3 + p^2y^3 + q^2z^3 + pqt^3 = 0 .$$

On obtient:

$$\nu = 1/pq \qquad \lambda\nu = p/q \qquad \mu = q^2$$
$$\lambda = p^2 \qquad \lambda/\nu = p^3q \sim q .$$

On peut prendre:

$$\varepsilon = \delta^2/p^2 \qquad \eta = 1 .$$

On a toujours $\mathcal{S}_{\mathcal{A}}=\mathcal{S}=\{3,p,q\}$ et on est évidemment toujours dans le *second cas*. On a le tableau suivant:

en 3	ν	cube	$G^v = \langle t \rangle$	$i_3 = 0$
en p	λ/ν	cube	$G^v = \langle r \rangle$	$i_p = [{}^q\varepsilon/\varepsilon,\nu] = [\theta,p]_p = 2$ car $p \equiv 2 \bmod 9$
en q	λ	cube	$G^v = \langle s \rangle$	$i_q = [\theta,\nu]_q = -[\theta,q]_q = 2$ car $q \equiv 5 \bmod 9$.

Finalement:

$$i = i_p + i_q = 1 \neq 0$$

et la conclusion est la même que précédemment. Du fait que l'ordre des coefficients de l'équation n'est pas le même que précédemment, il ne faut pas s'étonner qu'on trouve $i=1$ au lieu de 2 comme en (b), car on sait seulement que l'algèbre $\mathcal{A}$ engendre $\mathrm{Br}\, V/\mathrm{Br}\, k \simeq \mathbb{Z}/3$, groupe qui a 2 générateurs!

(d) *L'exemple de Cassels et Guy: variante avec* $i_3 \neq 0$.

Ecrivons l'équation (2) sous la forme:

$$x^3 + 30y^3 + 36z^3 + 15t^3 = 0\ .$$

Remplaçons même le dernier coefficient 15 par $15{\cdot}6^3 = 15{\cdot}216$ pour simplifier les calculs ultérieurs. On trouve alors:

$$\nu = 3 \qquad \lambda\nu = 90 \qquad \mu = 36$$
$$\lambda = 30 \qquad \lambda/\nu = 10\ .$$

On peut prendre:

$$\varepsilon = \gamma^2(\delta - 2)^2 \qquad \eta = 1\ .$$

Comme $N(\delta-2)=2$ et $N(\gamma)=3$, on a encore $\mathcal{S}_{\mathcal{A}}=\mathcal{S}=\{2,3,5\}$ et on est évidemment toujours dans le *second cas*. On a le tableau suivant:

en 2	ν	cube	$G^v = \langle t \rangle$	$i_2 = 0$
en 3	λ/ν	cube	$G^v = \langle r \rangle$	$i_3 = [{}^q\varepsilon/\varepsilon,\nu]_3 = [{}^q\varepsilon/\varepsilon,3]_3$
en 5	ν	cube	$G^v = \langle t \rangle$	$i_5 = 0$.

On a: ${}^q\varepsilon/\varepsilon = \theta^2{\cdot}[{}^q(2-\delta)^2/(2-\delta)^2]$. Comme $[\theta,3]_3=0$, voir par exemple (72), il vient:

$$i_3 = -[{}^q(2-\delta)/(2-\delta),3]_3 = -[2-\theta\delta_1,3]_3$$

où $\delta_1 = \sqrt[3]{10} \in \mathbb{Q}_3$, voir par exemple (110). Comme $2-\theta\delta_1$ est une unité, il suffit de connaître son reste mod 9, autrement dit mod λ^4, pour calculer $[2-\theta\delta_1,3]_3$. On a:

$$\begin{aligned} 2-\theta\delta_1 &\equiv 2(1-2\theta) \equiv (1+\lambda^2+\lambda^3)(1-2\lambda) \\ &\equiv 1-2\lambda+\lambda^2-\lambda^3 \equiv 1+\lambda+\lambda^2 \quad \mathrm{mod}\ \lambda^4 \end{aligned}$$

d'où, par application de la formule (71):

$$i = i_3 = -[1+\lambda+\lambda^2, 3]_3 = -1 \neq 0$$

et l'obstruction de Manin est donc "non nulle".

On peut traiter de manière tout à fait analogue une dizaine d'équations du début de la liste de Cassels et Guy, par exemple les trois figurant au début de ce § à la suite des deux premières, à condition de les écrire comme suit:

$$\begin{aligned} x^3 + 30y^3 + 6z^3 + 20t^3 &= 0 \\ x^3 + 30y^3 + 15z^3 + 50t^3 &= 0 \\ x^3 + 30y^3 + 60z^3 + 25t^3 &= 0\,. \end{aligned}$$

(e) *L'exemple de Cassels et Guy: variante avec* $\mathcal{S}_{\mathcal{A}} \neq \mathcal{S}$.

Ecrivons l'équation (2) sous la forme:

$$x^3 + 30y^3 + 15z^3 + 36t^3 = 0\,.$$

Remplaçons même le dernier coefficient 36 par $36\cdot 5^3 = 36\cdot 125$. On trouve alors:

$$\nu = 10 \qquad \lambda\nu = 300 \qquad \mu = 15$$
$$\lambda = 30 \qquad \lambda/\nu = 3\,.$$

Comme $N(\beta-6) = 84 = 4\cdot 3\cdot 7$ et que $N(10-\beta) = 700 = 4\cdot 25\cdot 7$, on peut prendre:

$$\varepsilon = 5(\beta-6)/(10-\beta) \qquad \eta = 1\,.$$

On a alors $\mathcal{S}_{\mathcal{A}} = \{2,3,5,7\} \neq \mathcal{S} = \{2,3,5\}$. On est évidemment toujours dans le *second cas*. On a le tableau suivant:

en 2	λ/ν cube	$G^\vee = \langle r\rangle$	$i_2 = 0$	car $\varepsilon = \varepsilon_\beta$
en 3	ν cube	$G^\vee = \langle t\rangle$	$i_3 = 0$	
en 5	λ/ν cube	$G^\vee = \langle r\rangle$	$i_5 = 0$	car $\varepsilon = \varepsilon_\beta$
en 7	$\lambda\nu$ cube	$G^\vee = \langle q\rangle$	$i_7 = [\theta\cdot{}^r\varepsilon/\varepsilon, \nu]_{7_1}$.	

On a:

$$^r\varepsilon/\varepsilon = \frac{\theta^2\beta-6}{\beta-6}\,\frac{\beta-10}{\theta^2\beta-10} = \frac{(\beta-6\theta)(\beta-10)}{(\beta-6)(\beta-10\theta)}\,.$$

En 7, les unités qui sont des cubes sont les nombres $\pm 1 \bmod 7$, par exemple 300, qui est congru à -1. Fixons par exemple θ_1 et $\beta_1 \in \mathbb{Q}_7$ tels que:

$$\theta_1 \equiv 2 \quad \text{et} \quad \beta_1 \equiv 3 \bmod 7 .$$

Comme ν est une unité en 7, seule compte, pour $\theta^r\varepsilon/\varepsilon$, sa valuation. Les 4 nombres $\beta-6, \beta-6\theta, \beta-10$ et $\beta-10\theta$ sont congrus $\bmod 7$ à $\beta+1, \beta+\theta, \beta-3$ et $\beta-3\theta$. Avec les choix ci-dessus de θ_1 et β_1, on voit que β_1-10 n'est pas une unité et que les trois autres nombres sont des unités. De plus, comme la valuation de $N(10-\beta)$ est 1, il en est de même pour β_1-10 - noter d'ailleurs que $\theta_1 \equiv 30$ et $\beta_1 \equiv 24 \bmod 49$. Finalement:

$$i_7 = [\theta^r\varepsilon/\varepsilon, 10]_{7_1} = [7,10]_{7_1} = -[3,7]_{7_1} = -[\theta_1^2,7]_{7_1} = [\theta_1,7]_{7_1} = 1 ,$$

et, comme $i = i_7$, l'obstruction de Manin est encore "non nulle".

Il est clair qu'on aura toujours $\mathcal{S}_{\mathcal{A}} \neq \mathcal{S}$ si $\varepsilon = \varepsilon_\beta$, puisqu'alors $i_v = 0$ pour $v \in \mathcal{S}$. On peut néanmoins avoir encore $\mathcal{S}_{\mathcal{A}} = \mathcal{S}$ en prenant par exemple:

$$\varepsilon = \delta(2-\delta)$$

avec $\eta=1$. Pour $v=2,5$, on trouve $i_v = [{}^q\varepsilon/\varepsilon, \nu]_v = [\theta(2-\theta\delta_1), 10]_v$, ce qui donne, avec $\delta_1 \equiv 1 \bmod 2$ et $\delta_1 \equiv 2 \bmod 5$:

$$i_2 = [\theta^2, 2]_2 = 1 \quad \text{et} \quad i_5 = [2(\theta-\theta^2), 5]_5 = [\theta-\theta^2, 5]_5 = 0$$

car $(\theta-\theta^2)^2 = -3$ et $(\theta-\theta^2)^8 \equiv 1 \bmod 5$. On trouve donc encore $i=1$, comme pour le choix précédent de ε, ce qui est normal.

(f) *L'exemple de Cassels et Guy: variante avec* $\eta \neq 1$.

Ecrivons l'équation (2) sous la forme:

$$x^3 + 15y^3 + 36z^3 + 30t^3 = 0 .$$

Ainsi:

$$\nu = 1/18 \qquad \lambda\nu = 5/6 \qquad \mu = 36$$
$$\lambda = 15 \qquad \lambda/\nu = 270 \sim 10 .$$

On peut prendre:

$$\varepsilon = (1+\alpha)/2\gamma \qquad \eta = 29 + 16\theta + 3\alpha - \alpha^2 .$$

On vérifie en calculant $N(\eta)$ que $\mathcal{S}_{\mathcal{A}} = \mathcal{S} = \{2,3,5\}$ et on a le tableau suivant:

en 2	λ	cube	$G^v = \langle s \rangle$	$i_2 = [\theta/\eta,\nu]_2$
en 3	λ/ν	cube	$G^v = \langle r \rangle$	$i_3 = [{}^q\varepsilon/\varepsilon\eta,\nu]_3$
en 5	ν	cube	$G^v = \langle t \rangle$	$i_5 = 0$.

Comme $\alpha = \sqrt[3]{15}$, on a $\alpha_1 \equiv 1 \bmod 2$ et par suite $\eta \equiv 1 \bmod 2$, ce qui donne:

$$i_2 = [\theta,1/18]_2 = -[\theta,2]_2 = 1 .$$

Comme $\varepsilon\eta/{}^q\varepsilon = 16(\theta - \theta^2)$, il vient:

$$i_3 = [16(\theta - \theta^2),18]_3 = -[3,18]_3 = 0$$

car $\theta - \theta^2 = (\theta-1)^3/3$. Finalement: $i = i_2 = 1$ et l'obstruction de Manin est "non nulle".

Il eût été évidemment plus habile de prendre:

$$\varepsilon = 1/(1-\beta)^2 \qquad \eta = 1 ,$$

car un tel choix donne aussitôt $i_2 = [\theta,\nu]_2 = 1$ comme ci-dessus et $i_3 = 0$ car $\varepsilon = \varepsilon_\beta$. On retrouve, comme il est naturel, la même valeur pour i, mais avec moins de peine!

(g) *Un exemple quelconque pris dans la liste.*

Il s'agit d'un exemple pris au hasard dans la liste établie par ordinateur et donnée à la fin du § suivant:

$$15x^3 + 21y^3 + 28z^3 + 50t^3 = 0 .$$

On trouve:

$$\nu = 125/98 \sim 1/98 \sim 28 \qquad \lambda\nu = 25/14 \sim 1/70 \qquad \mu = 28/15$$
$$\lambda = 7/5 \qquad \lambda/\nu = 686/625 \sim 2/5$$

On prend:

$$\varepsilon = 14/(5\delta(5 - 4\beta)) \qquad \eta = 1 .$$

On a $\mathcal{S}_\mathcal{A} = \mathcal{S} = \{2,3,5,7\}$. Comme $\varepsilon_\delta = 1/\delta$, on obtient le tableau suivant:

en 2	λ	cube	$G^v = \langle s \rangle$	$i_2 = [\theta,\nu]_2 = [\theta,4]_2 = 1$
en 3 et 5	ν	cube	$G^v = \langle t \rangle$	$i_3 = i_5 = 0$
en 7	λ/ν	cube	$G^v = \langle r \rangle$	$i_7 = [{}^q\varepsilon_\delta/\varepsilon_\delta,\nu]_7 = [\theta^2,7]_7 = 1$.

Ainsi, $i = 2 \neq 0$ et l'obstruction de Manin est "non nulle".

§8. La procédure

La procédure ci-après s'applique à une surface cubique diagonale V d'équation homogène

(1) $$ax^3 + by^3 + cz^3 + dt^3 = 0 \qquad a, b, c, d \in \mathbb{Q}^* .$$

Elle conclut de diverses manières:

(i) $\prod_p V(\mathbb{Q}_p) = \emptyset$

(ii) $\prod_p V(\mathbb{Q}_p) \neq \emptyset$ et l'obstruction de Manin est "non vide"

(iii) $\prod_p V(\mathbb{Q}_p) \neq \emptyset$ et l'obstruction de Manin est "vide"

(iv) $V(\mathbb{Q}) = \emptyset$

(v) $V(\mathbb{Q}) \neq \emptyset$.

On a évidemment les implications suivantes: (i) ⇒ (iv), (ii) ⇒ (iv) et (v) ⇒ (iii) et on peut donc se limiter aux conclusions (iii) et (iv). On espère naturellement que (iii) implique inversement (v). La conclusion (ii) correspond aux *contre-exemples au principe de Hasse* fournis par l'obstruction de Manin.

<u>La procédure</u>

p1 *équation réduite*

On se ramène, par multiplication des coordonnées et de l'équation par des rationnels convenables, au cas où l'équation (1) est "réduite", i.e.: a, b, c, d sont des entiers naturels et, pour tout nombre premier p, le quadruplet $(v_p(a), v_p(b), v_p(c), v_p(d))$ est, à permutation près, égal à (0,0,0,0) ou (0,0,0,1) ou (0,0,0,2) ou (0,0,1,1) ou (0,0,1,2).

p2 *cas trivial*

Si l'un des quotients a/b, a/c, a/d, b/c, b/d, c/d est un cube dans $\mathbb{Q}$,

alors (v) $V(\mathbb{Q}) \neq \emptyset$ *stop;*

sinon **p3**.

p3 *conditions locales*

On détermine $\mathcal{S}$:={p premier; p|3abcd} et $\mathcal{S}'$:={p∈$\mathcal{S}$; p≡1 ou 3 mod 3}.
S'il existe une permutation (a',b',c',d') de (a,b,c,d) telle que

(a',b',c',d') ≡ (1,2,4,0) mod 9 ,

ou s'il existe p∈$\mathcal{S}'$, p≠3, et une permutation de (a,b,c,d) de la forme (a',b',pc',pd') où a',b',c',d' sont des p-unités telles qu'aucun des quotients a'/b', c'/d' ne soit un cube mod p,

alors (i) $\prod_p V(\mathbb{Q}_p) = \emptyset$ *stop;*

sinon **p4**.

p4 *le cas de Selmer*

Si l'un des quotients ab/cd, ac/bd ou ad/bc est un cube dans $\mathbb{Q}$,

alors (v) $V(\mathbb{Q}) \neq \emptyset$ *stop;*

sinon **p5**.

p5 *le premier cas*

S'il existe p∈$\mathcal{S}$ tel qu'aucun des quotients ab/cd, ac/bd, ad/bc ne soit un cube dans $\mathbb{Q}_p$,

alors (iii) $\prod_p V(\mathbb{Q}_p) \neq \emptyset$ et l'obstruction de Manin est "vide" *stop;*

sinon p6.

p6 *conditionnement*

On se ramène, par permutation des coefficients, au cas où ad/bc est un cube dans $\mathbb{Q}_3$.
On définit (u~v signifie u/v∈$\mathbb{Q}^{*3}$)

λ :~ b/a μ :~ c/a ν :~ ad/bc

en imposant que λ, μ et ν soient "réduits" (u "réduit" signifie que, pour tout premier p, on a $v_p(u)$=0, 1 ou −1).
On définit

$\alpha := \sqrt[3]{\lambda}$ $\gamma := \sqrt[3]{\nu}$ $\beta := \alpha\gamma$ $\delta := \alpha/\gamma$.

p7 *calcul de* $\varepsilon=\varepsilon_\beta\varepsilon_\delta$

On détermine

$$\varepsilon_\beta = (u + v\beta)/(u' + v'\beta) \quad \text{et} \quad \varepsilon_\delta = (m + n\delta)/(m' + n'\delta) \qquad \begin{matrix} u,v,u',v' \in \mathbb{Z} \\ m,n,m',n' \in \mathbb{Z} \end{matrix}$$

tels que

$$N(\varepsilon_\beta)\cdot N(\varepsilon_\delta) = \mu$$

avec

$$N(\varepsilon_\beta) = (u^3 + \lambda\nu v^3)/(u'^3 + \lambda\nu v'^3) \quad \text{et} \quad N(\varepsilon_\delta) = (m^3 + n^3\lambda/\nu)/(m'^3 + n'^3\lambda/\nu).$$

On dira que ε est "simple" s'il est de la forme

$$\varepsilon = \zeta^j(u + v\beta)^{j'} \quad \text{ou} \quad \varepsilon = \zeta^j(m + n\delta)^{j'} \qquad j\,,j' = 0\,,1\,,-1$$

avec $\zeta=\alpha, \beta, \gamma$ ou δ.

p8 *calcul de* i

Si ε est "simple", on pose $\mathcal{S}_\varepsilon := \mathcal{S}$.
Si ε est quelconque, on pose

$$\mathcal{S}_\varepsilon = \{p;\ p \in \mathcal{S} \text{ ou } v_p(u^3 + \lambda\nu v^3) \neq 0 \text{ ou } v_p(u'^3 + \lambda\nu v'^3) \neq 0 \text{ ou } v_p(m^3 + n^3\lambda/\nu) \neq 0 \text{ ou } v_p(m'^3 + n'^3\lambda/\nu) \neq 0\}\,.$$

On pose $\mathcal{S}''_\varepsilon := \mathcal{S}_\varepsilon - \{3\}$ et on calcule la fonction

$$\text{"i"}:\ \mathcal{S}''_\varepsilon \longrightarrow \mathbb{Z}/3 \qquad p \longrightarrow i_p$$

à l'aide du tableau suivant qui se lit ainsi: "si ν est un cube dans $\mathbb{Q}_p$, alors $i_p=0$, si λ est un cube dans $\mathbb{Q}_p$, alors $i_p=[\theta,\nu]_p$, ..."

(114)

cube dans $\mathbb{Q}_p \longrightarrow$	ν	λ	λ/ν	$\lambda\nu$
$i_p \longrightarrow$	0	$[\theta,\nu]_p$	$[{}^q\varepsilon_\delta/\varepsilon_\delta,\nu]_p$	$-[\theta\cdot{}^r\varepsilon_\beta/\varepsilon_\beta,\nu]_p$

On pose $\quad i := \sum_{p\in\mathcal{S}''_\varepsilon} i_p \in \mathbb{Z}/3$

p9 *conclusion*

Si i=0,

alors (iii) $\prod_p V(\mathbb{Q}_p) \neq \emptyset$ et l'obstruction de Manin est "vide" *stop;*

sinon (ii) $\prod_p V(\mathbb{Q}_p) \neq \emptyset$ et l'obstruction de Manin est "non vide",

et alors V est un contre-exemple au principe de Hasse *stop.*

<u>*Commentaires*</u>

p1 Voir §4, pas n°0.
p3 Voir §4, pas n°0.
p4 Voir §1, lemme 1.
p5 Voir §5, proposition 2 et lemme 8.

p6 On pourrait aussi bien définir la propriété pour $u \in \mathbb{Q}^*$ d'être "réduit" par exemple par $v_p(u)=0$, 1 ou 2, pour tout p premier.

p7 D'après la proposition 4 du §6, on pourrait chercher ε uniquement sous la forme ε_β, ou uniquement sous la forme ε_δ. Néanmoins la recherche sous la forme $\varepsilon_\beta\varepsilon_\delta$ semble plus "rentable", en particulier, si l'on cherche une forme "simple" éventuelle. De fait, le calcul sur ordinateur a curieusement toujours fourni une solution ε "simple". La technique utilisée a consisté à faire des tables de nombres réduits $\sim u^3+\lambda\nu v^3$ ou $\sim m^3+n^3\lambda/\nu$, pour de petites valeurs des entiers u,v,m,n, et à chercher dans ces tables la réduction de l'un des nombres

$$\mu^{\pm1}, \mu^{\pm1}\alpha^{\pm1}, \mu^{\pm1}\beta^{\pm1}, \mu^{\pm1}\gamma^{\pm1}, \mu^{\pm1}\delta^{\pm1}.$$

p8 Voir §6. Le conditionnement fait en **p6** assure que ν est un cube dans $\mathbb{Q}_3$ et donc $i_3=0$, ce qui justifie qu'on se limite à $\mathcal{S}''_\varepsilon$, et même en pratique, si ε est simple, à $\mathcal{S}'':=\mathcal{S}-\{3\}$. La définition de $\mathcal{S}_\varepsilon$ pour ε quelconque inclut les premiers p au-dessus desquels il existe une place w_p de $\mathbb{Q}(\beta)$ telle que $w_p(\varepsilon_\beta)\neq0$, ou une place w_p de $\mathbb{Q}(\delta)$ telle que $w_p(\varepsilon_\delta)\neq0$. Si ε est "simple", il ne peut y avoir de "simplification" comme dans l'exemple (e) du §7 pour ε_β, et alors $\mathcal{S}_\varepsilon$ se réduit à $\mathcal{S}$.
Pour les symboles $[\ ,\ ]_p$ voir le formulaire du §4 de (64) à (77). Pour la signification et le mode de calcul des expressions figurant dans le tableau (114), voir le lemme 12 du §6. On trouvera néanmoins ci-après un formulaire abrégé indiquant les formules les plus courantes. Celles relatives à p=3 ne sont pas

reprises puisque la procédure ne les utilise pas. Néanmoins, on peut ne pas conditionner l'équation comme en **p6**, auquel cas il faut calculer i_3 qui peut être non nul et le faire intervenir dans la sommation définissant i. Plusieurs exemples ont été traités de la sorte au §7 et il en a été de même pour le calcul sur ordinateur.

Formulaire

(a) Une p-unité $u \in \mathbb{Z}_p^*$ est toujours un cube si $p \equiv 2 \bmod 3$. Si $p \equiv 1 \bmod 3$, u est un cube si et seulement si c'est un cube mod p. Si $p=3$, u est un cube si et seulement si $u \equiv \pm 1 \bmod 9$.

(b) Pour une équation (1) réduite, les seuls cas où $V(\mathbb{Q}_p)=\emptyset$ sont les suivants, à permutation près des coefficients a,b,c,d (a',b',c',d' sont des p-unités):

$p \equiv 1 \bmod 3$	(a',b',pc',pd')	et a'/b' et c'/d' non cubes mod p,
$p \equiv 1 \bmod 3$	(a',b',pc',p^2d')	et a'/b' non cube mod p ,
$p = 3$	$(a,b,c,d) \equiv (1,2,4,0) \bmod 9$.	

(c) Le symbole $[u,v]_p$ est défini pour $u,v \in \mathbb{Q}(\theta)^*$, il est biadditif, alterné et à valeurs dans $\mathbb{Z}/3$. Soit v_1, ou v_1 et v_2, le(s) prolongement(s) à $\mathbb{Q}(\theta)$ de la valuation p-adique de $\mathbb{Q}$:

$$[u,v]_p := [u,v]_{v_1} \qquad \text{si } p \equiv 2 \bmod 3$$

et

$$[u,v]_p := [u,v]_{v_1} + [u,v]_{v_2} \qquad \text{si } p \equiv 1 \bmod 3 .$$

Si $u=m^3+n^3v$ ou si $v=m^3+n^3u$ pour $m,n \in \mathbb{Q}(\theta)$, alors $[u,v]_p=0$. De même, si $u,v \in \mathbb{Q}^*$, alors:

$$[u,v]_p = 0 .$$

Les formules ci-dessous, où u et v sont des p-<u>unités</u>, déterminent $[\ ,\ ]_p$ si $p \neq 3$:

$$[up^i,vp^j]_p = j[u,p]_p - i[v,p]_p$$

$$[u,p]_p = -m \qquad \text{où } u^{(p^2-1)/3} \equiv \theta^m \bmod p \qquad \underline{\text{pour}}\ p \equiv 2 \bmod 3 ,$$

$$[u,p]_p = -m_1 - m_2 \qquad \text{où } u_i^{(p-1)/3} \equiv \theta_i^{m_i} \bmod p \qquad \underline{\text{pour}}\ p \equiv 1 \bmod 3 ,$$

où u_i et θ_i désignent les images de u et de θ par le plongement $\mathbb{Q}(\theta) \hookrightarrow \mathbb{Q}_p$ défini par v_i. En particulier, on a le tableau:

p mod 9	1	2	3	4	5	7	8
$[\theta,p]_p$	0	2	0	1	1	2	0

(d) Pour calculer les expressions du tableau (114), on fixe au début de chaque calcul, et à sa convenance, une valeur θ dans $\mathbb{Q}_p$ si $p \equiv 1 \bmod 3$, une valeur de β dans $\mathbb{Q}_p$ si $\lambda\nu$ est un cube dans $\mathbb{Q}_p$, une valeur de δ dans $\mathbb{Q}_p$ si λ/ν est un cube dans $\mathbb{Q}_p$. Si $p \equiv 1 \bmod 3$, on a

$$[u,v]_p = -[u,v]_{v_1}$$

si $[u,v]_p$ désigne l'expression figurant dans le tableau (114). Pour le calcul il est utile d'avoir le tableau suivant qui donne les valeurs de ${}^g\zeta/\zeta$:

$g \downarrow$ $\zeta \longrightarrow$	α	β	γ	δ
s	1	θ	θ	θ^2
t	θ	θ	1	θ
q	θ^2	1	θ	θ
r	θ	θ^2	θ	1

$\longleftarrow {}^g\zeta/\zeta$

Si λ/ν est un cube dans $\mathbb{Q}_p$, soit $\delta_1 \in \mathbb{Q}_p$ une racine cubique. Si alors

$$\varepsilon_\delta = m + n\delta \qquad m, n \in \mathbb{Z} \text{ et } (m,n) = 1,$$

on a les formules:

$$[{}^q\varepsilon_\delta/\varepsilon_\delta,\nu]_p = [m + n\theta\delta_1,\nu]_p \qquad \text{si } p \equiv 2 \bmod 3$$
$$= [m + n\delta_1,\nu]_{v_1} - [m + n\theta_1\delta_1,\nu]_v \qquad \text{si } p \equiv 1 \bmod 3$$

où $\theta_1 \in \mathbb{Q}_p$ correspond à v_1.

Si $\lambda\nu$ est un cube dans $\mathbb{Q}_p$, soit $\beta_1 \in \mathbb{Q}_p$ une racine cubique. Si alors

$$\varepsilon_\beta = u + v\beta \qquad u, v \in \mathbb{Z} \text{ et } (u,v) = 1,$$

on a les formules:

$$[{}^{\mathrm{r}}\varepsilon_\beta/\varepsilon_\beta,\nu]_p = [u + v\theta^2\beta_1,\nu]_p \qquad \text{si } p \equiv 2 \bmod 3$$

$$= [u + v\beta_1,\nu]_{v_1} - [u + v\theta_1^2\beta_1,\nu]_{v_1} \qquad \text{si } p \equiv 1 \bmod 3$$

où $\theta_1 \in \mathbb{Q}_p$ correspond à v_1.

<u>*Calcul sur ordinateur*</u> (M. Vallino)

Le calcul sur ordinateur a eu pour objet de tester numériquement, sur les surfaces cubiques diagonales, la conjecture énoncée dans l'introduction: *pour les surfaces rationnelles - propres et lisses - l'obstruction de Manin est la seule obstruction au principe de Hasse.* Autrement dit, si $\mathbb{Q}$ est le corps de base, si V est une surface rationnelle qui a des points dans chaque complété de $\mathbb{Q}$ et si l'obstruction de Manin est "vide" pour V, alors $V(\mathbb{Q}) \neq \emptyset$.

Le test a porté sur les équations

$$ax^3 + by^3 + cz^3 + dt^3 = 0 \qquad \begin{array}{r} a, b, c, d \text{ entiers} \\ 0 < a,b,c,d < 100\,. \end{array}$$

Les résultats numériques obtenus sont *en accord avec la conjecture* dans le domaine considéré et, bien que de portée limitée, ils confortent nettement celle-ci pour les surfaces cubiques diagonales.

Les calculs ont été effectués au centre de calcul de l'ENS, sur IBM 4341, par M. Vallino qui a écrit à cet effet trois programmes principaux:

(i) un programme qui détermine, pour une équation donnée, si elle a des solutions dans chaque complété $\mathbb{Q}_p$,

(ii) un programme qui recherche, pour une équation et un entier h donnés, une solution éventuelle à coordonnées entières (x,y,z,t) de hauteur <h,

(iii) un programme de calcul de l'obstruction de Manin suivant la procédure indiquée plus haut aux quelques modifications près signalées dans les commentaires.

De fait, l'expérimentation numérique et la partie théorique ont interagi à plusieurs reprises, les résultats numériques laissant prévoir des propositions qui, une fois établies, simplifiaient la procédure. C'est ainsi que les programmes (i) et (ii) ont largement précédé le programme (iii) qui n'est intervenu qu'en fin de calcul après un large débroussaillage par application des programmes (i) et (ii).

Avant d'indiquer les grandes lignes et les principaux résultats successifs de l'expérimentation numérique, il faut préciser - pour éviter toute erreur d'interprétation - que les programmes (i) et (ii) - sauf le premier programme (ii) pour h=200 - ont traité chaque équation

$$ax^3 + by^3 + cz^3 + dt^3 = 0$$

sous la forme "réduite" équivalente, forme qui minimise le produit des coefficients: voir le pas p1 de la procédure. Si l'équation donnée est réduite, on peut donc prendre les résultats indiqués à la lettre. Sinon, il faut prendre garde au fait que la solution éventuelle indiquée, et sa hauteur, sont relatives à la forme réduite et non à l'équation initiale. Précisons encore que le programme (ii) ne donne qu'une solution de l'équation dans le domaine considéré - lorsqu'il y en a - et non pas a priori une solution de la plus petite hauteur. Indiquons enfin qu'ont été éliminées toutes les équations dont la forme réduite appartient encore au domaine considéré $0<a,b,c,d<100$. Par exemple, les équations de Bremner, et de Cassels et Guy, de coefficients (1,4,10,25) et (5,9,10,12), sont réduites et ont des solutions dans chaque complété: elles ont donc été conservées, mais l'équation de coefficients (1,2,20,50) équivalente à celle de Bremner, a été éliminée. En revanche, les équations de coefficients (2,4,10,25) et (1,10,20,50) ont été maintenues quoique équivalentes, car leur forme réduite (1,2,5,100) sort du domaine considéré. Ceci explique la présence de "doublons" dans les résultats cités ci-après, "doublons" qui ont été éliminés dans la liste des contre-exemples au principe de Hasse.

Voici donc comment l'exploration numérique a eu lieu:

(a) grâce à (i) ont été stockées en fichiers les équations ayant des solutions dans chaque complété de $\mathbb{Q}$, en éliminant les équations non réduites dont la forme réduite avait des coefficients <100;

(b) grâce à (ii) pour h=200, ont été éliminées toutes les équations ayant une solution de hauteur <200; il restait alors environ 14000 équations;

(c) après application de (ii) pour h=1000, il resta 681 équations qui furent réparties en trois fichiers: 329 pour lesquelles il existait $p\equiv 2 \bmod 3$ divisant exactement 1 coefficient, 75 du même type pour $p\equiv 1$ ou 3 mod 3, enfin les autres (277);

(d) par application successive de (ii) pour h=2000, 4000, 8000, on trouva une solution pour chacune des équations du premier fichier, puis de même pour le second en appliquant (ii) pour h=4000, puis pour h=8000 pour la seule équation - réduite - de coefficients (43,44,77,86) qui a pour solution (4473,–3827,6493,–6392); ces 404 équations relevaient du *premier cas*;

(e) enfin, l'application de (iii) aux 277 dernières équations ne révéla que 8 équations pour lesquelles l'obstruction de Manin était "vide" et l'application de (ii) pour h=4000 fournit une solution pour chacune d'elles.

La liste ci-après ne comporte que 245 équations, et non 269 par suite de "doublons". Signalons enfin que le temps de calcul de (ii) est en général très nettement supérieur au temps de calcul de (iii) qui est de l'ordre de 1s.

Surfaces cubiques diagonales

Contre-exemples au principe de Hasse

245

Liste des équations $ax^3 + by^3 + cz^3 + dt^3 = 0$ à coefficients entiers, $0<a,b,c,d<100$, ayant des solutions dans chaque complété $\mathbb{Q}_p$ sans en avoir dans $\mathbb{Q}$.

a	b	c	d																
1	2	30	75	3	20	50	90	9	30	44	55	15	21	28	50	25	30	66	99
1	4	10	25	3	23	36	46	9	36	46	69	15	21	90	98	25	34	50	85
1	4	15	90	3	25	30	36	9	36	51	68	15	23	46	90	25	42	45	98
1	4	30	45	3	34	85	90	9	45	51	85	15	34	68	75	25	44	45	66
1	4	30	75	3	36	41	82	9	60	68	85	15	36	68	85	25	44	60	66
1	5	12	30	3	38	90	95	10	11	22	25	15	45	46	92	25	44	66	75
1	6	25	90	3	44	55	90	10	11	66	75	15	46	75	92	25	44	90	99
1	6	34	51	3	50	55	66	10	12	51	85	17	18	36	51	25	46	60	68
1	10	20	50	4	6	17	51	10	12	55	99	17	20	50	85	25	50	66	99
1	10	34	85	4	6	25	30	10	15	51	68	17	30	34	75	25	51	60	68
1	10	36	75	4	6	25	45	10	17	25	34	17	30	36	85	25	51	68	90
1	12	25	30	4	6	69	92	10	17	25	68	17	30	45	68	25	51	75	85
1	12	51	68	4	10	11	55	10	22	25	44	17	34	75	90	25	55	60	66
1	15	51	85	4	10	17	85	10	22	75	99	18	22	55	75	28	30	49	75
1	18	25	30	4	11	50	55	10	23	25	92	18	33	55	90	29	30	58	75
1	20	68	85	4	12	34	51	10	23	60	69	18	35	45	84	30	33	44	50
1	21	36	98	4	15	18	25	10	25	29	58	18	36	58	87	30	34	68	75
1	30	50	60	4	20	22	55	10	25	33	66	18	36	69	92	30	36	55	99
2	3	33	44	4	20	34	85	10	25	46	92	20	22	25	55	30	41	75	82
2	4	15	75	4	25	44	55	10	25	63	84	20	22	66	75	30	47	75	94
2	4	66	99	4	30	33	55	10	30	69	92	20	23	25	46	30	49	70	75
2	5	36	75	5	6	12	25	10	33	44	75	20	23	30	69	30	50	69	92
2	6	51	68	5	9	10	12	10	33	50	99	20	25	41	82	33	34	44	51
2	9	10	75	5	17	34	50	10	34	51	60	20	29	50	58	34	36	60	85
2	9	44	99	5	17	51	75	10	36	51	85	20	33	50	66	34	36	75	85
2	10	68	85	5	22	25	44	10	49	60	63	20	36	55	66	34	36	75	85
2	12	17	51	5	25	34	68	10	52	75	78	20	41	50	82	34	45	68	75
2	12	25	90	5	25	36	60	10	69	75	92	20	44	50	55	34	50	51	60

2 12 58 87	5 30 46 69	11 18 30 55	20 46 60 69	34 66 68 99
2 17 20 85	5 60 69 92	11 30 33 50	20 46 69 75	35 36 50 63
2 22 25 55	6 9 11 44	11 30 44 75	20 47 50 94	36 38 60 95
2 25 30 36	6 9 17 34	11 36 55 90	20 51 68 75	36 44 55 75
2 30 33 55	6 11 55 90	11 44 75 90	20 55 77 98	36 51 85 90
2 44 50 55	6 12 22 99	11 49 66 84	22 25 50 55	36 57 90 95
2 51 60 85	6 15 25 36	11 50 66 75	22 25 60 66	36 77 98 99
2 55 60 66	6 20 51 85	11 50 90 99	22 30 44 75	44 50 60 99
2 55 66 75	6 20 55 66	12 17 30 85	22 33 50 60	44 51 66 68
2 55 90 99	6 25 55 66	12 18 29 58	22 33 69 92	44 66 75 90
2 57 60 95	6 29 36 58	12 30 49 70	22 33 75 90	44 75 90 99
3 4 10 15	6 30 55 99	12 33 50 55	23 30 45 46	45 46 69 90
3 4 25 90	6 35 50 84	12 33 55 90	23 30 46 75	45 47 60 94
3 4 46 69	6 44 45 55	12 34 60 85	23 30 75 92	45 49 60 84
3 5 10 36	6 68 85 90	12 45 68 85	23 66 92 99	45 51 68 90
3 6 10 25	9 12 23 46	12 51 85 90	25 26 39 60	46 50 60 69
3 9 34 68	9 12 41 82	14 18 35 60	25 30 33 44	47 66 94 99
3 11 18 44	9 15 17 85	15 17 25 51	25 30 34 51	50 55 60 99
3 17 18 34	9 18 34 51	15 17 68 90	25 30 44 66	58 69 87 92
3 17 45 85	9 28 70 75	15 18 44 55	25 30 46 69	58 75 87 90
3 18 49 84	9 30 34 85	15 20 21 98	25 30 58 87	68 75 85 90

Dans cette liste ne figure qu'une équation à équivalence près: par exemple ne figure pas l'équation dont les coefficients sont (2,4,10,25), car équivalente à (1,10,20,50).

Liste établie par M. Vallino sur IBM 4341 au Centre de calcul de l'ENS le 13 mars 1985.

§9. Les hypersurfaces cubiques diagonales

Etant donné une classe $\mathcal{V}$ de variétés, on appellera *hypothèse* BM, pour abréger, l'hypothèse suivante: *l'obstruction de Brauer-Manin est la seule obstruction au principe de Hasse dans* $\mathcal{V}$.

Comme indiqué au § précédent, les calculs sur ordinateur ont confirmé cette hypothèse pour les équations (1) à coefficients entiers tels que $0<a,b,c,d<100$. Mais il peut être aussi tentant de contrôler indirectement cette hypothèse, pour une certaine classe $\mathcal{V}$, en considérant un certain nombre de ses conséquences. Par exemple, le corollaire de la proposition 2 du §5 se reformule comme suit:

Proposition 6. *Si l'hypothèse* BM *est vraie pour les surfaces cubiques diagonales sur* $\mathbb{Q}$, *alors le principe de Hasse vaut pour les équations*

$$ax^3 + by^3 + cz^3 + dt^3 = 0 \qquad \begin{matrix} abcd \neq 0 \\ a, b, c, d \in \mathbb{Z} \end{matrix}$$

pour lesquelles il existe un premier p *qui divise un, et un seul, des coefficients, supposés sans cube.*

Nous allons en déduire la conséquence suivante relative aux hypersurfaces cubiques diagonales:

Proposition 7. *Si l'hypothèse* BM *est vraie pour les surfaces cubiques diagonales sur* $\mathbb{Q}$, *alors le principe de Hasse vaut pour les hypersurfaces cubiques diagonales dans* $\mathbb{P}^n_{\mathbb{Q}}$ *pour* $n \geq 4$, *et celles-ci ont des points rationnels pour* $n \geq 6$.

Soit X une telle hypersurface cubique diagonale qui ait des points dans chaque complété de $\mathbb{Q}$. La méthode consiste à trouver une section linéaire $V \subset X$ sur $\mathbb{Q}$, qui soit une surface cubique diagonale ayant encore des points dans chaque complété et pour laquelle l'obstruction de Manin soit "vide". De fait, on va chercher V du type indiqué dans la proposition 6.

L'hypersurface X admet une équation du type

$$a_0 x_0^3 + \ldots + a_n x_n^3 = 0 \qquad n \geq 4$$

où les coefficients a_i sont entiers non nuls. Soit $\mathcal{S} = \{p \text{ premier};\ p \mid 3a_0 \cdot a_1 \ldots a_n\}$. Pour chaque $p \in \mathcal{S}$, choisissons $\xi_p = (x_{p,i})_{i=0,\ldots,n} \in X(\mathbb{Q}_p)$ de telle sorte que

$$(115) \qquad \sum_{i\geq 3} a_i x_{p,i}^3 \neq 0 .$$

En fait, un tel choix est possible pour tout p, par lissité, puisque $n \geq 3$ et que $X(\mathbb{Q}_p) \neq \emptyset$ par hypothèse. De plus, on prend les $x_{p,i}$ entiers p-adiques.

Soit alors $q \equiv 2 \bmod 3$ un nombre premier auxiliaire n'appartenant pas à $\mathcal{S}$. Comme $n \geq 4$ un tel choix de q permet de trouver $\xi_q = (x_{q,i})_{i=3,\ldots,n} \in \mathbb{Z}^{n-2}$ tel que:

$$(116) \qquad \sum_{i\geq 3} a_i x_{q,i}^3 \equiv 0 \bmod q \quad \text{et} \quad \not\equiv 0 \bmod q^2 .$$

Cherchons en effet ξ_q sous la forme $(u, -1, 0, \ldots, 0)$ ce qui est possible car $n \geq 4$. Comme $q \notin \mathcal{S}$, les a_i sont des q-unités, et comme $q \equiv 2 \bmod 3$, ce sont des cubes dans

$\mathbb{Z}_q$. Soit donc $\zeta := \sqrt[3]{a_4/a_3} \in \mathbb{Z}_q$. Choisissons alors

$$u \text{ entier} \equiv \zeta(1+q) \bmod q^2 .$$

Pour $\xi_q=(u,-1,0,\dots,0)$ on trouve alors:

$$\sum_{i\geq 3} a_i x_{q,i}^3 = a_3u^3 - a_4 \equiv 3a_4q \bmod q^2 ,$$

ce qui est $\equiv 0 \bmod q$, mais $\not\equiv 0 \bmod q^2$.

Si l'on prend alors $\xi = (\alpha_i)_{i\geq 3} \in \mathbb{Z}^{n-2}$ assez proche p-adiquement de ξ_p pour chaque $p \in \mathcal{S}$, et assez proche q-adiquement de ξ_q - ce qui est possible grâce au théorème des restes chinois - il résulte des conditions (115) et (116) que l'entier

$$d := \sum_{i\geq 3} a_i \alpha_i^3$$

vérifie:

(i) $d \neq 0$

(ii) pour chaque $p \in \mathcal{S}$, $(\sum_{i\geq 3} a_i x_{p,i}^3)/d$ est un cube dans $\mathbb{Q}_p$

(iii) q divise d, mais q^2 ne divise pas d.

Considérons alors la surface cubique diagonale $V \subset \mathbb{P}^3_{\mathbb{Q}}$ d'équation

(117) $$a_0x_0^3 + a_1x_1^3 + a_2x_2^3 + dt^3 = 0$$

et obtenue par la "section linéaire" de X définie par

(118) $$x_i = \alpha_i t \qquad i = 3, \dots, n .$$

Cette surface cubique a des points dans chaque complété $\mathbb{Q}_p$: pour $p \in \mathcal{S}$, cela résulte de la condition (ii) et du fait que $\xi_p = (x_{p,i})_{i=0,\dots,n}$ appartient à $X(\mathbb{Q}_p)$; pour $p \notin \mathcal{S}$, cela résulte simplement du fait que la courbe de genre 1 définie par

$$a_0x_0^3 + a_1x_1^3 + a_2x_2^3 = 0$$

a bonne réduction en p - grâce au choix de $\mathcal{S}$ - donc a des points dans $\mathbb{Q}_p$.

Par ailleurs, comme le premier q n'appartient pas à $\mathcal{S}$, il ne divise ni a_0, ni a_1, ni a_2. Mais, d'après (iii), il divise d et q^2 ne divise pas d, et a fortiori q^3 ne divise pas d.

Par suite, la surface cubique diagonale V d'équation (117) relève du *premier cas* et on peut lui appliquer la proposition 6. Comme elle a des points dans chaque complété de $\mathbb{Q}$, l'hypothèse BM pour les surfaces cubiques diagonales implique $V(\mathbb{Q})\neq\emptyset$ et donc, via la section (118), $X(\mathbb{Q})\neq\emptyset$.

Pour $n\geq 6$, les conditions locales $X(\mathbb{Q}_p)\neq\emptyset$ sont toujours vérifiées (cf. §4, pas n°0). L'énoncé pour $X\subset\mathbb{P}^n$ et $n\geq 6$ en résulte.

On trouvera d'autres conséquences de l'hypothèse BM - pour diverses classes $\mathcal{V}$ - dans l'exposé [11].

§10. Variétés de descente

La théorie développée dans [8] ramène la question de la validité de l'*hypothèse* BM pour une classe donnée $\mathcal{V}$ de variétés rationnelles à la question de la validité du *principe de Hasse* pour une classe auxiliaire $\mathcal{W}$, par exemple celle des *torseurs universels*, ou plus généralement celle des *variétés de descente* de certains types dits *admissibles*. Dans ce §, nous allons considérer des variétés de descente qui, sans être admissibles, jouent le même rôle pour la question de l'hypothèse BM. Avec les notations du §1, ce sont les variétés de descente de type

$$i: \mathrm{Pic}\, V_K \longrightarrow \mathrm{Pic}\, \overline{V} .$$

Après les avoir considérées et étudiées de deux points de vue différents, on en donne finalement des équations qui en fournissent une description directe.

(a) *Préliminaires algébriques*

On se propose d'étudier certaines suites exactes de tores attachées à une extension galoisienne bicubique K/k. On reprend les notations du §1, en particulier celles du diagramme (13).

On note $\mathbb{G}_m$ le groupe multiplicatif Spec $\mathbb{Z}[t,t^{-1}]$. On note $\mathbb{G}_{m,F}$ son extension à un corps F, dont le groupe des points à valeurs dans F est F*. On note enfin, pour F'/F séparable finie, $R_{F'/F}\mathbb{G}_m$ le F-tore descendu à la Weil de $\mathbb{G}_{m,F'}$: le groupe de ses points à valeurs dans F est F'*. Si F'/F est une extension galoisienne finie de groupe G, il y a une "dualité" entre la catégorie des F-tores déployés par F'/F et celle des G-modules $\mathbb{Z}$-libres de type fini: à un F-tore T elle associe le G-module $\hat{T}$ des caractères de T. Voir [25]. Ci-dessous, $\mathbb{G}_{m,k}$ est simplement noté $\mathbb{G}_m$.

Lemme 13. *Soit* K/k *une extension galoisienne bicubique dont on note* K_1, K_2, L_1, L_2 *les sous-corps cubiques. On a alors une suite exacte de* k*-tores:*

$$(119) \qquad 1 \longrightarrow \mathbb{G}_m \xrightarrow{\iota} R_{L_1/k}\mathbb{G}_m \times R_{L_2/k}\mathbb{G}_m \xrightarrow{\rho} \mathbb{G}_m \times R_{K/k}\mathbb{G}_m \xrightarrow{\upsilon} R_{K_1/k}\mathbb{G}_m \times R_{K_2/k}\mathbb{G}_m \xrightarrow{\pi} \mathbb{G}_m \longrightarrow 1$$

définie par:

$$(120) \qquad \begin{aligned} \iota(x) &= (x, x^{-1}) \\ \rho(x,y) &= (N_{L_1/k}(x) \cdot N_{L_2/k}(y), 1/xy) \\ \upsilon(x,y) &= (x \cdot N_{K/K_1}(x), x \cdot N_{K/K_2}(y)) \\ \pi(x,y) &= N_{K_1/k}(x)/N_{K_2/k}(y) . \end{aligned}$$

Soient $S := \ker(\upsilon)$ et $S_1 := \ker(\pi)$. Il est utile pour ce qui suit d'écrire les suites exactes de k-tores ci-après, qui sont des portions de la suite (119):

$$(121) \qquad 1 \longrightarrow S_1 \longrightarrow R_{K_1/k}\mathbb{G}_m \times R_{K_2/k}\mathbb{G}_m \xrightarrow{\pi} \mathbb{G}_m \longrightarrow 1$$

$$(122) \qquad 1 \longrightarrow \mathbb{G}_m \xrightarrow{\iota} R_{L_1/k}\mathbb{G}_m \times R_{L_2/k}\mathbb{G}_m \xrightarrow{\rho} \mathbb{G}_m \times R_{K/k}\mathbb{G}_m \xrightarrow{\upsilon} S_1 \longrightarrow 1$$

$$(123) \qquad 1 \longrightarrow \mathbb{G}_m \xrightarrow{\iota} R_{L_1/k}\mathbb{G}_m \times R_{L_2/k}\mathbb{G}_m \longrightarrow S \longrightarrow 1$$

$$(124) \qquad 1 \longrightarrow S \longrightarrow \mathbb{G}_m \times R_{K/k}\mathbb{G}_m \longrightarrow S_1 \longrightarrow 1 .$$

La démonstration de l'exactitude de la suite (119) se fait par dualité. On vérifie sans trop de peine qu'avec les notations du §1, spécialement du diagramme (13), la suite de G-modules ci-après est exacte:

$$(125) \qquad 0 \longrightarrow \mathbb{Z} \xrightarrow{\hat{\pi}} \mathbb{Z}[G/s] \oplus \mathbb{Z}[G/q] \xrightarrow{\hat{\upsilon}} \mathbb{Z} \oplus \mathbb{Z}[G] \xrightarrow{\hat{\rho}} \mathbb{Z}[G/t] \oplus \mathbb{Z}[G/r] \xrightarrow{\hat{\iota}} \mathbb{Z} \longrightarrow 0$$

$$\begin{aligned} \hat{\pi}(1) &= (N_{G/s}, -N_{G/q}) \\ \hat{\upsilon}(1,0) &= (1, N_s) \quad \text{et} \quad \hat{\upsilon}(0,1) = (1, N_q) \\ \hat{\rho}(1,0) &= (N_{G/r}, N_{G/t}) \quad \text{et} \quad \hat{\rho}(0,1) = (-1,-1) \\ \hat{\iota}(1,0) &= \hat{\iota}(0,1) = 1 . \end{aligned}$$

Si l'on prend les points à valeurs dans k de la suite exacte de k-tores (119), on obtient un complexe de groupes:

$$(126)\qquad 1 \longrightarrow k^* \xrightarrow{\ \iota\ } L_1^* \times L_2^* \xrightarrow{\ \rho\ } k^* \times K^* \xrightarrow{\ \upsilon\ } K_1^* \times K_2^* \xrightarrow{\ \pi\ } k^* \longrightarrow 1$$

qui n'est pas exact, mais si l'on prend les points à valeurs dans L_1 on obtient une suite exacte de σ-modules qui s'écrit (la suite (119) est déjà scindée sur L_1):

$$(127)\qquad 1 \longrightarrow L_1^* \xrightarrow{\ \iota\ } L_1^{*3} \times K^* \xrightarrow{\ \rho\ } L_1^* \times K^* \xrightarrow{\ \upsilon\ } K^{*2} \xrightarrow{\ \pi\ } L_1^* \longrightarrow 1$$

$$\begin{aligned}
\iota(x) &= (x, \sigma x, \sigma^2 x; x^{-1})\\
\rho(u,v,w;z) &= (u \cdot \sigma^2 v \cdot \sigma w \cdot N_t z; 1/(uz), 1/(v \cdot r(z)), 1/(w \cdot r^2(z)))\\
\upsilon(z;u,v,w) &= (zu \cdot s^2 v \cdot sw, zu \cdot q^2 v \cdot qw)\\
\pi(u,v) &= N_t(u)/N_t(v)\,,
\end{aligned}$$

moyennant les identifications:

$(k \otimes_k L_1)$	$\simeq L_1^*$	avec à droite l'action naturelle de σ sur L_1	
$(L_1 \otimes_k L_1)^*$	$\simeq L_1^* \times L_1^* \times L_1^*$	par $a \otimes b \longrightarrow (ab, a \cdot \sigma b, a \cdot \sigma^2 b)$	et $\sigma(u,v,w) = (v,w,u)$
$(L_2 \otimes_k L_1)^*$	$\simeq K^*$	par $a \otimes b \longrightarrow ab$	et σ agissant sur K via $\sigma(x) = r(x)$
$(K \otimes_k L_1)^*$	$\simeq K^* \times K^* \times K^*$	par $a \otimes b \longrightarrow (ab, a \cdot \sigma b, a \cdot \sigma^2 b)$	et l'action $\sigma(u,v,w) = (v,w,u)$
$(K_1 \otimes_k L_1)^*$	$\simeq K^*$	par $a \otimes b \longrightarrow ab$	et σ agissant sur K par $\sigma(x) = s(x)$
$(K_2 \otimes_k L_1)^*$	$\simeq K^*$	par $a \otimes b \longrightarrow ab$	et σ agissant sur K par $\sigma(x) = q(x)$.

La suite (126) n'est autre, évidemment, que la suite des invariants de (127) sous σ.

Soit F/k une extension quelconque. Les extensions (124) et (123) définissent des cobords:

$$(128)\qquad S_1(F) \xrightarrow{\ \partial\ } H^1(F,S) \hookrightarrow^{\partial} Br(F)$$

et on se propose d'expliciter l'application composée:

$$(129)\qquad S_1(F) \xrightarrow{\ \partial^2\ } Br(F)$$

qui se factorise en fait par $Br(F, F \otimes_k L_1) \cap Br(F, F \otimes_k L_2) \hookrightarrow Br(F)$. Via le choix de σ comme générateur du groupe de Galois de L_1/k, choix qui détermine l'identification $F^*/N((F\otimes_k L_1)^*)\simeq Br(F,F\otimes_k L_1)$, cette application s'interprète donc par exemple comme une application:

(129') $\quad S_1(F) \xrightarrow{\partial^2} F^*/N((F\otimes_k L_1)^*)$.

Pour simplifier les notations, nous traitons le cas $F=k$, et en fait cela n'est pas restrictif. Pour expliciter (129) pour $F=k$, il suffit d'exhiber un diagramme commutatif de σ-modules:

(130)
$$\begin{array}{ccccccccccc}
0 & \longrightarrow & \mathbb{Z} & \xrightarrow{N_\sigma} & \mathbb{Z}[\sigma] & \xrightarrow{\Delta_\sigma} & \mathbb{Z}[\sigma] & \xrightarrow{\varepsilon} & \mathbb{Z} & \longrightarrow & 0 \\
\downarrow & & \downarrow{\scriptstyle \omega_3} & & \downarrow{\scriptstyle \omega_2} & & \downarrow{\scriptstyle \omega_1} & & \downarrow{\scriptstyle \omega_0} & & \\
1 & \longrightarrow & L_1^* & \xrightarrow{\iota} & L_1^{*3}\times K^* & \xrightarrow{\rho} & L_1^*\times K^{*3} & \xrightarrow{"\upsilon"} & S_1(L_1) & \longrightarrow & 1\,.
\end{array}$$

Les lignes de ce diagramme sont exactes, la seconde est extraite de (127), où l'on a $S_1(L_1)=\ker \pi$, et s'obtient en prenant les points de (122) à valeurs dans L_1; elle est scindée comme suite de groupes abéliens.

Voici comment on peut procéder. D'après (121):

(131) $\quad S_1(k) = \{(u,v) \in K_1^*\times K_2^* \mid N_t(u) = N_t(v)\}$,

qui apparaît comme le sous-groupe des invariants sous σ agissant par $\sigma(u,v)=(su,qv)$ de $S_1(L_1)=\{(u,v)\in K^*\times K^* \mid N_t(u)=N_t(v)\}$ d'après (127).

Soit donc $(u,v)\in S_1(k)$, qui définit ω_0, par $\omega_0(1)=(u,v)$. D'après le théorème 90 appliqué à l'extension cyclique K/L_1, il existe

(132) $\quad w \in K^* \quad$ tel que $u/v = w/t(w)$.

On peut alors prendre

$$\omega_1(1) = \xi := (1;u\cdot t(w),1,1/q^2(w)))$$

car $\upsilon(\xi)=(u,v)$. On trouve:

$$\Delta_\sigma(\xi) = \xi/{}^\sigma\xi = (1;vw,q^2(w),1/(vw\cdot q^2(w)))\ .$$

D'après le théorème 90 appliqué à l'extension cyclique K/L_1, il existe

(133) $z \in K^*$ tel que $u/t(v) = z/t(z)$.

On peut alors prendre

$$\omega_2(1) = \xi' := (z/vw, r(z)/q^2(w), vw\cdot q^2(w)\cdot r^2(z); z)$$

car $\rho(\xi')=\xi/{}^{\sigma}\xi$. D'après (127), l'action de σ sur $(a,b,c;d)\in L_1^{*3}\times K^*$ est donnée par $\sigma(a,b,c;d)=(b,c,a;r(d))$ et on trouve effectivement:

$$N_\sigma(\xi') = \iota(N_r(z)).$$

On peut donc prendre:

$$\omega_3(1) = N_r(z).$$

En résumé, on obtient le résultat suivant:

Lemme 14. *Avec les hypothèses et notations du lemme* 13, *l'application composée* ∂^2:

$$S_1(k) \xrightarrow{\partial} H^1(k,S) \hookrightarrow^{\partial} k^*/N(L_1^*)$$

où $k^*/N(L_1^*)\simeq Br(k,L_1)$ *est vu comme un sous-groupe de* $Br(k)$ *via* σ, *et où*

$$S_1(k) = \{(u,v) \in K^*\times K^* \mid N_t(u) = N_t(v)\},$$

se décrit comme suit: si $(u,v)\in S_1(k)$, *il existe* $z\in K^*$ *tel que*

$$u/{}^t v = z/{}^t z,$$

et alors:

$$\partial^2(u,v) \textit{ est la classe de } N_r(z) \textit{ dans } k^*/N(L_1^*).$$

Le même description vaut évidemment pour l'application (128) où F remplace k: on remplace alors L_1 par $F\otimes_k L_1$ et K par $F\otimes_k K$. On notera d'autre part que le bord ∂: $H^1(F,S) \longrightarrow Br(F)$ déduit de la suite (123) est injectif par le théorème 90 qui assure que $H^1(F,R_{F'/F}\mathbb{G}_m)$=0 pour F'/F séparable finie, voir [25].

Lemme 15. *Avec les notations et hypothèses du lemme* 13, *les* k-*tores* S *et* S_1 *définis par* (123) *et* (121) *vérifient:*

(i) $Ш^1(G,\hat{S}) = \mathbb{Z}/3$

(ii) *si* k *est un corps des nombres,*

$$Ш^1(k,S) = Ш^1(k,S_1) = Ш^2(k,S) = 0 .$$

La notation $Ш^i(k,T)$ désigne le noyau de l'application $H^i(k,T) \longrightarrow \prod_v H^i(k_v,T)$. Le lemme assure donc que *le tore* S_1 *vérifie le principe de Hasse*, ce qui peut se démontrer d'autres façons.

L'égalité $Ш^1(k,S_1)=Ш^2(k,S)$ est donnée par le cobord déduit de la suite (124) et résulte des nullités $H^1(k,T)=Ш^2(k,T)=0$ lorsque T est isomorphe à un produit de tores du type $R_{k_1/k}\mathbb{G}_m$ (théorème 90 et injectivité de $\mathrm{Br}\, k_1 \longrightarrow \bigoplus \mathrm{Br}\, k_{1,w}$).

Par dualité, la suite (123) donne la suite exacte de G-modules:

$$(134) \qquad 0 \longrightarrow \hat{S} \longrightarrow \mathbb{Z}[G/t]\oplus\mathbb{Z}[G/r] \xrightarrow{\hat{\iota}} \mathbb{Z} \longrightarrow 0 .$$

En prenant la suite de cohomologie modifiée à la Tate, on obtient:

$$(135) \qquad \hat{H}^0(t,\mathbb{Z})\oplus\hat{H}^0(r,\mathbb{Z}) \simeq (\mathbb{Z}/3)^2 \longrightarrow \hat{H}^0(G,\mathbb{Z}) \simeq \mathbb{Z}/9 \longrightarrow H^1(G,\hat{S}) \longrightarrow 0$$

d'où (i). Si l'on considère l'application naturelle $(135) \longrightarrow \prod_v (135)_v$, il y a une infinité de places v telles que $G^v=\langle s\rangle$, auquel cas la suite $(135)_v$ s'écrit simplement $0 \longrightarrow \mathbb{Z}/3 \longrightarrow H^1(G^v,\hat{S}) \longrightarrow 0$. On voit donc qu'un générateur de $\hat{H}^0(G,\mathbb{Z}) \simeq \mathbb{Z}/9$ ne donne pas 0 dans $H^1(G^v,\hat{S})$ pour un tel v. Par suite $Ш^1(G,\hat{S})=0$ et, par la dualité de Tate-Nakayama (voir [25]): $Ш^2(k,S) \simeq Ш^1(G,\hat{S})=0$.

Le fait que $Ш^1(k,S)=0$ résulte aussitôt de la suite exacte (123) qui définit une injection $Ш^1(k,S) \hookrightarrow Ш^2(k,\mathbb{G}_m)=0$.

(b) *Construction d'un torseur sur* V *sous* S

On revient aux hypothèses et notations du §1. On considère une surface cubique diagonale V d'équation homogène

$$(1) \qquad ax^3 + by^3 + cz^3 + dt^3 = 0 \qquad\qquad a, b, c, d \in \mathbb{Q}^*$$

qu'on écrit encore:

(1') $x^3 + \lambda y^3 + \mu z^3 + \lambda\mu\nu t^3 = 0$

et on considère l'extension $K/k=k(\sqrt[3]{\lambda},\sqrt[3]{\nu})/k$ associée. On se place dans le *cas générique* - par exemple on fait l'hypothèse (24) - auquel cas K/k est une extension bicubique de $k=\mathbb{Q}(\theta)$. On peut alors considérer les suites exactes de k-tores (119), (121) à (124) introduites en (a). On fait l'hypothèse supplémentaire:

$$\prod_p V(\mathbb{Q}_p) \neq \emptyset .$$

Nous allons appliquer les préliminaires de la partie (a) à F=k(V) et considérer le cobord itéré:

(136) $S_1(k(V)) \xrightarrow{\partial} H^1(k(V),S) \hookrightarrow{\partial} Br(k(V))$.

L'équation (1') s'écrit encore:

(1") $N_t(x+\alpha y) = -\mu \cdot N_t(z+\beta t)$.

Considérons les fonctions:

(137) $\varphi = (x+\alpha y)/t \in K_1(V)^*$

(138) $\psi = (z+\beta t)/t \in K_2(V)^*$.

Comme par hypothèse V a des points dans chaque complété de $\mathbb{Q}$ et que l'équation de V s'écrit:

(1") $N_t(\varphi/\psi) = -\mu$ $\varphi \in K_1(V)^*,\ \psi \in K_2(V)^*$,

la k-variété d'équation

$$N_t(\xi_1/\xi_2) = -\mu \qquad \xi_1 \in K_1^*, \quad \xi_2 \in K_2^* ,$$

a des points dans chaque complété de k. Elle en a donc dans k car c'est un espace principal homogène sous le k-tore S_1 et celui-ci vérifie le principe de Hasse (cf. lemme 15). Il existe donc

(139) $\varepsilon_\alpha \in K_1^*$ et $\varepsilon_\beta \in K_2^*$ tels que $N_t(\varepsilon_\alpha/\varepsilon_\beta) = -\mu$.

En fait, on pourrait opérer sur $\mathbb{Q}$ et trouver $\varepsilon_\alpha \in \mathbb{Q}(\alpha)^*$ et $\varepsilon_\beta \in \mathbb{Q}(\beta)^*$ ayant les mêmes propriétés. On déduit alors de (1") et (139) l'égalité $N_t(\varphi/\varepsilon_\alpha)=N_t(\psi/\varepsilon_\beta)$ avec

$\varphi/\varepsilon_\alpha \in K_1(V)^*$ et $\psi/\varepsilon_\beta \in K_2(V)^*$. Autrement dit:

(140) $\quad (\varphi/\varepsilon_\alpha, \psi/\varepsilon_\beta) \in S_1(k(V))$.

Lemme 16. *Sous les hypothèses, et avec les notations, ci-dessus, le* k(V)*-torseur*

$$\partial(\varphi/\varepsilon_\alpha, \psi/\varepsilon_\beta) \in H^1(k(V),S)$$

où ∂ *est le cobord défini par l'extension* (124), *se prolonge en un torseur*

$$\mathcal{T} \in H^1(V,S)$$

sur V *sous le tore* S.

Dans cet énoncé on note V la variété sur k, au lieu de V_k, et on identifie un torseur et sa classe dans le H^1, mais ces abus ne sont pas gênants ici.

Soit $T := \mathbb{G}_m \times R_{K/k}\mathbb{G}_m$ le tore médian de l'extension (124). On a le diagramme commutatif, à lignes et colonnes exactes (cf. $[8]_{\mathrm{II}}$):

(141)
$$\begin{array}{ccccc} & & & & \mathrm{Hom}_k(\hat{T}, \mathrm{Div}\, V_K) \\ & & & & \Big\downarrow \pi \\ & & S_1(k(V)) & \xrightarrow{\ \mathrm{div}\ } & \mathrm{Hom}_k(\hat{S}_1, \mathrm{Div}\, V_K) \\ & & \Big\downarrow \partial & & \Big\downarrow \partial \\ H^1(V,S) & \xrightarrow{\ j\ } & H^1(k(V),S) & \longrightarrow & \mathrm{Ext}^1_k(\hat{S}, \mathrm{Div}\, V_K). \end{array}$$

Les colonnes sont tirées de la suite exacte de G-modules duale de la suite exacte de tores (124), celle de droite est donc exacte. Le morphisme π est induit par le morphisme $\hat{\upsilon}$ qui s'écrit:

$$\begin{array}{ccc} \mathbb{Z}[G/s] \oplus \mathbb{Z}[G/q] & \longrightarrow & \mathbb{Z} \oplus \mathbb{Z}[G] = \hat{T} \\ (u,v) & \longrightarrow & (\varepsilon(u) + \varepsilon(v), uN_s + vN_q) . \end{array}$$

Il s'écrit donc lui-même:

$$\begin{array}{ccc} \mathrm{Div}(V) \oplus \mathrm{Div}(V_K) & \longrightarrow & \mathrm{Div}(V_{K_1}) \oplus \mathrm{Div}(V_{K_2}) \\ (D,D') & \longrightarrow & (D + N_s(D'), D + N_q(D')) . \end{array}$$

Pour établir le lemme, il faut donc montrer que $\mathrm{div}(\varphi/\varepsilon_\alpha,\psi/\varepsilon_\beta)$ appartient à l'image de π. Comme l'équation (1) s'écrit encore:

$$(1''') \qquad N_q(x+\alpha y) = -\mu\cdot N_s(z+\beta t)\,,$$

on trouve:

$$\mathrm{div}(\varphi/\varepsilon_\alpha,\psi/\varepsilon_\beta) = (\mathrm{div}(\varphi),\mathrm{div}(\psi)) = (N_s(L(0)) - \mathrm{div}(t), N_q(L(0)) - \mathrm{div}(t))$$

où

$$L(0) = \begin{cases} x+\alpha y = 0 \\ z+\beta t \;= 0\,. \end{cases}$$

Ceci achève la preuve du lemme.

<u>Remarque 6:</u> $\partial(\varphi/\varepsilon_\alpha,\psi/\varepsilon_\beta)\in H^1(k(V),S)$ dépend du choix de $(\varepsilon_\alpha,\varepsilon_\beta)$, mais il est immédiat qu'il n'en dépend qu'à un élément de $H^1(k,S)$ près. Quant au prolongement $\mathcal{T}$ à V tout entière, il n'est a priori pas unique, j n'étant pas nécessairement injective, mais, V étant lisse, la valeur de $\mathcal{T}$ en un point de V ne dépend en fait que de l'image de $\mathcal{T}$ dans $H^1(k(V),S)$, voir $[8]_{\mathrm{II}}$.

(c) *Interprétation comme torseur de type* i

On conserve les hypothèses et notations de la partie (b). Pour la notion de torseur d'un type donné, voir [8]. Ici i: $\mathrm{Pic}\, V_K \longrightarrow \mathrm{Pic}\, \overline{V}$ est l'injection canonique.

Lemme 17. *Avec les hypothèses et notations antérieures, on a le diagramme commutatif de* G*-modules suivant, dont les lignes sont exactes:*

$$\begin{array}{ccccccc}
1 & \longrightarrow & (-\mu)^{\mathbb{Z}} & \longrightarrow & \varphi^{\mathbb{Z}[G/s]}\cdot\psi^{\mathbb{Z}[G/q]} & \xrightarrow{\text{div}} & \\
 & & \big\downarrow & & \big\downarrow & & \\
1 & \longrightarrow & K^* & \longrightarrow & K(V)^* & \xrightarrow{\text{div}} &
\end{array}$$

$$\begin{array}{ccccc}
\mathbb{Z}\cdot H_t \oplus \mathbb{Z}[G]\cdot L(0) & \xrightarrow{\text{cl}} & \text{Pic}(V_K) & \longrightarrow & 0 \\
\big\downarrow & & \big\| & & \\
\text{Div}(V_K) & \xrightarrow{\text{cl}} & \text{Pic}(V_K) & \longrightarrow & 0\,.
\end{array}$$

En particulier, les G*-modules* $\text{Pic}(V_K)$ *et* $\hat{S}$ *sont isomorphes.*

Dans ce diagramme, les flèches verticales sont les plongements naturels évidents.

La première ligne s'écrit de manière "abstraite":

$$0 \longrightarrow \mathbb{Z} \longrightarrow Q := \mathbb{Z}[G/s] \oplus \mathbb{Z}[G/q] \longrightarrow P := \mathbb{Z} \oplus \mathbb{Z}[G] \longrightarrow \hat{S} = \text{Pic}(V_K) \longrightarrow 0\,,$$

et on vérifie aisément que ce n'est autre que la suite exacte de G-modules (125).

Pour la démonstration, on reprend les notations du §1 et on en utilise certains résultats. Tout d'abord, d'après le tableau (17), le sous-G-module de $\text{Pic}(V_K)$ engendré par la classe de $L(0)$ contient les classes des $L(i)$, $L'(i)$ et $L''(i)$, donc ℓ d'après (19) et M d'après (20). C'est donc $\text{Pic}(V_K)$ tout entier.

Soit P le sous-G-module de $\text{Div}(V_K)$ engendré par le diviseur H_t de la section hyperplane $t=0$ et par $L(0)$. On voit aisément, cf. (17), que $P \simeq \mathbb{Z} \oplus \mathbb{Z}[G]$ comme G-modules. D'autre part, le noyau de l'application "classe": $P \xrightarrow{\text{cl}} \text{Pic}(V_K)$ est engendré par les diviseurs des fonctions φ et ψ:

$$\text{(142)}\qquad \begin{aligned} \text{div}(\varphi) &= N_s(L(0)) - H_t = L(0) + L'(0) + L''(0) - H_t \\ \text{div}(\psi) &= N_q(L(0)) - H_t = L(0) + L'(2) + L''(1) - H_t\,. \end{aligned}$$

Enfin, le sous-G-module Q de $K(V)^*$ engendré par les fonctions φ et ψ a pour trace sur K^* le G-module trivial $(-\mu)^{\mathbb{Z}}$, compte tenu de la relation

$$\text{(1'')}\qquad N_t(\varphi) = -\mu N_t(\psi)\,,$$

et du fait que le morphisme div: $Q \longrightarrow P$ s'identifiant de façon abstraite au morphisme $\hat{\upsilon}: \mathbb{Z}[G/s]\oplus\mathbb{Z}[G/q] \longrightarrow \mathbb{Z}\oplus\mathbb{Z}[G]$ - comme il résulte des formules (142) - à condition de prendre $-H_t$ comme générateur du module $\mathbb{Z}\cdot H_t$, on sait que son noyau est engendré dans $\mathbb{Z}[G/s]\oplus\mathbb{Z}[G/q]$ par $\hat{\pi}(1)=(N_t,-N_t)$.

<u>Remarque 7:</u> Ce lemme donne une présentation du G-module Pic V_K, alors que les calculs du §1 avaient évité, pour plus de simplicité, l'étude de la structure de G-module de Pic V_K. Les résultats plus forts ci-dessus - lemmes 17 et 15 (i) - redonnent le lemme 2, à savoir $Br(V)/Br(k)\simeq\mathbb{Z}/3$.

Avec les notations et hypothèses antérieures, S est donc le k-tore dual du G-module Pic V_K. Si i désigne l'injection naturelle Pic $V_K \hookrightarrow$ Pic $\overline{V}$, un *torseur* $\mathcal{T}$ *sur* V *de type* i est un torseur sur V sous S dont l'image par l'application naturelle

$$H^1(V,S) \longrightarrow \mathrm{Hom}_k(\hat{S},\mathrm{Pic}\,\overline{V}) = \mathrm{Hom}_k(\mathrm{Pic}\,V_K,\mathrm{Pic}\,\overline{V})$$

est i.

Proposition 8. *On garde les hypothèses et notations de* (b), *lemme* 16. *Le torseur* $\partial(\varphi/\varepsilon_\alpha,\psi/\varepsilon_\beta)\in H^1(k(V),S)$ *se prolonge en un torseur* $\mathcal{T}\in H^1(V,S)$ *de type* i *et un seul.*

Pour établir ce résultat, on fait appel à la théorie exposée dans [8], et plus particulièrement $[8]_{II}$, §II (description locale des torseurs). Soit Y le fermé de V réunion des 9 droites L(i), L'(i), L''(i) et de la section hyperplane H_t définie par t=0. Soit U l'ouvert complémentaire. D'après le lemme 17, et avec les notations du lemme 13, si $\mathfrak{g}$ désigne le groupe de Galois de $\bar{k}/k$, on a le diagramme commutatif suivant de $\mathfrak{g}$-modules:

$$(143)\qquad \begin{array}{ccccccccc} 1 & \longrightarrow & \bar{k}[U]^*/\bar{k}^* & \xrightarrow{\mathrm{div}} & \mathrm{Div}_{\overline{Y}}\overline{V} & \xrightarrow{\mathrm{cl}} & \mathrm{Pic}\,\overline{V} & \longrightarrow & 0 \\ & & \uparrow\chi & & \uparrow\kappa & & \uparrow i & & \\ 0 & \longrightarrow & \hat{S}_1 & \longrightarrow & \hat{P} & \longrightarrow & \hat{S} & \longrightarrow & 0 \end{array}$$

dont les lignes sont exactes; celle du bas est la suite de G-modules, donc de $\mathfrak{g}$-modules, duale de la suite exacte de tores (124), et est extraite de (125); celle du haut est la suite naturelle évidente. On a $P=\mathbb{Z}\oplus\mathbb{Z}[G]$ et κ est définie par $\kappa(1,0)=-H_t$ et $\kappa(0,1)=L(0)$. Enfin, toujours d'après le lemme 17, si l'on considère la suite exacte de

G-modules duale de la suite de tores (121), voir le début de (125), l'application χ est induite par $\tilde{\chi}$:

(144)
$$\begin{array}{ccccccccc} & & & & & & \bar{k}[U]^*/\bar{k}^* & & \\ & & & & & \overset{\tilde{\chi}}{\nearrow} & \uparrow \chi & & \\ 0 & \longrightarrow & \mathbb{Z} & \xrightarrow{\hat{\pi}} & \mathbb{Z}[G/s] \oplus \mathbb{Z}[G/q] & \longrightarrow & \hat{S}_1 & \longrightarrow & 0, \end{array}$$

où $\tilde{\chi}$ est définie par $\tilde{\chi}(1,0)$=classe de $\varphi \bmod \bar{k}^*$ et $\tilde{\chi}(0,1)$=classe de $\psi \bmod \bar{k}^*$. De plus, si l'on considère $\varepsilon_\alpha \in K_1^*$ et $\varepsilon_\beta \in K_2^*$ tels que $N_t(\varepsilon_\alpha/\varepsilon_\beta)=-\mu$, cf. (139), et le $\mathfrak{g}$-morphisme $\tilde{\xi}: \mathbb{Z}[G/s] \oplus \mathbb{Z}[G/q] \longrightarrow \bar{k}[U]^*$ défini par $\tilde{\xi}(1,0)=\varphi/\varepsilon_\alpha$ et par $\tilde{\xi}(0,1)=\psi/\varepsilon_\beta$, c'est un relèvement de $\tilde{\chi}$ qui est trivial sur $\hat{\pi}(\mathbb{Z})$ et induit donc un relèvement $\xi: \hat{S}_1 \longrightarrow \bar{k}[U]^*$ de χ. On obtient donc le diagramme commutatif suivant:

(145)
$$\begin{array}{ccccccccc} & & & & \bar{k}[U]^* & \longrightarrow & \bar{k}[U]^*/\bar{k}^* & & \\ & & & & \tilde{\xi}\uparrow & \overset{\xi}{\nwarrow} & \uparrow \chi & & \\ 0 & \longrightarrow & \mathbb{Z} & \xrightarrow{\hat{\pi}} & \mathbb{Z}[G/s] \oplus \mathbb{Z}[G/q] & \longrightarrow & \hat{S}_1 & \longrightarrow & 0. \end{array}$$

De fait, les $\mathfrak{g}$-morphismes χ, ξ et $\tilde{\xi}$ aboutissent dans $K[U]^*/K^*$ et $K[U]^*$ respectivement.

Comme l'algèbre des fonctions régulières sur la variété affine S_1 est l'algèbre des $\mathfrak{g}$-invariants de l'algèbre de groupe $\bar{k}[\hat{S}_1]$, le $\mathfrak{g}$-morphisme $\xi: \hat{S} \longrightarrow \bar{k}[U]^*$ définit un k-morphisme $\Phi: U \longrightarrow S_1$ qui n'est autre que $(\varphi/\varepsilon_\alpha, \psi/\varepsilon_\beta) \in S_1(U)$ - où l'on considère $S_1(U)$ comme plongé dans $\mathbb{G}_m(U_{K_1}) \times \mathbb{G}_m(U_{K_2})$ par définition même, cf. (121). La suite exacte de k-tores (124) fait du tore médian $\mathbb{G}_m \times R_{K/k}\mathbb{G}_m$ un torseur sur la k-variété S_1 sous le k-tore S. On en déduit, par pull-back via Φ, un torseur $\mathcal{T}_U$ sur U sous le k-tore S:

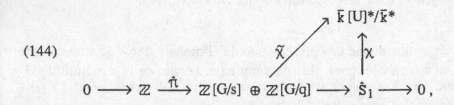

$$
\begin{array}{ccc}
\mathcal{T}_U & \longrightarrow & \mathbb{G}_m \times R_{K/k}\mathbb{G}_m \\
\downarrow & & \downarrow \upsilon \\
U & \xrightarrow{\ \Phi\ } & S_1 .
\end{array}
\tag{146}
$$

Or, d'après $[8]_{II}$, § II (description locale des torseurs), la construction du torseur $\mathcal{T}_U$ à partir d'un diagramme tel que (143) et d'un relèvement ξ de χ, fournit un torseur qui est la restriction à U d'un torseur sur V de type i, et celui-ci est d'ailleurs unique comme on le verra ci-après.

Par construction même de $\mathcal{T}_U$, sa fibre générique $(\mathcal{T}_U)_{k(V)} \in H^1(k(V),S)$ n'est autre que $\partial(\Phi_{k(V)}) = \partial(\varphi/\varepsilon_\alpha, \psi/\varepsilon_\beta)$ où $\partial: S_1(k(V)) \longrightarrow H^1(k(V),S)$ désigne le cobord déduit de la suite exacte de k-tores (124) qui définit précisément le torseur-type υ sur S_1 sous S.

D'autre part, comme $\prod_v V(k_v) \neq \emptyset$, les applications $H^1(k_v,S) \longrightarrow H^1(k_v(V),S)$ sont injectives. Comme, d'après le lemme 15, $Ш^1(k,S)=0$, on en déduit que l'application naturelle $H^1(k,S) \longrightarrow H^1(k(V),S)$ est elle-même injective. Comme un torseur de type i est défini à un élément de $H^1(k,S)$ près, on voit donc qu'un *torseur de type* i *est déterminé par sa restriction au point générique de* V, et a fortiori par sa restriction à U.

Désormais, on désignera par $\mathcal{T} \in H^1(V,S)$ *le torseur de type* i *qui prolonge à* V *le torseur* $\mathcal{T}_U$ *obtenu par pull-back de* υ *via* $\Phi = (\varphi/\varepsilon_\alpha, \psi/\varepsilon_\beta) \in S_1(U)$. C'est aussi le torseur de type i qui prolonge le torseur $\partial(\varphi/\varepsilon_\alpha, \psi/\varepsilon_\beta) \in H^1(k(V),S)$.

On notera que cette proposition améliore et précise le lemme 16. Elle fournit en outre une autre preuve du fait que le torseur $\partial(\varphi/\varepsilon_\alpha, \psi/\varepsilon_\beta)$ se prolonge à V.

Avant d'énoncer le corollaire qui suit, rappelons en quoi consiste l'*obstruction* $\mathcal{O}_\omega$ *définie par un type donné* $\omega: \hat{T} \longrightarrow \mathrm{Pic}\,\overline{V}$ *de torseur sur* V : "pour tout torseur T de type ω il existe une place v de k telle que $T(k_v)=\emptyset$ ". Lorsqu'il existe un torseur T de type ω, cette obstruction équivaut à la suivante: "pour tout $z \in H^1(k,T)$, il existe une place v de k telle que $T^z(k_v)=\emptyset$ ". Enfin, si l'on introduit, pour chaque place v de k, le morphisme composé

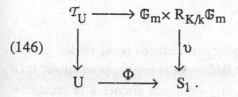

(147) $\tau_v : H^1(k_v,T) \xrightarrow{\simeq} H^1(k_v,\hat{T})^{\sim} \longrightarrow H^1(k,\hat{T})^{\sim}$

de la dualité de Tate-Nakayama locale et de l'application duale de la restriction de k à k_v - ~ désignant la dualité de Pontrjagin Hom(·,$\mathbb{Q}/\mathbb{Z}$) - et si l'on suppose encore qu'il existe un torseur T de type ω, l'obstruction Θ_ω équivaut encore à la condition suivante: "pour tout $(x_v) \in \prod_v V(k_v)$, on a

(148) $\sum_v \tau_v(T(x_v)) \neq 0$."

Pour tout ce qui précède, voir $[8]_{II}$, § III. Dans ce qui suit, on prend pour k-tore T le tore S et pour ω l'injection naturelle i: Pic $V_K \longrightarrow$ Pic $\overline{V}$. Comme on l'a déjà vu il existe bien alors un torseur de type i, ce qui résulte d'ailleurs a priori, d'après $[8]_{II}$, de $Ш^2(k,S)=0$, cf. lemme 15.

Corollaire. *On garde les hypothèses et notations de* (b), *lemme* 16. *On désigne par* $\mathcal{T}$ *le torseur de type* i *qui prolonge à* V *le torseur* $\mathcal{T}_U$ *défini par* (146), *ou, ce qui revient au même, le torseur* $\partial(\varphi/\varepsilon_\alpha,\psi/\varepsilon_\beta) \in H^1(k(V),S)$. *Alors:*

(i) *l'obstruction de Manin pour* V *équivaut à celle définie par le torseur* $\mathcal{T}$*;*

(ii) $\partial(\mathcal{T})$ *engendre* Br(V)/Br(k), *où* ∂ *désigne le cobord défini par la suite* (123)*;*

(iii) *si l'obstruction de Manin est "vide" pour* V, *on peut choisir* $(\varepsilon_\alpha,\varepsilon_\beta) \in S_1(k)$ *de telle sorte que le torseur* $\mathcal{T}$ *de type* i *associé ait des points dans chaque complété* k_v, *et alors l'obstruction de Manin pour* $\mathcal{T}$ *est encore vide.*

Dans (i), on appelle *obstruction définie par le torseur* $\mathcal{T}$ *la condition* (148). C'est une obstruction au fait que V possède un point rationnel, et elle équivaut à l'obstruction Θ_i définie de diverses façons plus haut. Les divers torseurs de type i s'obtiennent par torsion de l'un d'eux par un élément $z \in H^1(k,S)$, dont il est utile de rappeler la présentation qu'en donnent les suites (124) et (121):

(149) $k^* \times K^* \xrightarrow{\upsilon} \ker(K_1^* \times K_2^* \xrightarrow{\pi} k^*) \longrightarrow H^1(k,S) \longrightarrow 0$.

Les divers torseurs de type i sont donc les prolongements à V des divers torseurs $\partial(\varphi/\varepsilon_\alpha,\psi/\varepsilon_\beta) \in H^1(k(V),S)$ pour $(\varepsilon_\alpha,\varepsilon_\beta) \in S_1(k) = \ker(K_1^* \times K_2^* \xrightarrow{\pi} k^*)$.

Comme V est K-rationnelle, l'assertion (i) vaut pour les torseurs de type i, aussi bien que pour les torseurs *universels*, d'après $[8]_{II}$, § III. On en donnera néanmoins une autre démonstration s'appuyant sur (ii), grâce à (ii) $\Rightarrow$ (i).

Nous allons établir l'implication (ii) $\Rightarrow$ (i). Elle résulte aussitôt de la formule:

$$(150) \qquad \sum_v \mathrm{inv}_v((\partial\mathcal{T})(x_v)) = (\sum_v \tau_v(\mathcal{T}(x_v)))(\hat{\partial}(1))$$

où $\partial\mathcal{T}\in\mathrm{Br}(V)$ est défini par le cobord de la suite exacte de k-tores (123), où $\hat{\partial}(1)\in H^1(k,\hat{S})$ est l'image de $1\in H^0(k,\mathbb{Z})=\mathbb{Z}$ par le cobord $\hat{\partial}$ de la suite exacte de $\mathfrak{g}$-modules duale de (123):

$$(151) \qquad 0 \longrightarrow \hat{S} \longrightarrow \mathbb{Z}[G/t]\oplus\mathbb{Z}[G/r] \xrightarrow{\ \iota\ } \mathbb{Z} \longrightarrow 0,$$

et où $(x_v)\in\prod_v V(k_v)$. Montrons d'abord la formule (150). Soit $z_v:=\mathcal{T}(x_v)\in H^1(k_v,S)$. Par fonctorialité, $\partial(z_v)=(\partial\mathcal{T})(x_v)\in\mathrm{Br}(k_v)$. La définition du cup-produit et des cobords définis par les suites exactes (123) et (151) donnent par un calcul standard utilisant l'accouplement $M\times\hat{M} \longrightarrow \mathbb{G}_m$ où $M:=R_{L_1/k}\mathbb{G}_m\times R_{L_2/k}\mathbb{G}_m$ est le terme médian de (123) et $\hat{M}$ celui de (151):

$$(152)_v \qquad z_v \cup \hat{\partial}_v(1) = \partial(z_v) \qquad \text{dans } \mathrm{Br}(k_v).$$

Dans cette formule, $\hat{\partial}_v$ désigne le cobord $H^0(k_v,\mathbb{Z}) \longrightarrow H^1(k_v,\hat{S})$ défini par (151). Or, par définition même, $(\tau_v(z_v))(\hat{\partial}(1))=\mathrm{inv}_v(z_v\cup\hat{\partial}(1))$. D'où la formule (150) en prenant la somme $\sum_v \mathrm{inv}_v((152)_v)$.

Comme le $\mathfrak{g}$-module $\hat{M}$ est de permutation, la suite (151) définit une surjection $\mathbb{Z}=H^0(\mathfrak{g},\mathbb{Z}) \xrightarrow{\ \hat{\partial}\ } H^1(\mathfrak{g},\hat{S})=:H^1(k,\hat{S})$. Ainsi, $\hat{\partial}(1)$ engendre $H^1(k,\hat{S})\simeq\mathbb{Z}/3$ (lemme 15) et la condition (148) pour $\mathcal{T}$ équivaut donc à la non-nullité du membre de droite dans le formule (150). Si $\partial(\mathcal{T})$ engendre $\mathrm{Br}(V)/\mathrm{Br}(k)\simeq\mathbb{Z}/3$, on peut calculer l'obstruction de Manin au moyen de $\mathcal{A}:=\partial(\mathcal{T})$, cf. lemme 4, et elle équivaut donc à la non-nullité du membre de gauche de la formule (150). Ceci achève de prouver l'implication (ii) $\Rightarrow$ (i).

Nous allons démontrer (ii). La suite exacte (123) fournit le diagramme commutatif suivant, dont les lignes et colonnes sont exactes:

$$
(153)\qquad
\begin{array}{ccccccc}
 & & H^1(k,S) & \xrightarrow{\partial} & Br(k) & \longrightarrow & H^2(k,M) \\
 & & \big\downarrow & & \big\downarrow & & \big\downarrow \\
H^1(V,M) & \longrightarrow & H^1(V,S) & \xrightarrow{\partial} & Br(V) & \longrightarrow & H^2(V,M) \\
\big\downarrow & & \big\downarrow{\scriptstyle\omega} & & & & \\
Hom_{\mathfrak{g}}(\hat{M},Pic\,\overline{V}) & \longrightarrow & Hom_{\mathfrak{g}}(\hat{S},Pic\,\overline{V}) & \xrightarrow{\partial} & Ext^1_{\mathfrak{g}}(\mathbb{Z},Pic\,\overline{V}) & & \\
 & & & & \| & & \\
 & & & & H^1(k,Pic\,\overline{V}) & &
\end{array}
$$

et dont les trois verticales du haut sont injectives - pour la première c'est un fait général, pour les deux autres c'est vrai au niveau de chaque k_v car $V(k_v)\neq\emptyset$ et $Ш^2(k,\mathbb{G}_m)=Ш^2(k,M)=0$. Supposons que $\partial(\mathcal{T})$ appartienne à l'image de Br(k) dans Br(V). Une chasse facile dans la partie supérieure du diagramme montre qu'on peut alors modifier $\mathcal{T}$ par $z\in H^1(k,S)$ de telle sorte que $\partial(\mathcal{T}^z)=0$, ce qui implique, par une chasse dans la partie gauche, que $\partial\omega(\mathcal{T}^z)=0$. Or, $\omega(\mathcal{T}^z)=\omega(\mathcal{T})=i$ et $\partial(i)$ n'est autre que l'extension de $\mathfrak{g}$-modules obtenue par push-out de (151) par i. C'est aussi l'image de $\hat{\partial}(1)\in H^1(k,Pic\,V_K)=H^1(\mathfrak{g},\hat{S})$ par l'application induite par i:

$$H^1(\mathfrak{g},Pic\,V_K)\longrightarrow H^1(\mathfrak{g},Pic\,\overline{V})\,.$$

Or, comme V est K-rationnelle, cette application s'identifie d'après les suites exactes d'inflation-restriction à l'application $H^1(G,Pic\,V_K)\longrightarrow H^1(G,(Pic\,\overline{V})^{\mathfrak{h}})$ où $\mathfrak{h}$ est le groupe de Galois de $\bar{k}/K$, et cette application est l'identité car évidemment $(Pic\,\overline{V})^{\mathfrak{h}}=Pic\,V_K$. Comme $\hat{\partial}(1)\neq0$ - voir plus haut - $\partial(i)\neq0$, ce qui montre que l'hypothèse initiale est contradictoire, et établit donc (ii). On obtient ainsi également une autre démonstration de (i) grâce à l'implication (ii) $\Rightarrow$ (i) établie ci-dessus.

Si l'obstruction de Manin est "vide" pour V, il existe d'après $[8]_{II}$, §III, un torseur *universel* qui a des points dans chaque complété k_v de k. Il existe donc aussi un torseur de type i ayant la même propriété. De plus, il est établi dans $[8]_I$, §II que, pour une k-compactification lisse $\mathcal{T}^c$ d'un torseur $\mathcal{T}$ de type i, on trouve, comme pour les torseurs universels: $H^1(k,Pic\,\overline{\mathcal{T}}^c)=0$. L'obstruction de Manin pour une telle variété est donc automatiquement vide lorsqu'elle a des points dans chaque complété de k. Par définition même, l'*obstruction de Manin pour* $\mathcal{T}$, qui n'est pas complète, est l'obstruction de Manin pour $\mathcal{T}^c$. Ceci achève donc de prouver le corollaire.

Remarque 8: Les torseurs de type i ne sont pas *admissibles* au sens de $[8]_{II}$, car le conoyau de i n'est pas de permutation. En effet, avec les notations du §1, on a la suite exacte de $\mathfrak{G}$-modules

$$(154) \qquad 0 \longrightarrow \operatorname{Pic} V_K \xrightarrow{\;i\;} \operatorname{Pic} \overline{\nabla} \longrightarrow \mathbb{Z}[w]/\mathbb{Z} \longrightarrow 0$$

où $\operatorname{Pic} \overline{\nabla} = \operatorname{Pic} V_{K'}$ et où le groupe de Galois G' de K'/k est isomorphe à $G \times \langle w \rangle$. Néanmoins, pour le *principe de Hasse* et l'*obstruction de Manin*, les torseurs de type i se comportent exactement comme les torseurs admissibles.

(d) $\partial(\mathcal{T})$ *comme algèbre cyclique*

On va écrire $\partial(\varphi/\varepsilon_\alpha, \psi/\varepsilon_\beta) \in \operatorname{Br}(k(V))$ sous la forme d'une algèbre cyclique $(h', L_1(V)/k(V), \sigma)$ de la même manière qu'on a déjà écrit $\mathcal{A}_{\varepsilon,\eta}$ au §4, remarque 4, sous la forme $(h, L_1(V)/k(V), \sigma)$ avec $h \in k(V)^*$ défini par

$$(88) \qquad h = N_s(\omega) \cdot f/\eta$$

$$(86') \qquad \Delta_t(\omega) = g/\varepsilon \qquad\qquad \omega \in K(V)^* .$$

Le théorème 90 appliqué à l'extension cyclique $K(V)/L_1(V)$ assure l'existence de $\omega' \in K(V)^*$ telle que

$$(155) \qquad \Delta_t(\omega') = (\varphi/\varepsilon_\alpha)/{}^t(\psi/\varepsilon_\beta) .$$

D'après le lemme 14,

$$(156) \qquad \begin{aligned} &\partial(\varphi/\varepsilon_\alpha, \psi/\varepsilon_\beta) = (h', L_1(V)/k(V), \sigma) \in \operatorname{Br}(k(V), L_1(V)) \\ &h' = N_r(\omega') \in k(V)^* . \end{aligned}$$

Considérons alors

$$\varepsilon = {}^q\varepsilon_\alpha / {}^t\varepsilon_\beta .$$

On a bien $N_t(\varepsilon) = -\mu$. Soit alors $\eta \in K^*$ vérifiant $\Delta_t(\eta) = -\mu/N_s(\varepsilon)$. Un calcul facile montre que

(157) $\quad {}^{q}((\varphi/\varepsilon_\alpha)/{}^{t}(\psi/\varepsilon_\beta)) = g/\varepsilon$,

car déjà ${}^{q}(\varphi/{}^{t}\psi)=g$. Considérons alors:

$$\omega = {}^{q}\omega' .$$

Les relations (155) et (157) donnent aussitôt

(86') $\quad \Delta_t(\omega) = g/\varepsilon$.

Il en résulte, comme on l'a déjà vu au §4, remarque 4, que

$$h := N_s(\omega)\cdot f/\eta \quad \text{appartient à } k(V)^*$$

et que, si, avec les notations du §1, cf. remarque 2, on pose

$$D := L'(2) - L''(0) = \Delta_r L'(2) \in \mathrm{Div}(V_K) ,$$

alors:

$$D' := D + \mathrm{div}(\omega) \quad \text{appartient à } \mathrm{Div}(V_{L_1}) \quad \text{et} \quad \mathrm{div}(h) = N_\sigma(D') .$$

Comme $h'=N_r(\omega')\in k(V)^*$, on a $N_r(\omega)=N_r(\omega')=h'$. Comme en outre $D=\Delta_r L'(2)$, on obtient

$$\mathrm{div}(h) = N_\sigma(D') = N_r(D') = N_r(\mathrm{div}(\omega)) = \mathrm{div}(N_r(\omega)) = \mathrm{div}(h') ,$$

d'où finalement:

(158) $\quad h'/h \in k^*$.

Les algèbres cycliques $(h',L_1(V)/k(V),\sigma)$ et $(h,L_1(V)/k(V),\sigma)\in \mathrm{Br}(k(V),L_1(V))$ définissent donc le même élément dans $\mathrm{Br}(k(V))/\mathrm{Br}(k)$. Or, d'après la remarque 4, l'algèbre $\mathcal{A}_{\varepsilon,\eta}$ s'écrit au point générique de V comme l'algèbre $(h,L_1(V)/k(V),\sigma)$, et d'après ce qui précède, l'algèbre $\partial(\mathcal{T})$ s'écrit au point générique de V comme l'algèbre $(h',L_1(V)/k(V),\sigma)$. Il en résulte que les algèbres $\mathcal{A}_{\varepsilon,\eta}$ et $\partial(\mathcal{T})$ définissent le même élément dans $\mathrm{Br}(V)/\mathrm{Br}(k) \hookrightarrow \mathrm{Br}(k(V))/\mathrm{Br}(k)$. En particulier, comme $\mathcal{A}_{\varepsilon,\eta}$ engendre $\mathrm{Br}(V)/\mathrm{Br}(k)$, il en est de même de $\partial(\mathcal{T})$, ce qui prouve de nouveau l'assertion (ii) du corollaire précédent: $\partial(\mathcal{T})$ *engendre* $\mathrm{Br}(V)/\mathrm{Br}(k)$.

On peut résumer les résultats ci-dessus dans l'énoncé suivant:

Proposition 9. *On reprend les hypothèses et notations du §4. L'obstruction de Manin pour* V *peut se calculer au moyen de l'algèbre cyclique*

$$(h',L_1(V)/k(V),\sigma)$$

où $h'=N_r(\omega')\in k(V)^*$, *la fonction* $\omega'\in K(V)^*$ *étant déterminée, après le choix préalable de* $(\varepsilon_\alpha,\varepsilon_\beta)\in K_1^*\times K_2^*$ *tel que* $N_t(\varepsilon_\alpha/\varepsilon_\beta)=-\mu$, *par la relation*

$$(155)\qquad \Delta_t(\omega')=(\varphi/\varepsilon_\alpha)/{}^t(\psi/\varepsilon_\beta)\,.$$

Une fois h' déterminée, le calcul est très simple. L'obstruction de Manin est "vide" si, et seulement si, il existe $(x_v)\in\prod_v V(k_v)$ tel que

$$(159)\qquad \sum_v [h'(x_v),\nu]_v=0\,.$$

Néanmoins, cette méthode de calcul achoppe sur le calcul de ω', cf. remarque 4.

(e) *Des équations pour* $\mathcal{T}$

Les *variétés de descente* que nous avons associées à V comme auxiliaires pour l'étude de l'obstruction de Manin pour V sont les *torseurs* $\mathcal{T}$ *de type* i. Elles sont paramétrées par

$$H^1(k,S)=\ker((K_1^*\times K_2^*)/(\Delta(k^*)\cdot N(K^*))\xrightarrow{\ \pi\ }k^*)\,,$$

où Δ est l'application diagonale et $N=(N_{K/K_1},N_{K/K_2})$, alors que π est le quotient des normes $N_{K_1/k}/N_{K_2/k}$. A tout $(\varepsilon_\alpha,\varepsilon_\beta)\in K_1^*\times K_2^*$ tel que $N_t(\varepsilon_\alpha/\varepsilon_\beta)=-\mu$ est ainsi associée une variété de descente $\mathcal{T}$ qui est un torseur de type i. D'après le corollaire de la proposition 8, l'obstruction de Manin disparaît pour une telle variété, supposée avoir des points dans chaque complété de k. Il paraît alors raisonnable de poser la question suivante:

Question: Les torseurs $\mathcal{T}$ de type i vérifient-ils le principe de Hasse?

L'intérêt de cette question - et plus précisément d'une réponse affirmative - tient à l'énoncé suivant:

Proposition 10. *Si les torseurs de type* i *vérifient le principe de Hasse, l'obstruction de Manin est la seule obstruction au principe de Hasse pour* V.

Autrement dit, si $\prod V(k_v) \neq \emptyset$ et si l'obstruction de Manin est "vide", alors $V(k) \neq \emptyset$. Sinon, $V(k) = \emptyset$. En cas de réponse affirmative à la question ci-dessus, on disposerait d'une méthode effective pour décider si $V(\mathbb{Q}) = \emptyset$ ou non.

La proposition ci-dessus est une conséquence immédiate de l'assertion (iii) du corollaire de la proposition 8: si l'obstruction de Manin est "vide" pour V, il existe un torseur $\mathcal{T}$ de type i qui a des points dans chaque complété de k.

En vue de traiter la question ci-dessus il peut être utile de connaître des équations pour les torseurs $\mathcal{T}$. C'est ce que nous allons faire pour conclure. Il résulte aussitôt de la définition même de $\mathcal{T}$ que cette variété admet au-dessus de l'ouvert U de V les équations suivantes, cf. (146):

$$(160) \quad \begin{aligned} &ax^3 + by^3 + cz^3 + dt^3 = 0 \\ &\varphi(x,y,z,t) = \varepsilon_\alpha \cdot u \cdot N_s(\xi) \\ &\psi(x,y,z,t) = \varepsilon_\beta \cdot u \cdot N_q(\xi) \end{aligned}$$

où

$$(x,y,z,t;u;\xi) \in \mathbb{P}^3 \times \mathbb{G}_m \times R_{K/k}\mathbb{G}_m \hookrightarrow \mathbb{P}^3 \times \mathbb{A}^{10} .$$

De fait, on peut écrire ces équations plus simplement. L'équation (1) s'écrit encore:

$$(1'') \qquad N_q(x + \alpha y) = -\mu \cdot N_s(z + \beta t) .$$

C'est donc une conséquence des deux autres équations:

$$\begin{aligned} N_q(x + \alpha y) &= N_q(\varepsilon_\alpha) \cdot u^3 t^3 N_{K/k}(\xi) \\ N_s(z + \beta t) &= N_s(\varepsilon_\beta) \cdot u^3 t^3 N_{K/k}(\xi) \end{aligned}$$

compte tenu de la relation:

$$N_q(\varepsilon_\alpha)/N_s(\varepsilon_\beta) = N_t(\varepsilon_\alpha/\varepsilon_\beta) = -\mu .$$

Faisons alors le changement de variables défini par:

$$X = x/ut \qquad Y = y/ut \qquad Z = z/ut \qquad U = 1/u .$$

On obtient pour nouvelles équations dans $\mathbb{A}^3_k \times \mathbb{G}_m \times R_{K/k}\mathbb{G}_m$ où l'on note les coordonnées $(X,Y,Z;U;\xi)$:

$$X + \alpha Y = \varepsilon_\alpha N_s(\xi) \neq 0$$

(161)

$$Z + \beta U = \varepsilon_\beta N_q(\xi) \neq 0 .$$

On a évidemment:

$$\varepsilon_\alpha \cdot N_s(\xi) = Q_0(\xi) + Q_1(\xi)\alpha + Q_2(\xi)\alpha^2$$

(162)

$$\varepsilon_\beta \cdot N_q(\xi) = R_0(\xi) + R_1(\xi)\beta + R_2(\xi)\beta^2$$

où les Q_j et R_j sont des <u>formes cubiques</u> sur $R_{K/k}\mathbb{G}_a \simeq \mathbb{A}^9_k$ à valeurs dans $\mathbb{A}^1_k$ dépendant de ε_α et ε_β.

Proposition 11. *Avec les notations ci-dessus, le torseur $\mathcal{T}$ possède un ouvert k-isomorphe au cône épointé intersection dans l'espace affine $R_{K/k}\mathbb{G}_a \simeq \mathbb{A}^9_k$ des deux hypersurfaces cubiques d'équations*

$$Q_2(\xi) = 0$$

(163)

$$R_2(\xi) = 0 .$$

On peut expliciter davantage ces équations sur $\bar{k}$ pour mieux apprécier la géométrie de cette variété. Ecrivons:

$$\xi = \sum_{i,j} W_{i,j}\alpha^i\beta^j \qquad i, j = 0, 1, 2$$

où les $W_{i,j}$ sont 9 formes linéaires coordonnées de ξ. On peut écrire:

$$\varepsilon_\alpha \cdot N_s(\xi) = \varepsilon_\alpha \xi_{\alpha,\beta_0}\xi_{\alpha,\beta_1}\xi_{\alpha,\beta_2}$$

où les β_j sont les trois valeurs possibles de β dans $\bar{k}$. Posons:

$$Q(\xi) = \varepsilon_\alpha \cdot N_s(\xi) = Q_0(\xi) + Q_1(\xi)\alpha + Q_2(\xi)\alpha^2 .$$

On trouve aisément que sur $\bar{k}$ l'équation $Q_2(\xi)=0$ s'écrit encore:

$$(\alpha_1 - \alpha_2)Q_{\alpha_0}(\xi) + (\alpha_2 - \alpha_0)Q_{\alpha_1}(\xi) + (\alpha_0 - \alpha_1)Q_{\alpha_2}(\xi) = 0 ,$$

où $Q_{\alpha_i}(\xi)$ désigne l'image de Q par $\alpha \longrightarrow \alpha_i$, les α_i désignant les trois valeurs possibles de α dans $\bar{k}$. Si ε_{1i} est l'image de ε_α par $\alpha \longrightarrow \alpha_i$ et ε_{2j} celle de ε_β par $\beta \longrightarrow \beta_j$, on trouve donc pour équations sur $\bar{k}$, en prenant pour variables coordonnées $X_{ij}:=\xi_{\alpha_i,\beta_j}$,

$$(164)\quad \begin{aligned} &(\alpha_1-\alpha_2)\varepsilon_{10}X_{00}X_{01}X_{02}+(\alpha_2-\alpha_0)\varepsilon_{11}X_{10}X_{11}X_{12}+(\alpha_0-\alpha_1)\varepsilon_{12}X_{20}X_{21}X_{22}=0 \\ &(\beta_1-\beta_2)\varepsilon_{20}X_{00}X_{10}X_{20}+(\beta_2-\beta_0)\varepsilon_{21}X_{01}X_{11}X_{21}+(\beta_0-\beta_1)\varepsilon_{22}X_{02}X_{12}X_{22}=0\,. \end{aligned}$$

Les X_{ij} sont des formes linéaires en les W_{ij}. Elles forment un autre système de coordonnées dans $\mathbb{A}^9_{\bar{k}}$.

Remerciements

L'un d'entre nous - D. Kanevsky - remercie la fondation Alexander von Humboldt de son soutien lors de son séjour à l'Institut Max-Planck de Bonn. Il remercie aussi l'Institute for Advanced Study de Princeton pour le long séjour qu'il a pu y faire.

Nous remercions G. Wüstholz qui a su faciliter à plusieurs reprises l'élaboration commune de cet article.

Enfin, nous remercions tout particulièrement M. Vallino pour son rôle décisif dans la forme finale de cet article: l'expérimentation intensive sur ordinateur a souvent mis en évidence des phénomènes que nous n'avions pas soupçonnés, et c'est à ce rapport dialectique entre expérimentation et théorie qu'est dû le caractère performant de l'algorithme que nous avons exposé ici, algorithme bien meilleur que celui que nous avions envisagé initialement.

BIBLIOGRAPHIE

[1] E. Artin, J.T. Tate, *Class field theory*, Harvard 1961.

[2] H.-J. Bartels, Über Normen algebraischer Zahlen, Math. Ann. **251** (1980), 191-212.

[3] A. Bremner, Some cubic surfaces with no rational points, Math. Proc. Camb. Phil. Soc. **84** (1978), 219-223.

[4] J.W.S. Cassels, M.J.T. Guy, On the Hasse principle for cubic surfaces, Mathematika **13** (1966), 111-120.

[5] F. Châtelet, Points rationnels sur certaines surfaces cubiques, *Colloque intern. CNRS, Les tendances géométriques en algèbre et théorie des nombres,* Clermont-Ferrand (1964), 67-75, CNRS, Paris 1966.

[6] J.-L. Colliot-Thélène, Hilbert's theorem 90 for K_2, with application to the Chow groups of rational surfaces, Invent. Math. **71** (1983), 1-20.

[7] J.-L. Colliot-Thélène, Surfaces cubiques diagonales, *Séminaire de théorie des nombres, Paris 1984-85*, éd. C. Goldstein, Progress in Math. **63** (1986), 51-66, Birkhäuser, Boston 1986.

[8] J.-L. Colliot-Thélène, J.-J. Sansuc, La descente sur les variétés rationnelles, in *Journées de géométrie algébrique d'Angers* (1979), 223-227, éd. A. Beauville, Sijthoff & Noordhoff, Alphen aan den Rijn 1980; La descente sur les variétés rationnelles, II, Duke Math. J. **54** (1987) (à paraître).

[9] A. Grothendieck, J. Dieudonné, *Eléments de géométrie algébrique,* Publi. math. IHES **24** & **28**, Paris 1965 & 1966.

[10] K. Iwasawa, On explicit formulas for the norm residue symbol, J. Math. Soc. Japan **20** (1968), 151-165.

[11] D. Kanevsky, Application of the conjecture on the Manin obstruction to various diophantine problems, *Journées arithmétiques de Besançon, 24-28 juin 1985, Astérisque* **147-148** (1987), 307-314.

[12] S. Lichtenbaum, Duality theorems for curves over p-adic fields, Invent. Math. **7** (1969), 120-136.

[13] Yu.I. Manin, Le groupe de Brauer-Grothendieck en géométrie diophantienne, *Actes du congrès intern. math. Nice,* **1** (1970), 401-411.

[14] Yu.I. Manin, *Formes cubiques* (en russe), Nauka, Moscou 1972 (trad. anglaise: *Cubic forms*, North Holland, Amsterdam 1974, 2ème éd. 1986).

[15] L.J. Mordell, Rational points on cubic surfaces, Publ. math. Debrecen **1** (1949), 1-6.

[16] L.J. Mordell, On the conjecture for rational points on a cubic surface, J. London Math. Soc. **40** (1965), 149-158.

[17] E.S. Selmer, Sufficient congruence conditions for rational points on a cubic surface, Math. Scand. **1** (1953), 113-119.

[18] J.-P. Serre, *Corps locaux*, Actualités sci. et ind. 1296, Hermann, Paris 1962.

[19] C.L. Siegel, Normen algebraischer Zahlen, Nachr. Akad. Wiss. Göttingen Math. Phys. Kl. II, Nr **11** (1973), 197-215.

[20] Th. Skolem, Einige Bemerkungen über die Auffindung der rationalen Punkte auf gewissen algebraischen Gebilden, Math. Z. **63** (1955), 295-312.

[21] H.P.F. Swinnerton-Dyer, Two special cubic surfaces, Mathematika **9** (1962), 54-56.

[22] H.P.F. Swinnerton-Dyer, Applications of algebraic geometry to number theory, Proc. Symp. Pure Math. AMS **XX** (1971), 1-52.

[23] S. Takahashi, Cohomology groups of finite abelian groups, Tohôku Math. J. **4** (1952), 294-302.

[24] J.T. Tate, Global class field theory, in *Algebraic number theory*, Brighton (1965), 162-203 & (avec J.-P. Serre) 348-364, ed. J. W. S. Cassels & A. Fröhlich, Academic Press, London-New York 1967.

[25] V.E. Voskresenskiĭ, *Tores algébriques* (en russe), Nauka, Moscou 1977.

J.-L. C.-T.:
Université de Paris-Sud,
C.N.R.S. Mathématique, bât. 425,
91405 ORSAY CEDEX
France

D. K.:
IBM, Thomas J. Watson Research Center,
P.O. Box 218,
Yorktown Heights,
NEW YORK 10598
USA

J.-J. S.:
Université de Paris VII,
Mathématiques,
2, place Jussieu,
75251 PARIS CEDEX 05
France

SMALL VALUES OF HEIGHTS ON FAMILIES OF ABELIAN VARIETIES

by

D. W. MASSER

Department of Mathematics
University of Michigan
Ann Arbor
Michigan 48104
U. S. A.

1. INTRODUCTION

In this paper we continue our study of small values of the Néron-Tate height on abelian varieties. In general terms the problem is as follows. Suppose we have an abelian variety defined over a number field k, and let q be the Néron-Tate height corresponding to an ample divisor. There is then a positive minimum for the values of q at all non-torsion points defined over k, and we wish to find non-trivial lower bounds for this minimum.

In our earlier paper [Ma 2] we treated such problems for a varying field k and a fixed abelian variety. Now in the present paper k will be considered fixed and so we have to measure the dependence on the varying abelian variety. This will be done by the traditional method of taking a family of abelian varieties parametrized by some variety V.

Thus let V be a variety, also defined over k, and let A be an abelian variety defined over the function field k(V). Let D be an ample divisor on A also defined over k(V). We shall assume that for each v in V the corresponding specialization from k(V) to k(v) yields an abelian variety A_v defined over k(v) and an ample divisor D_v defined over k(v) (this can always be ensured by throwing away a proper closed subset of V). Thus if v is in $V(\bar{k})$, there is an associated Néron-Tate height q_v on $A_v(\bar{k})$, defined relative to k for convenience, which is positive definite modulo torsion. Our lower bounds for q_v will be obtained, following some remarks of Daniel Bertrand [Be], through the study of a certain distance function on $A_v(\bar{k})$. Let T_v be the tangent space of A_v at the origin, and write $\exp_v$ for the associated exponential map from T_v to A_v. Since the divisor D_v is ample, its associated Riemann form gives rise to a positive definite Hermitian form H_v on T_v. For P in $A_v(\bar{k})$ we define

$$r_v(P) = \inf_t H_v(t)\,,$$

where the infimum is taken over all t in T_v with $P=\exp_v(t)$. We shall loosely refer to this as a distance function, although it is more properly the square root $r_v^{1/2}$ that is a norm on $A_v(\bar{k})$. In our Theorem below we give lower bounds for $\max(q_v,r_v)$ in terms of v.

To measure v let us fix a projective embedding of V itself defined over k. Then for v in $V(\bar{k})$ there is a (logarithmic) Weil height $h(v)$ relative to k.

Theorem . *There is a non-empty open subset* V_0 *of* V *with the following property. For each* $d\geq 1$ *there exists* $C>0$, *depending only on* k, A, D, d *and the embedding of* V, *such that for any extension* K *of* k *of relative degree at most* d *and any* v *in* $V_0(K)$ *we have*

$$\max(q_v(P),r_v(P)) > C^{-1}(\max(1,h(v)))^{-1}$$

for all non-zero P *in* $A_v(K)$.

Our first corollary gives the required lower bound for the Néron-Tate height. Let n be the dimension of the generic abelian variety A, and let V_0 be as above.

Corollary 1 . *For each* $d\geq 1$ *there exists* $C_1>0$, *depending only on* k, A, D, d *and the embedding of* V, *such that for any extension* K *of* k *of relative degree at most* d *and any* v *in* $V_0(K)$ *we have*

$$q_v(P) > C_1^{-1}(\max(1,h(v)))^{-2n-1}$$

for all non-torsion P *in* $A_v(K)$.

The second corollary provides an estimate for the torsion group of $A_v(K)$. While similar results may be obtained by more elementary methods, we include it because it comes out of the present work with hardly any extra effort.

Corollary 2. *For each* $d \geq 1$ *there exists* $C_2 > 0$, *depending only on* k, A, D, d *and the embedding of* V, *such that for any extension* K *of* k *of relative degree at most* d *and any* v *in* $V_0(K)$ *the torsion group of* $A_v(K)$ *has cardinality at most* $C_2(\max(1,h(v)))^n$.

Paula Cohen [Co] was the first to obtain results in the style of the corollaries above. She treats the case of elliptic curves (n=1), and she considers the Weierstrass family over $k(g_2,g_3)$, embedded in $\mathbb{P}_2$ by the standard equation

$$y^2z = 4x^3 - g_2xz^2 - g_3z^3 .$$

Thus V is affine space $\mathbb{A}^2$ without the discriminant locus defined by $g_2^3=27g_3^2$. In this case she proved in place of Corollary 2 that every torsion point has order at most $Ch^{3/2}\log h$ for $h=\max(e,h(v))$ (so the torsion group itself has cardinality at most $C^2h^3(\log h)^2$). In fact in her results the dependence on d is also made explicit, and they apply in principle also to Néron-Tate heights in the style of Corollary 1. Moreover V_0 can be taken to be V itself.

The methods of [Co], as well those of [Ma 2], are taken from transcendental number theory, and the methods of the present paper are similar. It seems interesting to ask if the same sort of results can be obtained with more elementary techniques. In fact Coates, Ogg, and Silverman have independently suggested ways of obtaining bounds for torsion using reduction theory (see also [Frey] for the case of elliptic curves); these all give estimates depending polynomially on h(v). I don't know if such techniques can lead to lower bounds for Néron-Tate heights (see however Silverman [Sil 2]).

Of course there are very simple counting arguments (see [Ma 1] p. 217) that do give such bounds, but in general the estimates are much weaker; for example in the present situation they depend exponentially on h(v).

As explained in [Ma 1], Corollaries 1 and 2 have applications to problems of linear dependence on abelian varieties. These arise in the determination of the Mordell-Weil group (at least theoretically), but also in other contexts; see for example a paper of Singer and Davenport [SD] on differential equations, and also the work of Mason [M] on diophantine equations over function fields. Further, in the article [Ma 3] we have applied our results to give estimates for the exceptional v in $V(\overline{k})$ for which the specialization homomorphism from $A(k(V))$ to $A_v(k(v))$ is not injective. The estimates are non-trivial (and indeed nearly best possible) precisely because of the polynomial dependence on $h(v)$ in Corollaries 1 and 2. So for this last application the elementary exponential bounds referred to above are certainly not adequate (in contrast to the other applications, for which any bounds suffice).

Our paper is divided into several sections. We actually prove the Theorem for a certain "universal family" defined by standard theta-series, and in section 2 we describe this family and record some of its properties. The proof itself occupies sections 3, 4, 5, 6, 7 and 8.

Section 3 contains some technical estimates for the associated elements of the Siegel upper half space. These are applied in section 4 to prove some auxiliary results on distance functions, in particular a Box Principle for abelian varieties that may have independent interest. They are also applied to section 5 to give some fairly sharp analytic growth estimates for theta-functions.

Then, before starting on the construction of the auxiliary function, we give the necessary zero estimate in section 6. Because of our descent argument in section 7, we need the rather precise results recently established by Philippon [Ph]. In fact we need an easy modification, in which a large set of consecutive integers is replaced by a subset whose density is close to one in an obvious sense.

In section 7 we construct a suitable auxiliary function. In the paper [Ma 2] it sufficed to consider only simple abelian varieties, but in the present context such restrictions apparently lead to difficult questions of estimating isogenies (even if we suppose A itself is simple, the set of v in V for which A_v is simple will probably not contain a non-empty open subset of V). So we have to use a descent argument of the type introduced by Philippon and Waldschmidt in their recent work [PW] on linear forms. The inductive step is summarized in the Proposition.

Then in section 8 we use the Proposition to prove the Theorem, at least for the universal family described in section 2. This is extended in section 9 to arbitrary families using the "maximal fibre systems" constructed by Shimura [Sh].

Finally in section 10 we deduce Corollaries 1 and 2 from the Theorem and the Box Principle, and we make some further remarks; in particular we prove that the Theorem is the best possible with regard to its dependence on the height $h(v)$ of the parameter point.

We shall use the following notation throughout this paper. First $\varepsilon(x)=\exp(2\pi i x)$. Then $\mathbb{Z}^n$, $\mathbb{Q}^n$, $\mathbb{R}^n$ and $\mathbb{C}^n$ will denote sets of row vectors. The transpose of a matrix will be expressed by a dash. The Siegel upper half space of order n will be denoted by $\mathfrak{S}$ or $\mathfrak{S}_n$, and its quotient by the action of a group Γ will be written $\Gamma\backslash\mathfrak{S}$. Finally for a field K and a subset V' of V we denote by V'(K) the set of points of V' defined over K, and for h≥1 we write $V'_h(K)$ for the set of points v in $V'(K)$ with h(v)≤h.

In writing this paper I have had the benefit of much correspondence with Joe Silverman, and also many conversations with Gisbert Wüstholz, and I wish to thank them very much for their patience. I am also grateful to Alice Silverberg for discussions about [Sh]. The work was essentially finished at the Institute for Advanced Study (Princeton) while I held a Sloan Fellowship. It was supported throughout by the National Science Foundation.

2. A UNIVERSAL FAMILY

Fix a dimension n≥1, and let e be a positive integer to be chosen later. We shall describe a number field k, a variety V defined over k, and an abelian variety A defined over k(V), all depending only on n and e. This is essentially due to Igusa and Mumford, and it provides explicit coordinates for the moduli space of principally polarized abelian varieties of dimension n with so-called level (e,2e)-structure.

We start by defining V; this is best done analytically, and we obtain at the same time a projective embedding. For τ in $\mathfrak{S}$, z in $\mathbb{C}^n$ and m, m* in $\mathbb{R}^n$ write

$$\theta_{mm*}(\tau,z) = \sum_t \varepsilon\{\frac{1}{2}(t+m)\tau(t+m)' + (t+m)(z+m*)'\} ,$$

where the sum is over all t in $\mathbb{Z}^n$. Note that this depends only on the class of m in $\mathbb{R}^n/\mathbb{Z}^n$. We define a map $\Theta(\tau,z)$ from the product $\mathfrak{S}\times\mathbb{C}^n$ to the projective space $\mathbb{P}_N$ with $N+1=e^n$ by fixing some ordering of the N+1 elements of $e^{-1}\mathbb{Z}^n/\mathbb{Z}^n$ and taking the projective coordinates of $\Theta(\tau,z)$ as the functions

$$\theta_{m0}(\tau,z) \qquad\qquad (m\ in\ e^{-1}\mathbb{Z}^n/\mathbb{Z}^n) .$$

In section 8 we define a certain modular group $\hat{\Gamma}^e$ acting on $\mathfrak{S}$. If e is divisible by 4, it is known that the map $\Theta(e\tau,0)$ gives an analytic isomorphism from the quotient $\hat{\Gamma}^e\backslash\mathfrak{S}$ to a "modular variety" V in $\mathbb{P}_N$ (see [Ig] p.189). If e is divisible by 8, V is even quasiprojective (see [Mu 1] p.83). Since the functions $\theta_{m0}(e\tau,0)$ lie in the Laurent ring generated over $\mathbb{Z}$ by the $\varepsilon(\frac{1}{2}e^{-1}\tau_{ij})$, where the τ_{ij} are the entries of τ (1≤i,j≤n), it follows easily that V is defined over k=$\mathbb{Q}$.

Next fix τ in $\mathfrak{S}$. If e≥3, the map $\Theta(e\tau,ez)$ gives an analytic isomorphism from the quotient of $\mathbb{C}^n$ by the lattice $\mathbb{Z}^n+\mathbb{Z}^n\tau$ to an abelian variety $A(\tau)$ in $\mathbb{P}_N$ with origin $\Theta(e\tau,0)$. If e is divisible by 4, it is known that $A(\tau)$ is defined in $\mathbb{P}_N$ by homogeneous equations of degree 2, and further that the coefficients of these equations are homogeneous forms in the $\theta_{m0}(e\tau,0)$ (m in $e^{-1}\mathbb{Z}^n/\mathbb{Z}^n$) with coefficients in $\mathbb{Z}$ (see [Ig] pp. 167, 170, where the equations are given explicitly). Taking τ to correspond to a generic (or even arbitrary) point of V, we obtain an abelian variety A, whose defining equations are over the function field $k(V)$ (and even over the coordinate ring $\mathbb{Z}[V]$).

Since the origin of A is obviously defined over $k(V)$, it follows by general arguments that there exists a complete set of additional laws on A also defined over $k(V)$ (some indications of how to write these explicitly are in [Bai 1] pp. 345-346). Actually the main Theorem (p.603) of Lange and Ruppert [LR] yields bihomogeneous forms of bidegree (2,2); the argument as it stands works only over the algebraic closure of $k(V)$, but it is easy to descend to $k(V)$ using traces. It turns out that for the purpose of section 6 we need only to know that for each τ in $\mathfrak{H}$ there is a complete set of addition laws of bidegree (2,2) defined over $\mathbb{C}$; and this of course does follow immediately from [LR].

Finally for each τ in $\mathfrak{H}$ the divisor on $A(\tau)$ corresponding to the function $\theta_{00}(e\tau,ez)$ is symmetric, positive and very ample; putting these together, we obtain a divisor D on A, defined over $k(V)$, which is also symmetric, positive and very ample.

In sections 3 to 8 of this paper we shall prove the Theorem for the special family A over $k(V)$ with respect to D. Since a general parameter variety will end up mapped into V, it is essential not to omit any closed subset of V. In fact for any v in V we have the specialized objects A_v, D_v defined over $k(v)$. So if v is in $V(\bar{k})$ we can define the Néron-Tate height q_v on $A_v(\bar{k})$, and also the (logarithmic) Weil height h_v on $A_v(\bar{k})$, relative to k, corresponding to our explicit embedding in $\mathbb{P}_N$.

It is well-known that the difference between these two heights is bounded on $A_v(\bar{k})$. In section 7 we shall need the following estimate for the difference in terms of v.

Difference Lemma. *Given* $d\geq 1$ *there is a constant* c, *depending only on* n, e *and* d, *such that for any extension* K *of* k *of relative degree at most* d, *any* $h\geq 1$, *and any* v *in* $V_h(K)$ *we have*

$$|q_v(P) - h_v(P)| \leq ch$$

for all P *in* $A_v(\bar{k})$.

Proof. This is a special case of Theorem A (p. 201) of Silverman and Tate [Sil 1], expressed in a slightly different language. In fact for the special family considered here it was proved by Zarhin and Manin [ZM] in a very explicit form.

In section 5 we shall obtain a partial analogue of the above result for the distance function r_v on $A_v(\bar{k})$. For the moment we just note that there is a simple expression for the corresponding Hermitian form. Namely, take any τ_v in $\mathfrak{H}$ with $\vartheta_{00}(e\tau_v,0)=v$. Then the Hermitian form with respect to the exponential map $\Theta(e\tau_v,ez)$ on $A_v=A(\tau_v)$ is given by

$$H_v(z) = e\bar{z}y_v^{-1}z' ,$$

where $y_v=\mathrm{Im}\tau_v$ is the imaginary part of τ_v. This is an abuse of notation since the matrix of H_v depends on the choice of representative τ_v; however, in the course of the next section we shall improve matters by specifying τ_v in a certain fundamental region.

3. MATRIX ESTIMATES

From now on we fix an integer e so that all the properties in section 2 hold (for example e=8 will do). We retain the notation k, V, A, D for the special family constructed there. Recall that for each v in V we have assigned an element τ_v of the Siegel space $\mathfrak{H}$. This is defined only up to the action of elements $\sigma=\begin{pmatrix}\alpha & \beta\\ \gamma & \delta\end{pmatrix}$ of a certain proper subgroup $\hat{\Gamma}^e$ of the full modular group Γ. We remind the reader that Γ itself is the set of all such σ with α, β, γ, δ integer matrices of order n with $\alpha\beta'$, $\gamma\delta'$ symmetric and $\alpha\delta'-\beta\gamma'=\iota$ the identity matrix; and it acts on τ in $\mathfrak{H}$ by

$$\sigma(\tau) = (\alpha\tau + \beta)(\gamma\tau + \delta)^{-1} .$$

There is a standard closed fundamental region $\mathfrak{F}$ in $\mathfrak{H}$ for this action; see for example [Fr] (p. 36) or [Ig] (p. 194). For technical reasons it is convenient to work with the set V' of all $\Theta(e\tau,0)$ as τ ranges over $\mathfrak{F}$; since V is isomorphic to $\hat{\Gamma}^e\backslash\mathfrak{H}$ this is a proper subset of V. The main result of this section is the following.

Matrix Lemma. *For each* $d\geq1$ *there is a constant* c, *depending only on* n, e *and* d, *such that for any extension* K *of* k *of relative degree at most* d, *any* $h\geq1$, *any* v *in* $V_h'(K)$, *and any* τ_v *in* $\mathfrak{F}$ *with* $\Theta(e\tau_v,0)=v$ *the entries of* $\mathrm{Im}\tau_v$ *have absolute values at most* ch.

The proof requires the use of modular forms and cusp forms on $\mathfrak{H}=\mathfrak{H}_n$. Let M_n, $M_n^{(0)}$ denote the set of modular forms and cusp forms respectively, regarded as graded algebras over $\mathbb{C}$, and for $r\geq 1$ let $M_n(r)$, $M_n^{(0)}(r)$ be the finite-dimensional vector spaces corresponding to weight r. We shall use Siegel's map Φ from $M_n(r)$ to $M_{n-1}(r)$ (see [Fr] p. 45 or [Ig] p. 204), so that $M_n^{(0)}(r)$ is by definition the kernel of Φ.

Lemma 3.1. *There exists* $s=s_n\geq 1$ *with the following property. For any generators* $g_1,\ldots,g_Q$ *of* $M_n^{(0)}(s)$ *as a vector space over* $\mathbb{C}$, *we have*

$$\max_{1\leq b\leq Q} |g_b(\tau)| > 0$$

for all τ *in* $\mathfrak{H}_n$.

Proof. Since M_n is finitely generated as an algebra, it follows that the ideal $M_n^{(0)}$ of cusp forms is finitely generated as an M_n-module; let $h_1,\ldots,h_R$ be generators. By Theorem 4.9 (p. 63) of [Fr] the elements of $M_n^{(0)}$ have no common zero in $\mathfrak{H}_n$; so the $h_1,\ldots,h_R$ have no common zero in $\mathfrak{H}_n$. The lemma follows at once, with s for example as the lowest common multiple of the weights of $h_1,\ldots,h_R$.

Our next lemma is a sharper version holding for arbitrary modular forms.

Lemma 3.2. *There exists* $r=r_n\geq 1$ *with the following property. Let* $f_1,\ldots,f_P$ *be any generators of* $M_n(r)$ *as a vector space over* $\mathbb{C}$. *Then we can find* $c>0$, *depending only on* $f_1,\ldots,f_P$, *such that*

$$\max_{1\leq a\leq P} |f_a(\tau)| \geq c$$

for all τ *in* $\mathfrak{H}_n$.

Proof. It is convenient to interpret this for $n=0$ by defining $M_0(r)=\mathbb{C}$ and $\mathfrak{H}_0$ as a single point. Also write $S_0=2$, $S_n=2s_1\cdots s_n$, where the s's are as in the preceding lemma. We shall then prove by induction on n the slightly more precise proposition that Lemma 3.2 holds for any $r>2n$ divisible by S_n. With the above conventions this is clear for $n=0$.

So assume $n \geq 1$ and that the above proposition has been verified for all m satisfying $0 \leq m < n$. We will deduce it for n. We first suppose that the element τ lies in the fundamental region $\mathfrak{F}$.

If the inequality of the lemma is false in this case, we can find a sequence of τ in $\mathfrak{F}$ with

$$f_a(\tau) \longrightarrow 0 \qquad (1 \leq a \leq P). \qquad (3.1)$$

Let y_{ij} $(1 \leq i,j \leq n)$ be the entries of each $y = \mathrm{Im}\,\tau$. Suppose for the moment that there exists m with $1 \leq m \leq n$ such that y_{mm} remains bounded. Since ([Ig] p. 192)

$$0 \leq y_{11} \leq \ldots \leq y_{nn}, \; y_{ij}^2 \leq y_{ii}\, y_{jj} \qquad (1 \leq i, j \leq n) \quad (3.2)$$

it follows that we can write

$$x = \mathrm{Re}\,\tau = \begin{pmatrix} x' & ? \\ ? & ? \end{pmatrix}, \; y = \begin{pmatrix} y' & ? \\ ? & y'' \end{pmatrix}$$

with x', y' of order m and y' bounded. But x' is already bounded ([Ig] p. 194). Thus by selecting a subsequence we can assume that $x'+iy'$ converges to some limit τ_0'. Now each y' is M-reduced ([Ig] p. 191), so its determinant is bounded away from zero; thus $\mathrm{Im}\,\tau_0'$ is positive definite and so τ_0' lies in the Siegel space $\mathfrak{S}_m$.

If the above actually holds with $m=n$, then we have $\tau \longrightarrow \tau_0'$ and so $f_a(\tau_0')=0$ $(1 \leq a \leq P)$ by (3.1). But since s divides r we see easily from Lemma 3.1 that $f_1, \ldots, f_P$ have no common zero in $\mathfrak{S}_n$. So this is not possible.

Therefore either the above holds with some $m \neq n$ or not at all; in which latter case we interpret it as holding for $m=0$. In either case we can appeal to Theorem 6 (p. 201) of [Ig]. Using this and the formulae for the Fourier expansions of $\Phi^{n-m}f$, we conclude that

$$f_a(\tau) \longrightarrow \Phi^{n-m} f_a(\tau_0') \qquad (1 \leq a \leq P).$$

Thus we deduce

$$\Phi^{n-m} f_a(\tau_0') = 0 \qquad (1 \leq a \leq P).$$

However, using Theorem 5.1 (p. 64) of [Fr] we see that the map Φ^{n-m} from $M_n(r)$ to $M_m(r)$ is surjective. Since $f_1, \ldots, f_P$ generate $M_n(r)$, it follows that all the elements of $M_m(r)$ vanish at τ_0'. But this contradicts our inductive hypothesis. So the inequality of our lemma is established, at least for τ and $\mathfrak{F}$.

Finally since

$$|f_a(\sigma(\tau))| = |\det(\gamma\tau + \delta)|^r \, |f_a(\tau)| \qquad (1 \le a \le P)$$

and $|\det(\gamma\tau+\delta)| \ge 1$ on $\mathfrak{F}$ ([Fr] p. 36), we see that it follows immediately for all τ in $\mathfrak{H}$. This completes the proof. Note that for convenience we have expressed the argument in "ineffective" form; however it would not be difficult to obtain effective estimates for the constant c.

Next we need some properties of the group $\hat{\Gamma}^e$. It is known that the functions $f_{pq}(\tau)=\theta_{p0}(e\tau,0)\theta_{q0}(e\tau,0)$ (p,q in $e^{-1}\mathbb{Z}^n/\mathbb{Z}^n$) are modular forms of weight 1 with respect to $\hat{\Gamma}^e$, and that the graded ring of all modular forms with respect to $\hat{\Gamma}^e$ is the integral closure of the ring generated over $\mathbb{C}$ by the f_{pq}. For all this see [Ig] Theorem 9 (p. 222) (although the reader should note that the discrepancy between our $e\tau$ and Igusa's τ is due to a difference in defining the group action).

Proof of Matrix Lemma. It is known (see for example [Bai 2]) that the graded algebra M_n has a finite set of generators with Fourier expansions over $\mathbb{Q}$. Fix s satisfying the conditions of Lemma 3.1. It follows that we can choose corresponding generators $g_1,\dots,g_Q$ with Fourier expansions over $\mathbb{Q}$. Also fix r satisfying the conditions of Lemma 3.2. Since the map Φ simply sets to zero certain Fourier coefficients, it follows easily that we can choose corresponding generators $f_1,\dots,f_P$ also with Fourier expansions over $\mathbb{Q}$.

For $d \ge 1, h \ge 1$ let K be an extension of k of relative degree at most d, let v be in $V_h'(K)$, and let τ_v in $\mathfrak{F}$ correspond to v. We use $c_1,c_2,\dots$ for positive constants depending only on n, d and e (and the choice of generators above). Then we can find g among $g_1,\dots, g_Q$ and f among $f_1,\dots,f_P$ such that

$$g(\tau_v) \ne 0\,, \quad |f(\tau_v)| \ge c_1\,. \qquad (3.3)$$

Our proof proceeds by comparing estimates for $g(\tau_v)$. The upper bound is provided by Lemma 20 (p. 204) of [Ig], and since τ_v is in $\mathfrak{F}$ it shows that

$$|g(\tau_v)| \le c_2\exp(-c_3\mathrm{tr}(\mathrm{Im}\tau_v))\,, \qquad (3.4)$$

where the presence of the trace in $\mathrm{tr}(\mathrm{Im}\tau_v)$ is crucial.

We next prove the lower bound

$$|g(\tau_v)| \ge \exp(-c_4h)\,. \qquad (3.5)$$

Let φ denote either of the functions f^s or g^r. Since φ is a modular form with respect to the group Γ containing $\hat{\Gamma}^e$, it is integral over the ring generated over $\mathbb{C}$ by the f_{pq} (p,q in $e^{-1}\mathbb{Z}^n/\mathbb{Z}^n$). Now these f_{pq}, as well as φ, lie in the Laurent ring generated over $\mathbb{Q}$ by the $\varepsilon(\frac{1}{2}e^{-1}\tau_{ij})$, where the τ_{ij} $(1\le i,j\le n)$ are the entries of τ. It follows that in fact φ is integral over the ring generated over $\mathbb{Q}$ by the f_{pq}.

In particular since $g(\tau_v)\neq 0$, not all of the $f_{pq}(\tau_v)$ are zero. This implies that we can find m in $e^{-1}\mathbb{Z}^n/\mathbb{Z}^n$ with

$$\mu = \theta_{m0}(e\tau_v,0) \neq 0 .$$

Then

$$(\varphi(\tau))^2(\theta_{m0}(e\tau,0))^{-2rs}$$

is a modular function, and it is integral over the ring generated over $\mathbb{Q}$ by the ratios

$$f_{pq}(\tau)/f_{mm}(\tau) = (\theta_{p0}(e\tau,0)/\theta_{m0}(e\tau,0))\,(\theta_{q0}(e\tau,0)/\theta_{m0}(e\tau,0)) .$$

But since $\Theta(e\tau_v,0)=v$, the values at $\tau=\tau_v$ of the quotients on the right are algebraic numbers of degree at most c_5 and heights at most h, and we find without difficulty that $\mu^{-2rs}(\varphi(\tau_v))^2$ is an algebraic number of degree at most c_6 and height at most c_6h. Taking $\varphi=f^s$ and g^r, we deduce similar estimates for the algebraic number

$$\xi_v = (g(\tau_v))^{2r}/(f(\tau_v))^{2s} .$$

Since $\xi_v\neq 0$, Liouville's inequality yields

$$|\xi_v| \geq \exp(-c_7h) .$$

Thus by (3.3)

$$|g(\tau_v)|^{2r} \geq \exp(-c_7h)|f(\tau_v)|^{2s} \geq \exp(-c_8h) ,$$

and this gives (3.5).

Finally on comparing (3.4) with (3.5) we find that

$$\mathrm{tr}(\mathrm{Im}\tau_v) \leq c_9h .$$

The general inequalities (3.2) now imply that all the entries of $\mathrm{Im}\tau_v$ have absolute values at most c_9h. This completes the proof of the Matrix Lemma.

4. Box Principle

In this section we prove two auxiliary results about the distance functions r_v defined in section 1. For v in V choose any τ_v in $\mathfrak{S}$ with $\Theta(e\tau_v,0)=v$. Then if

$$H_v(z) = e\bar{z}y_v^{-1}z' ,$$

for $y_v = \mathrm{Im}\,\tau_v$ we have for any P on A_v the definition

$$r_v(P) = \inf_z H_v(z) ,$$

where the infimum is taken over all z with $\Theta(e\tau_v,ez)=P$. Now it is easily verified that $H_v(\alpha+i\beta)=H_v(\alpha)+H_v(\beta)$ for α, β in $\mathbb{R}^n$. It will be useful to write $z=\lambda+\mu\tau_v$ for λ, μ in $\mathbb{R}^n$; thus

$$H_v(z) = H_v(\lambda + \mu x_v) + H_v(\mu y_v) \tag{4.1}$$

for $x_v = \mathrm{Re}\,\tau_v$.

For the present section we shall consider only v in the subset V' of V, so we may choose τ_v in the fundamental region $\mathfrak{F}$. Let $y_1,\ldots,y_n$ be the diagonal elements of y_v, so that from (3.2) and [Fr] (p. 37) or [Ig] (p. 195) we have

$$\frac{1}{2}\sqrt{3} \le y_1 \le \ldots \le y_n . \tag{4.2}$$

Let η_v denote the diagonal matrix with diagonal entries $y_1,\ldots,y_n$. Then there exists $c>0$ depending only on n such that

$$c^{-1}\eta_v \le y_v \le c\eta_v , \tag{4.3}$$

in the sense that $c\eta_v - y_v$ and $y_v - c^{-1}\eta_v$ are positive semi-definite (see [Fr] p. 35). It is not difficult to deduce that also

$$c^{-1}\eta_v^{-1} \le y_v^{-1} \le c\eta_v^{-1} . \tag{4.4}$$

For the rest of this section $c_1,c_2,\ldots$ will denote positive constants depending only on n and the usual integers d, e.

Lemma 4.1 (Torsion Lemma). *For each* $d\geq 1$ *there exists* $C_1>0$*, depending only on* n, d *and* e*, such that for any extension* K *of* k *of relative degree at most* d*, any* $h\geq 1$*, any* v *in* $V_h'(K)$ *and any integer* $t>1$ *we have*

$$r_v(P)\geq C_1^{-1}t^{-2}h^{-1}$$

for all P *in* $A_v(\bar{k})$ *of order* t.

Proof. There is a $z=\lambda+\mu\tau_v$ as above with $\lambda=(\lambda_1,\ldots,\lambda_n)$, $\mu=(\mu_1,\ldots,\mu_n)$ and

$$\Theta(e\tau_v,ez)=P\,,\ r_v(P)=H_v(z)\,;$$

and since P has order t, the elements $t\lambda$, $t\mu$ are in $\mathbb{Z}^n$ and not both zero.

Suppose first $\mu\neq 0$. Then by (4.1)

$$r_v(P)\geq H_v(\mu y_v)=e\mu y_v\mu'\,.$$

By (4.3) this is at least $c^{-1}e\mu\eta_v\mu'$, and so we have

$$t^2r_v(P)\geq c^{-1}e\sum_{i=1}^{n}y_i(t\mu_i)^2\,.$$

The right-hand side is at least $c^{-1}ey_1\geq c_1$, by (4.2), and so in this case the lemma holds (even without the h^{-1}).

Next suppose $\mu=0$. Then by (4.1)

$$r_v(P)=H_v(\lambda)=e\lambda y_v^{-1}\lambda'\,.$$

By (4.4) this is at least $c^{-1}e\lambda\eta_v^{-1}\lambda'$, and so we have

$$t^2r_v(P)\geq c^{-1}e\sum_{i=1}^{n}y_i^{-1}(t\lambda_i)^2\,.$$

Now the right-hand side is at least $c^{-1}ey_n^{-1}$, by (4.2). The Matrix Lemma gives $y_n\leq c_2h$, and so the present lemma follows in this case too. This completes the proof.

We now state the main result of this section.

Lemma 4.2 (Box Principle). *For each* $d\geq 1$ *there exists* $C_2>0$*, depending only on* n, d *and* e*, such that for any extension* K *of* k *of relative degree at most* d*, any* $h\geq 1$*, and any* v *in* $V_h'(K)$ *the following holds. If for some integer* $B\geq C_2h^n$ *we are* *given points* $P_0,\ldots,P_B$ *on* $A_v(\bar{k})$, *then we can find* a, b, *with* $0\le a<b\le B$, *such that*

$$r_v(P_a - P_b) \le C_2 B^{-1/n}\,.$$

Proof. For each b with $0\le b\le B$ there is $z_b=\lambda_b+\mu_b\tau_v$ as above with $\lambda_b=(\lambda_{b1},\ldots,\lambda_{bn})$, $\mu_b=(\mu_{b1},\ldots,\mu_{bn})$ and $\Theta(e\tau_v,ez_b)=P_b$. Define

$$L_i = 1+[\tfrac{1}{2}B^{1/2n}y_i^{-1/2}]\,,\ M_i = 1+[\tfrac{1}{2}B^{1/2n}\,y_i^{1/2}] \qquad (1\le i\le n)\,,$$

and $N=L_1\cdots L_nM_1\cdots M_n$. By considering the reductions of the (λ_b,μ_b) in $(\mathbb{R}^n/\mathbb{Z}^n)^2$ and dividing into N boxes, we see from the conventional Box Principle that as soon as $B\ge N$ there must exist a, b with $0\le a<b\le B$ and $\ell=(\ell_1,\ldots,\ell_n)$, $m=(m_1,\ldots,m_n)$ in $\mathbb{Z}^n$ with

$$|\lambda_{ai}-\lambda_{bi}-\ell_i|\le L_i^{-1}\,,\ |\mu_{ai}-\mu_{bi}-m_i|\le M_i^{-1} \qquad (1\le i\le n)\,. \qquad (4.5)$$

Now if B/h^n is sufficiently large we have by the Matrix Lemma

$$B^{1/2n}\,y_i^{\pm 1/2}\ge c_3B^{1/2n}h^{-1/2}\ge 1 \qquad (1\le i\le n)\,.$$

In this case, since $1+[\frac{1}{2}x]\le x$ for $x\ge 1$, we deduce that indeed

$$N=\prod_{i=1}^{n}L_iM_i\le\prod_{i=1}^{n}(B^{1/2n}y_i^{-1/2})\,(B^{1/2n}\,y_i^{1/2})=B$$

as required.

Thus (4.5) does hold, and we proceed to deduce the inequality of the lemma. Clearly

$$r_v(P_a-P_b)\le H_v(z)\,,$$

where

$$z=(\lambda_a-\lambda_b-\ell)+(\mu_a-\mu_b-m)\tau_v\,.$$

So by Cauchy-Schwarz

$$(r_v(P_a-P_b))^{1/2}\le H_1^{1/2}+H_2^{1/2}+H_3^{1/2}\,, \qquad (4.6)$$

where

$$H_1 = H_v(\lambda_a - \lambda_b - \ell)\,, \quad H_3 = H_v((\mu_a - \mu_b - m)y_v)\,.$$

and

$$H_2 = H_v((\mu_a - \mu_b - m)x_v)$$

Using (4.3), (4.4) and (4.5) we find that

$$H_1 \le ce\sum_{i=1}^{n} y_i^{-1}L_i^{-2} \le c_4B^{-1/n}\,, \quad H_3 \le ce\sum_{i=1}^{n} y_iM_i^{-2} \le c_4B^{-1/n}\,.$$

Finally, since all the entries of x_v have absolute values at most $\frac{1}{2}$ ([Ig] p. 194), the supremum norms satisfy

$$|(\mu_a - \mu_b - m)x_v| \le c_5|\mu_a - \mu_b - m| \le c_6B^{-1/2n}\,.$$

Therefore $H_2 \le c_7B^{-1/n}$, and these inequalities, together with (4.6), complete the proof of the Box Principle.

5. Analytic Growth

For v in V' and τ_v in $\mathcal{F}$ with $\Theta(e\tau_v,0)=v$ we recall the theta-functions

$$\theta_{m0}(e\tau_v,ez) \qquad\qquad (m \text{ in } e^{-1}\mathbb{Z}^n/\mathbb{Z}^n)$$

defined in section 2. We would like analytic upper and lower bounds for these functions. The upper bound is not difficult to obtain, and once again it can be expressed in terms of the Hermitian form

$$H_v(z) = e\bar{z}y_v^{-1}z'\,,$$

for $y_v = \mathrm{Im}\,\tau_v$. But we are unable to prove a corresponding lower bound, except at the origin $z=0$. This fact is responsible for some complications in the Zeroes Lemma of section 6 and the main proof of section 7.

Analytic Lemma. *For each* $d\geq 1$ *there is a constant* $c>0$, *depending only on* n, d *and* e, *such that for any extension* K *of* k *of relative degree at most* d, *any* $h\geq 1$, *any* v *in* $V_h'(K)$ *and any* τ_v *in* $\mathfrak{F}$ *with* $\Theta(e\tau_v,0)=v$ *the following hold:*

(i) *We have*

$$\max_m |\theta_{m0}(e\tau_v,0)| \geq c^{-1}$$

(ii) *we have*

$$\max_m |\theta_{m0}(e\tau_v,ez)| \leq \exp(ch + 8\pi H_v(z)) .$$

Proof. Let $c_1,c_2,\ldots$ denote positive constants independent of h and K. We do (i) first. Fix v and $f_1,\ldots,f_P$ satisfying the conditions of Lemma 3.2. Then

$$\max_{1\leq a\leq P} |f_j(\tau_v)| \geq c_1 .$$

Now each $f_j(\tau)$ is integral over the ring generated over $\mathbb{C}$ by the

$$f_{pq}(\tau) = \theta_{p0}(e\tau,0)\theta_{q0}(e\tau,0) \qquad\qquad (p,q \text{ in } e^{-1}\mathbb{Z}^n/\mathbb{Z}^n) ,$$

as remarked earlier. From this (i) follows without difficulty.

For (ii) we start by proving that for any λ, μ and m in $\mathbb{R}^n$ we have

$$|\theta_{m0}(e\tau_v,e(\lambda + \mu\tau_v))| \leq \exp(c_2h + 8\pi eM)| \tag{5.1}$$

where

$$M = M(\mu) = \mu y_v \mu' \geq 0 .$$

Without loss of generality we may suppose that the supremum norm $|m|$ of m does not exceed 1. Then from the series definition of θ_{m0} we find easily that

$$|\theta_{m0}(e\tau_v,e(\lambda + \mu\tau_v))| \leq \sum_t \exp(-2\pi eQ(t)) , \tag{5.2}$$

where the sum is taken over all t in $\mathbb{Z}^n$ and

$$Q(t) = \frac{1}{2}(t + m)y_v(t + m)' + (t + m)y_v\mu' .$$

Let

$$T = T(t) = t y_v t' \geq 0 .$$

From the Cauchy-Schwarz inequality, the Matrix Lemma, and the inequality of the arithmetic and geometric means we get

$$|m y_v t'| \leq (m y_v m')^{1/2} T^{1/2} \leq c_3 h^{1/2} T^{1/2} \leq c_4 h + T/16 ,$$
$$|t y_v \mu'| \leq M^{1/2} T^{1/2} \leq 2M + T/8$$

and

$$|m y_v \mu'| \leq (m y_v m')^{1/2} M^{1/2} \leq c_5 h^{1/2} M^{1/2} \leq c_6 h + M .$$

Thus we find that

$$Q(t) \geq \frac{1}{4} T - (Ch + 3M)$$

for a certain $C = c_7$.

We now divide the t's into two classes. Suppose first

$$T(t) \geq 8(Ch + 3M) .$$

Then $Q(T) \geq T/8$. Now $T \geq c_8 |t|^2$ by (4.2) and (4.3), so this part of the sum (5.2) is bounded above by

$$\sum_t \exp(-c_9 |t|^2) \leq c_{10} .$$

Next suppose

$$T(t) < 8(Ch + 3M) . \tag{5.3}$$

Then $Q(t) \geq -(Ch+3M)$. Since $T \geq c_8 |t|^2$ we see that $|t| \leq c_{11} T^{1/2}$, and so the number of t satisfying (5.3) is at most

$$c_{12}(h + M)^{n/2} \leq c_{13} \exp(h + M)$$

So this part of the sum (5.2) is bounded above by $c_{14}\exp(c_{15}h + 8\pi e M)$. Combining the two parts, we deduce (5.1).

Finally any z in $\mathbb{C}^n$ may be written as $\lambda + \mu\tau_v$ for λ, μ in $\mathbb{R}^n$. Therefore by (4.1)

$$H_v(z) \geq H_v(\mu y_v) = eM ,$$

and this together with (5.1) gives (ii). The proof of the Analytic Lemma is thereby completed.

It is interesting to note that well-known classical arguments (see for example Lemma 4 ((p. 69) of [Ig]) give an upper bound of the shape $C\exp(\frac{1}{2}\pi H_v(z))$ in (ii) above, and a lower bound of exactly the same shape may be derived by similar methods. Thus (ii) is best possible in z apart from absolute constants. But of course C here depends on h, and the classical arguments apparently do not suffice to determine it explicitly. Nevertheless I would conjecture a precise analogue of the Difference Lemma in section 2; namely that

$$\left|\frac{1}{2}\pi H_v(z) - \log \max_m |\theta_{m0}(e\tau_v, ez)|\right| \leq ch$$

for some c independent of z and h.

6. ZERO ESTIMATES

We take our basic result from Philippon's article [Ph], but it has to be modified to deal with sets of density close to 1. Let $\mathcal{A}$ be an abelian variety of dimension $m\geq 1$ embedded in projective space $\mathbb{P}_N$ with coordinates $x_0,\ldots,x_N$. Let a be a positive integer such that all translation formulae on $\mathcal{A}$ can be completely described by homogeneous polynomials of degree at most a. Let δ be the degree of $\mathcal{A}$ in $\mathbb{P}_N$. We identify the additive group $\mathbb{G}_a$ with the affine line $\mathbb{A}$, with coordinate z.

Zeroes Lemma. *There exist* $\varepsilon>0$ and c, *depending only on* m *and* a, *with the following property. Suppose for integers* $L\geq 1$, $M\geq\delta$ *we have a polynomial* F, *of degree at most* L *in* z *and homogeneous of degree at most* M *in* $x_0,\ldots,x_N$, *that does not vanish identically on the product* $\mathbb{G}_a\times\mathcal{A}$. *Suppose further that for some point* $\mathcal{P}$ *in* $\mathcal{A}$ *and some integer*

$$S \geq c\delta LM^m \tag{6.1}$$

we can find at least $(1-\varepsilon)S$ *different integers* s *with* $1\leq s\leq S$ *such that* F *vanishes at* $(s,s\mathcal{P})$ *on* $\mathbb{G}_a\times\mathcal{A}$. *Then there is a connected abelian subvariety* $\mathcal{A}'$ *of* $\mathcal{A}$, *with dimension* m' *satisfying* $0\leq m'<m$, *such that* $s_0\mathcal{P}$ *lies in* $\mathcal{A}'$ *for some integer* $s_0\geq 1$. *Further if* δ' *is the degree of* $\mathcal{A}'$ *in* $\mathbb{P}_N$ *we have* $s_0\delta'\leq c\delta M^m$.

Proof. We define the constants b, ε by

$$b = 2(m+1)^2(2a)^m, \quad b\varepsilon(m+1)! = \frac{1}{2},$$

and then we put

$$c = \max(b, \varepsilon^{-1}).$$

We start by filling in the "missing zeroes". We will construct a new polynomial F^*, also not vanishing identically on $\mathbb{G}_a \times \mathcal{A}$, that vanishes at $(s, s\mathcal{P})$ for all integers s with $0 \leq s \leq S$. For there exists T with

$$1 \leq T \leq \varepsilon S + 1$$

and integers $s_1, \ldots, s_T$ such that F vanishes at all such points except those with $s = s_1, \ldots, s_T$. Define

$$H = \delta M^m/(m+1)!$$

and

$$L_0 = [T/H], \tag{6.2}$$

so that

$$(L_0 + 1)H > T. \tag{6.3}$$

Now since $M \geq \delta$, the estimate for Hilbert functions given as Proposition 1 (p. 29) of Nesterenko's paper [N] enables us to find at least H monomials in $x_0, \ldots, x_N$, of degree M, that are linearly independent on $\mathcal{A}$. It follows from (6.3) that we can construct a polynomial F_0, of degree at most L_0 in z and homogeneous of degree M in $x_0, \ldots, x_N$, that vanishes at $(s, s\mathcal{P})$ for $s = s_1, \ldots, s_T$ but not identically on $\mathbb{G}_a \times \mathcal{A}$. We now take $F^* = F_0 F$; it clearly has the required vanishing properties, and it also has degree at most $L^* = L_0 + L$ in z and degree at most 2M in $x_0, \ldots, x_N$.

We are going to apply Theorem 2.1 of [Ph] with T=0 and $G_1 = \mathbb{G}_a$, $G_2 = \mathcal{A}$, so that $c_1 = 1$, $c_2 = a$. We find that there is a proper connected algebraic subgroup G' of $G = \mathbb{G}_a \times \mathcal{A}$ such that the set Σ of points $(s, s\mathcal{P})$ $(0 \leq s \leq S/(m+1))$ satisfies

$$|(\Sigma + G')/G'| \, H(G'; L^*, 2M) \leq H(G; L^*, 2aM). \tag{6.4}$$

Since $\mathbb{G}_a$ and $\mathcal{A}$ are disjoint in the sense of [MW], we have $G' = H \times \mathcal{A}'$ for H=0 or $\mathbb{G}_a$ and some abelian subvariety $\mathcal{A}'$ of $\mathcal{A}$. Let m', δ' be the dimension and the degree of $\mathcal{A}'$.

We deal with the case H=0 first. It is then clear that

$$|(\Sigma + G')/G'| = |\Sigma| > S/(m+1) .$$

Using Lemma 3.4 of [Ph] to calculate the Hilbert term on the right of (6.4) we find that

$$S < \frac{1}{2} b\delta L^* M^m = \frac{1}{2} b\delta (L_0 + L) M^m . \tag{6.5}$$

Now (6.1) and the definition of c give $S \geq \varepsilon^{-1}$, so $T \leq 2\varepsilon S$. Hence by (6.2) and the definition of ε we have

$$\frac{1}{2} b\delta L_0 M^m \leq \frac{1}{2} b(m+1)!T \leq b\varepsilon(m+1)!S = \frac{1}{2} S .$$

This ensures that (6.1) and (6.5) contradict each other.

Thus (6.4) must hold for $G'=\mathbb{G}_a \times \mathcal{A}'$, and in particular $0 \leq m' < m$. In this case $|(\Sigma+G')/G'|$ is the number $\mathcal{N}$ of distinct points $s\mathcal{P}$ $(0 \leq s \leq S/(m+1))$ modulo $\mathcal{A}'$, and using Lemma 3.4 of [Ph] now to calculate both Hilbert terms in (6.4), we find that

$$(m+1)\mathcal{N}\delta' \leq \frac{1}{2} b\delta M^m . \tag{6.6}$$

In particular (6.1) and (6.6) imply that

$$S/(m+1) \geq \mathcal{N} ,$$

and so at least two among the points $s\mathcal{P}$ $(0 \leq s \leq \mathcal{N})$ must be congruent modulo $\mathcal{A}'$. Hence $s_0\mathcal{P}$ lies in $\mathcal{A}'$ for some positive integer $s_0 \leq \mathcal{N}$, and the required estimates for $s_0\delta'$ now follow from (6.6). This completes the proof.

We note that this result improves the Proposition (p. 159) of [Ma 2] in two respects: firstly the bound for s_0 is independent of L, and secondly there is a descent from $\mathcal{A}$ to $\mathcal{A}'$ with the dependence on degrees made explicit.

Lastly we remark that in our particular application it would have been possible to avoid the "missing zeroes" in the Zeroes Lemma simply by giving them to the auxiliary function in Lemma 7.5 using Siegel's Lemma. But the present exposition seems clearer.

7. DESCENT

We recall the family A over k(V) introduced in section 2, with the specialized abelian variety A_v defined for each v in V. In this section we give the main argument from transcendence, which leads to a descent on connected abelian subvarieties of A_v. This prompts the following makeshift definition. For P in A_v and a connected abelian subvariety $\mathcal{A}$ of A_v write $t_v(P,\mathcal{A})=\infty$ if no positive integer multiple of P lies in $\mathcal{A}$; else let σ be the least positive integer with σP in $\mathcal{A}$, and write $t_v(P,\mathcal{A})$ for the maximum of σ and the degree of $\mathcal{A}$ (probably it is more natural to consider the product, but the maximum suffices for our present purposes).

The main result of this section is the following Proposition. Again for technical reasons it refers only to points v of V'.

Proposition. *For each* $d\geq 1$ *there exists* $C_0>0$, *depending only on* n, d *and* e, *such that for any extension* K *of* k *of relative degree at most* d, *any* $h\geq 1$, *and any* v *in* $V_h'(K)$ *the following holds. Assume that for some* t *with* $1\leq t\leq h^{1/n}$ *and some* P *in* $A_v(K)$ *we have the inequality*

$$\max(q_v(P),r_v(P)) \leq C_0^{-1}t^{-4}h^{-1} .$$

Further suppose that

$$t_v(P,\mathcal{A}) \leq t$$

for some $\mathcal{A}$ *of positive dimension. Then we also have*

$$t_v(P,\mathcal{A}') \leq C_0 t^{n+2}$$

for some $\mathcal{A}'$ *of dimension strictly less than the dimension of* $\mathcal{A}$.

The proof proceeds by a sequence of lemmas and the construction of a suitable auxiliary function. We use positive constants $c_0,c_1,\ldots$ depending only on n, d and e.

Suppose in fact σP is in $\mathcal{A}$, where $1\leq\sigma\leq t$ and $\mathcal{A}$ has dimension $m\geq 1$ and degree $\delta\leq t$. We shall deduce the existence of $\mathcal{A}'$ assuming that

$$q_v(P) \leq C^{-6m-8}\sigma^{-2}\delta^{-2}h^{-1} , \qquad (7.1)$$
$$r_v(P) \leq C^{-6m-8}\sigma^{-2}\delta^{-2}h^{-1} \qquad (7.2)$$

for some sufficiently large integer $C \leq c_0$. We define the parameters

$$L = [C^3 \delta^{-m} h], \quad M = C^3 \delta$$
$$S = C^{2m+3} \delta h, \quad S = C^{3m+4} \delta h$$

(note that $\delta^{-m} h \geq \delta^{-n} h \geq t^{-n} h \geq 1$ due to our upper bound on t; so $L \geq 1$). We choose any τ_v in the fundamental region $\mathcal{F}$ with $\Theta(e\tau_v, 0) = v$. For brevity we change notation and we suppose the $N+1 = e^n$ functions $\theta_{m0}(e\tau_v, ez)$ (m in $e^{-1}\mathbb{Z}^n/\mathbb{Z}^n$) ordered in some way as $\theta_0(z), \ldots, \theta_N(z)$; thus we may rewrite $\Theta(e\tau_v, ez)$ as $\Theta(z)$.

We recall the Hermitian form

$$H_v(z) = e\bar{z} y_v^{-1} z', \qquad (y_v = \operatorname{Im} \tau_v).$$

Since this induces the function r_v, we can pick u in $\mathbb{C}^n$ such that

$$\Theta(u) = \sigma P, \quad H_v(u) = r_v(\sigma P) \leq \sigma^2 r_v(P).$$

Recall also the Weil height h_v of section 2.

Lemma 7.1. *For any integer* s *with* $0 \leq s \leq S$ *we have*

$$h_v(\Theta(su)) \leq c_1 h.$$

Proof. By the Difference Lemma and (7.1) we have

$$h_v(\Theta(su)) \leq q_v(\Theta(su)) + c_2 h = \sigma^2 s^2 q_v(P) + c_2(h) \leq c_1 h.$$

Lemma 7.2. *For any complex* z *with* $|z| \leq 11S$ *we have*

$$\max_{0 \leq i \leq N} |\Theta_i(zu)| \leq \exp(c_3 h).$$

Proof. By (ii) of the Analytic Lemma and (7.2) we have

$$\max_{0 \leq i \leq N} |\theta_i(zu)| \leq \exp(c_4 h + 8\pi H_v(zu)) = \exp(c_4 h + 8\pi |z|^2 H_v(u))$$
$$\leq \exp(c_4 h + 8\pi |z|^2 \sigma^2 r_v(P)) \leq \exp(c_3 h).$$

Lemma 7.3. *There are at most* $C^{-1}S$ *integers* s *with* $1 \le s \le S$ *such that*

$$\max_{0 \le i \le N} |\theta_i(su)|^M < 2^{-S_0}.$$

Proof. Suppose this is false. Then we can certainly find $T=[C^{-1}S_0]$ different integers between 1 and S, say $s=s_1,\dots,s_T$, for which the inequality holds. Fix any integer i with $0 \le i \le N$; then the function

$$g(z) = (\theta_i(zu))^M$$

satisfies

$$|g(s_t)| < 2^{-S_0} \qquad (1 \le t \le T).$$

Write

$$p(z) = \sum_{t=1}^{T} \prod_{\substack{j=1 \\ j \ne t}}^{T} ((z - s_j)/(s_t - s_j))g(s_t),$$

so that the Lagrange Interpolation Formula gives

$$h(s_t) = 0 \qquad (1 \le t \le T)$$

for the difference

$$h(z) = g(z) - p(z).$$

Thus by the usual Schwarz Lemma

$$|h(0)| \le 4^{-T}Y \tag{7.3}$$

where

$$Y = \sup_{|z|=9S} |h(z)|.$$

But for any z with $|z|=9S$ we see from Lemma 7.2 that

$$|g(z)| \le \exp(c_5Mh).$$

And with $T_1=[\frac{1}{2}(T-1)]$ we have

$$|p(z)| \le T(10\,S)^T(T_1!)^{-2}2^{-S_0} \le c_6^T(S/T)^T 2^{-S_0}\,.$$

Since $T=[C^{-1}S_0]$ we find that

$$|p(z)| \le \exp(C^{1/2}T)2^{-S_0} \le 2^{-S_0/2}\,. \tag{7.4}$$

Putting these together gives

$$Y \le \exp(c_5Mh) + 2^{-S_0/2} \le \exp(c_7Mh)\,,$$

and it follows from (7.3) that $|h(0)|\le 3^{-T}$. But (7.4) implies $|p(0)|\le 2^{-S_0/2}$, so we conclude that $|g(0)|\le 2^{-T}$. On recalling the definition of g and varying the subscript i we find that

$$\max_{0\le i\le N} |\theta_i(0)|^M \le 2^{-T}\,.$$

On the other hand the estimate (i) of the Analytic Lemma leads to the contradictory lower bound $\exp(-c_8M)$. This completes the proof of the lemma.

We use the familiar Δ-polynomials

$$\Delta_0(z) = 1\,,\ \ \Delta_\lambda(z) = (z+1)\cdots(z+\lambda)/\lambda! \qquad (\lambda \ge 1)$$

to avoid extra logarithmic terms in our final estimates.

Lemma 7.4. *For any integer* λ *with* $0\le\lambda\le L$ *and any complex* z *with* $|z|\le 11S$ *we have*

$$|\Delta_\lambda(z)| \le \exp(C^{7/2}h)\,.$$

Proof. Since $\Delta_\lambda(z)$ has non-negative coefficients we have $|\Delta_\lambda(z)|\le\Delta_\lambda(11S)$. Also for fixed $x\ge 0$ the values $\Delta_\lambda(x)$ increase as λ increases; so $\Delta_\lambda(11S)\le\Delta_L(11S)$. Finally

$$\Delta_L(11S) \le (11S+L)^L/L! \le c_9^L((S+L)/L)^L \le c_{10}^L(S/L)^L\,,$$

and the required estimate follows, since

$$S/L \le C^{3m+2}\delta^{m+1} \le \exp(C^{1/2}\delta)\,.$$

It is interesting to note that this estimate does not follow from Lemma 1 (p. 7) of [Bak]; we use only the binomial coefficients near the edges of the Pascal triangle.

We now construct the auxiliary function. We use z for the coordinate on $\mathbb{A}$ and $x_0,\ldots,x_N$ for the coordinates on $\mathbb{P}_N$ corresponding to the functions $\theta_0,\ldots,\theta_N$. Let $\mu=(\mu_0,\ldots,\mu_N)$ be a subscript in $\mathbb{Z}^{N+1}$.

Lemma 7.5. *There exist rational integers*

$$p_{\lambda\mu} \qquad (0 \le \lambda \le L;\ \mu_0 \ge 0,\ldots,\mu_N \ge 0,\ \mu_0+\ldots+\mu_N = M)$$

with absolute values at most $\exp(C^4\delta h)$, *such that the polynomial*

$$F(z,x_0,\ldots,x_N) = \sum_\lambda \sum_\mu p_{\lambda\mu} \Delta_\lambda(z) x_0^{\mu_0} \cdots x_N^{\mu_N}$$

is not identically zero on $\mathbb{G}_a \times \mathcal{A}$ *but the function*

$$f(z) = F(z,\theta_0(zu),\ldots,\theta_N(zu))$$

satisfies

$$f(s) = 0$$

for all integers s *with* $1 \le s \le S_0$.

Proof. Since $M \ge \delta$, Nesterenko's lower bound in [N] again shows that we can find

$$H \ge \delta M^m/(m+1)!$$

monomials $\mathcal{M}_1,\ldots,\mathcal{M}_H$, homogeneous of degree M in $x_0,\ldots,x_N$, that are linearly independent on $\mathcal{A}$ (compare section 6). Further for each s there exists $j=j(s)$ with $0 \le j \le N$ such that $\Theta_j(su) \ne 0$. Now it suffices to solve the equations

$$0 = \sum_\lambda \sum_\mu{}^{*} p_{\lambda\mu} \Delta_\lambda(s) (\theta_0(su)/\theta_j(su))^{\mu_0} \cdots (\theta_N(su)/\theta_j(su))^{\mu_N}$$

with $1 \le s \le S_0$, and where this time the sum over μ involves only those subscripts corresponding to the monomials $\mathcal{M}_1,\ldots,\mathcal{M}_H$. The number of unknowns is at least

$$(L+1)H > C^{3m+3}\delta h/(m+1)! = c_{11} C^m S_0,$$

and the equations are defined over the field K of degree at most c_{12}. So we can use Siegel's Lemma to solve the system. By Lemma 7.4 the $\Delta_\lambda(s)$ are rational integers of absolute values at most $\exp(C^{7/2}h)$, and so by Lemma 7.1 the projective heights of the linear forms are at most

$$\exp(C^{7/2}h + c_{13}Mh) \leq \exp(c_{14}C^{7/2}\delta h).$$

Hence, using for example Lemma 6 (p. 161) of [Ma 2], we obtain a non-trivial solution with the required properties. Since the $\Delta_\lambda(z)$ are linearly independent polynomials, it is clear that the resulting polynomial F is not identically zero on $\mathbb{G}_a \times \mathcal{A}$.

Lemma 7.6. *We have*

$$|f(s)| \leq 4^{-S_0}$$

for all integers s *with* $1 \leq s \leq S$.

Proof. The usual Schwarz Lemma gives

$$|f(s)| \leq 5^{-S_0}X$$

where

$$X = \sup_{|z|=11S} |f(z)| .$$

Now Lemmas 7.2, 7.4 and 7.5 yield

$$X \leq \exp(C^4\delta h + C^{7/2}h + c_{15}Mh) \leq \exp(c_{16}C^4\delta h).$$

These imply the inequalities of the present lemma.

Lemma 7.7. *There are at least* $(1-C^{-1})S$ *different integers* s *with* $1 \leq s \leq S$ *such that*

$$f(s) = 0 .$$

Proof. By Lemma 7.3, it will suffice to prove that $f(s)=0$ whenever $1 \leq s \leq S$ and

$$\max_{0 \leq i \leq N} |\theta_i(su)|^M \geq 2^{-S_0} .$$

Thus let $j=j(s)$ be an integer with $0 \leq j \leq N$ such that $|\theta_j(su)|^M \geq 2^{-S_0}$. By Lemma 7.6 the number

$$\xi = (\theta_j(su))^{-M} f(s)$$

satisfies $|\xi| \leq 2^{-S_0}$. But since

$$\xi = \sum_\lambda \sum_\mu p_{\lambda\mu} \Delta_\lambda(s) \, (\theta_0(su)/\theta_j(su))^{\mu_0} \cdots (\theta_N(su)/\theta_j(su))^{\mu_N} ,$$

calculations as in the proof of Lemma 7.5 show that ξ has height at most $c_{17} C^4 \delta h$. It follows that $\xi = 0$, and this proves the lemma.

We can now deduce our Proposition using the Zeroes Lemma. For we remarked in section 2 that the addition laws on A_v could be completely described by polynomials bihomogeneous of bidegree (2,2). This therefore holds for the subvariety $\mathcal{A}$ as well, and in particular we may choose the constant a of section 6 as a=2. Hence the constants ε, c of the Zeroes Lemma depend only on m. We may thus suppose that indeed $S \geq c\delta L M^m$ and that $f(s)=0$ for at least $(1-\varepsilon)S$ different integers s with $1 \leq s \leq S$.

Taking $\mathcal{P} = \sigma P$ in the Zeroes Lemma, we find a connected abelian subvariety $\mathcal{A}'$, strictly contained in $\mathcal{A}$, and a positive integer s_0 with $s_0 \mathcal{P} = s_0 \sigma P$ in $\mathcal{A}'$. Thus if δ' is the degree of $\mathcal{A}'$, we have

$$t_v(P, \mathcal{A}') \leq \max(s_0 \sigma, \delta') .$$

But $\sigma \leq t$ and

$$\max(s_0, \delta') \leq s_0 \delta' \leq c_{18} \delta M^m \leq c_{18} C^{3m} t^{m+1} ,$$

so we end up with

$$t_v(P, \mathcal{A}') \leq C^{3n+1} t^{n+2} .$$

This completes the proof of the Proposition.

8. Proof Of The Theorem (I)

In this section we complete the proof of the Theorem for the special family A over k(V) of section 2. We do this first for v in the subset V' of V.

Let C_0 be a sufficiently large constant (in particular so that the Proposition of section 7 holds), and put $\nu = (n+3)^n$. For $d \geq 1$, $h \geq 1$ let K be an extension of k of relative degree at most d, and let v be in $V'_h(K)$. We may assume without loss of generality that

$$h \geq C_0^{\nu}. \tag{8.1}$$

Suppose there exists non-zero P in $A_v(K)$ with

$$\max(q_v(P), r_v(P)) \leq C_0^{-2\nu-2} h^{-1}. \tag{8.2}$$

We shall deduce a contradiction.

We start by applying the Proposition with $\mathcal{A}=A_v$. In this case we can take

$$t = \max(1, \delta_v) \leq C_0,$$

where δ_v is the degree of A_v (in fact $\delta_v=n!(N+1)=n!e^n$, but we don't need this). It is easily checked that (8.1) and (8.2) imply the conditions of the Proposition, and so we get the existence of a connected abelian subvariety $\mathcal{A}'$, strictly contained in $\mathcal{A}$, with $t_v(P,\mathcal{A}')\leq C_0^{n+3}$.

If $\mathcal{A}'\neq 0$ we apply the Proposition again; and so on repeatedly. It is readily verified that at each stage (8.1) and (8.2) imply the desired conditions. Eventually, after at most n applications, we must arrive at the zero-dimensional abelian variety 0, and then we find that $t_v(P,0)\leq C_0^{\nu}$. But this means simply that P is a torsion point of order at most C_0^{ν}. So by Lemma 4.1 we have

$$\max(q_v(P), r_v(P)) = r_v(P) \geq C_0^{-2\nu-2} h^{-1}.$$

This contradicts (8.2). Hence (8.2) is false, and this completes the proof of the Theorem for v in V'.

To extend to all of V we need the following observation.

Lemma 8.1. *There exists a finite set* Λ *of automorphisms* λ *of* $\mathbb{P}_N$, *defined over* $\bar{k}$, *with the following property. Given any* v *in* V, *we can find* v' *in* V' *and* λ *in* Λ *such that* λ *induces an isomorphism of abelian varieties from* A_v *to* $A_{v'}$; *and furthermore* D_v *and the pullback* $\lambda^* D_{v'}$ *are linearly equivalent divisors on* A_v.

Proof. We use an explicit transformation formula for theta-functions. Fix $\sigma=\begin{pmatrix}\alpha & \beta\\ \gamma & \delta\end{pmatrix}$ in the modular group Γ with $\det\gamma\neq 0$, and for variables τ in $\mathfrak{H}$, z in $\mathbb{C}^n$ write

$$\tau' = \sigma(\tau) = (\alpha\tau + \beta)(\gamma\tau + \delta)^{-1},\ z' = (\gamma\tau + \delta)^{-1}z$$

(*not* transposes). Then for any m in $\mathbb{Q}^n$ we have

$$\Delta(\tau)E(z)\theta_{m0}(e\tau',ez') = \sum_p G(p,m)\theta_{m(p)m^*}(e\tau,ez)\,, \tag{8.3}$$

where the sum is over all p in $e^{-1}\mathbb{Z}^n/\mathbb{Z}^n$ and

$$m(p) = p + m\alpha + \frac{1}{2}\xi\,,\ m^* = em\beta + \frac{1}{2}e\xi^*.$$

Here ξ, ξ^* in $\mathbb{Z}^n$ depend only on σ, the $G(p,m)$ are algebraic numbers (essentially Gauss sums) depending only on p, m, e and σ, and $\Delta(\tau)$, $E(z)$ are non-vanishing functions of τ, z respectively which also depend on e and σ but not on m. This formula follows in a straightforward way from Satz 6 (p. 393) of [Sie] (compare also Theorem 6 (p. 84) of [Ig]).

Now recall that e is even. Then m^* is in $\mathbb{Z}^n$ and so by equation (θ.2) (p. 49) of [Ig] we may replace m^* in (8.3) by 0 at the cost of modifying the algebraic numbers $G(p,m)$. Also $m(p)$ is in $e^{-1}\mathbb{Z}^n/\mathbb{Z}^n$ whenever m is, and so (8.3) yields an automorphism $\lambda=\lambda_\sigma$ of $\mathbb{P}_N$, defined over $\bar{k}$, which induces an isomorphism of abelian varieties, also denoted by λ, from $A(\tau)$ to $A(\tau')$ in the notation of section 2 (note that the determinant formed by the modified $G(p,m)$ cannot vanish, since the θ_{m0} are well-known to be linearly independent functions on $\mathbb{C}^n$).

To specify the σ's we have at last to know the definition of the congruence subgroup $\hat{\Gamma}^e$ introduced in section 2; it consists of all $\rho=\left(\begin{smallmatrix}\alpha & \beta\\ \gamma & \delta\end{smallmatrix}\right)$ in Γ with

$$\alpha \equiv \delta \equiv \iota \pmod e \text{ and } \beta \equiv \gamma \equiv 0 \pmod e$$

and such that the diagonal elements of β, γ are divisible by $2e$. It is of finite index in Γ ([Ig] p. 177). So we can find a finite set Σ of elements σ of Γ such that the cosets $\sigma\hat{\Gamma}^e$ cover Γ. It is not difficult to see that we can assume $\det\gamma\neq 0$ for each such $\sigma=\left(\begin{smallmatrix}\alpha & \beta\\ \gamma & \delta\end{smallmatrix}\right)$; for example by replacing each σ by $\sigma\left(\begin{smallmatrix}\alpha_a & \beta_a\\ \gamma_a & \delta_a\end{smallmatrix}\right)$ for a suitable integer a with $0\leq a\leq n$, where $\alpha_a=(1+2ea)\iota$, $\delta_a=(1-2ea)\iota$, $\beta_a=\gamma_a=2ea\iota$ clearly defines an element of $\hat{\Gamma}^e$ (compare [Sie] p. 399).

To deduce the lemma take any v in V, and choose any τ in $\mathfrak{H}$ with $\Theta(e\tau,0)=v$. We can find σ_0 in Γ with $\tau_v'=\sigma_0(\tau)$ in $\mathfrak{F}$. Then $\sigma_0=\sigma\rho$ for some σ in Σ and ρ in $\hat{\Gamma}^e$, so that $\tau_v=\rho(\tau)$ also satisfies $\Theta(e\tau_v,0)=v$, and $\tau_v'=\sigma(\tau_v)$.

We take $v'=\Theta(e\tau_v',0)$ in V'. Then the map λ_σ constructed above goes from $A(\tau_v)=A_v$ to $A(\tau_v')=A_{v'}$. It is clear that D_v and $\lambda_\sigma^* D_{v'}$ are linearly equivalent divisors, and the lemma is thereby proved.

We are now able to extend the Theorem to all v in V. For $d \geq 1, h \geq 1$ let K be an extension of k of relative degree at most d and let v be in $V_h(K)$. Let $c_1, c_2, \ldots$ be positive constants depending only on n, d and e. Since Λ depends only on n and e, the field K' obtained by adjoining all the coefficients of all the λ in Λ is an extension of K of relative degree at most c_1. By Lemma 8.1 we can find v' in V' and λ in Λ inducing an isomorphism of A_v to $A_{v'}$ defined over K'. Looking at the origins we see that $v'=\lambda(v)$. It follows easily that v' is in $V(K')$ and that the height of v' is at most $h'=c_2h$, i.e., v' is in $V'_{h'}(K')$.

Let P be a non-zero point of $A_v(K)$. Then $P'=\lambda(P)$ is in $A_{v'}(K')$, and since $P' \neq 0$ we can apply the case of the Theorem already proved. We find that

$$\max(q_{v'}(P'), r_{v'}(P')) \geq c_3 h'^{-1} \geq c_4 h^{-1} . \tag{8.4}$$

But since the divisors D_v, $\lambda^* D_{v'}$ are linearly equivalent and λ is an isomorphism, we see by functoriality that

$$q_{v'}(P') = q_v(P) , \quad r_{v'}(P') = r_v(P) ;$$

and these together with (8.4) complete the proof of the Theorem for the special family A.

For later purposes we note that exactly similar arguments enable us to extend Lemmas 4.1 and 4.2 from V' to V. We leave the details to the reader.

Finally we observe that we have established these results for all algebraic points of V and not just those in some open subset; as remarked in section 2, this is crucial for the extension to arbitrary families, which will be carried out in the following section.

9. PROOF OF THE THEOREM (II)

In this section we extend the Theorem (and simultaneously the results of section 4) to arbitrary families. This would be rather easy if we knew that our special family in section 2 was "maximal" in a sense similar to that of Theorem 5.5 (p. 324) of Shimura's article [Sh]. Presumably this is true with suitable assumptions about the level structure; however, our family is associated with the so-called (e,2e)-level structure,which is somehow between the levels e and 2e, and it appears that this does not correspond to any "PEL structure" in [Sh]. So we have to work a bit harder. Nevertheless the arguments of this section are purely technical, and to save space we do not refer to [Sh] in the detailed way it deserves, but instead we leave it to the reader to do the necessary checking.

We use standard notations for divisors D, D'; thus $D \geq D'$ means that D-D' is positive and $D \sim D'$ means that D, D' are linearly equivalent. We shall later need the following result, which implies that the statement of the Theorem is essentially independent of the choice of the divisor D.

Lemma 9.1. *Let* V *be a variety defined over a number field* k, *let* A *be an abelian variety defined over* k(V), *and let* D, D' *be ample divisors on* A *also defined over* k(V). *Then there exists* m *and a non-empty open subset* V_0 *of* V *with the following property. For all* v *in* $V_0(\bar{k})$ *the height and distance functions* q_v, r_v *(with respect to* D*) and* q'_v, r'_v *(with respect to* D'*) are defined, and they satisfy*

$$q_v(P) \leq m q'_v(P) \ , \ r_v(P) \leq m r'_v(P)$$

for all P *in* $A_v(\bar{k})$.

Proof. Since D is ample, 3D is very ample ([Mu 2] p. 163), and in particular there exists a function f on A, defined over k(V), whose divisor (f) satisfies $(f) \geq -3D$. Replacing D by (f)+3D, we may therefore assume D itself is positive and very ample. This has the effect of multiplying q_v, r_v by 3. Similar remarks apply to D'.

We can thus find embeddings $\varepsilon, \varepsilon'$ of A into projective spaces $\mathbb{P}_N, \mathbb{P}_{N'}$ for the initial coordinates correspond to D, D'. We can further suppose that $\varepsilon, \varepsilon'$ and their inverses are defined over k(V). There is then a morphism φ, also defined over k(V), from $X'=\varepsilon'(A)$ to $X=\varepsilon(A)$ making a commutative triangle. Therefore it can be described completely by finitely many polynomial maps

$$x = (P_{\alpha 0}(x'),\ldots,P_{\alpha N}(x'))$$

with coefficients in k(V) (using the obvious notation for coordinates on X, X'). If now m_α denotes the common degree of the polynomials $P_{\alpha 0},\ldots,P_{\alpha N}$, we show that the present lemma holds with

$$m = \max_\alpha m_\alpha .$$

For we can suppose that all of the above continues to hold under specialization, at least for v in some non-empty open subset V_0 of V. Then for v in $V_0(\bar{k})$ the inequality $q_v(P) \leq m q'_v(P)$ is a well-known consequence (see e.g. [Ma 2] p. 168). To deduce the same for r_v, r'_v requires a bit more work.

Let D_α be the pullback by ε' of the divisor corresponding to $P_{\alpha 0}(x')$. The definition of φ then gives the inequality $D \leq \sup_\alpha D_\alpha$ between divisors. Also $D_\alpha \sim m_\alpha D'$. Using subscripts to indicate specializations, we find that

$$D_v \leq \sup_\alpha D_{\alpha v'}, \quad D_{\alpha v} \sim m_\alpha D'_v,$$

at least for all v in some non-empty open subset. But it is easy to see by considering the decomposition into irreducibles (see [Ig] p. 127) that the Hermitian form of $\sup_\alpha D_{\alpha v}$ is mH'_v. Thus $H_v \leq mH'_v$, and the desired inequality $r_v \leq mr'_v$ follows at once. This completes the proof.

Note that in general the Hermitian form of a supremum of divisors is not easily related to the Hermitian forms of the individual divisors.

We return now to the family A over $k(V)$ of section 2, which we rewrite as $\hat{A}^e$ over $k(\hat{V}^e)$ to remind ourselves of its special nature; also let $\hat{D}^e$ be the divisor D. We define another special system A^e over $k(V^e)$ as that corresponding to the "maximal fibre system" of level e constructed in Theorem 5.3 (p. 324) of [Sh] (take the principal polarization class and ignore endomorphisms). Here V^e is analytically isomorphic to $\Gamma^e \backslash \mathfrak{S}$, where Γ^e is the principal congruence subgroup of level e, and it is defined over $k = \mathbb{Q}(\varepsilon(1/e))$. Also A^e is defined over $k(V^e)$. Let D^e be a divisor on A^e, also defined over $k(V^e)$, corresponding to the principal polarization.

Our first step is to deduce the Theorem for A^e over $k(V^e)$ with respect to D^e from the Theorem for $\hat{A}^e$ over $k(\hat{V}^e)$ with respect to $\hat{D}^e$. We start by noting that the latter family already has a natural level e-structure. This follows from the formulae

$$\theta_{m0}(e\tau, a + b\tau) = \varepsilon(ma' - \frac{1}{2}e^{-1}b\tau b')\theta_{m+e^{-1}b,0}(e\tau, 0)$$

for a, b in $\mathbb{Z}^n$ (see [Ig] pp. 49, 50) which show that the points of order dividing e on $\hat{A}^e$ are already defined over $k(\hat{V}^e)$.

After reformulating in terms of fibre systems (which we leave to the reader) we may apply Theorem 5.5 (p. 324) of [Sh]. Denote specializations (or fibres) by subscripts. We obtain a morphism λ from $\hat{V}^e$ to V^e, defined over k, such that for each w in $\hat{V}^e$ there is an isomorphism Λ_w from $\hat{A}^e_w$ to $A^e_{\lambda(w)}$, defined over k(w), which respects the level structure and polarization class. We call such an isomorphism an e-isomorphism. In particular, the pullback $\Lambda^*_w D^e_{\lambda(w)}$ corresponds to the principal polarization on $\hat{A}^e_w$ From this it is easy to see that there exist positive integers s_w, t_w, bounded independently of w, such that

$$s_w \Lambda^*_w D^e_{\lambda(w)} \sim t_w \hat{D}^e_w . \tag{9.1}$$

We next prove that λ is surjective and that $\lambda^{-1}(v)$ is finite for all v in V^e. Let φ be the analytic isomorphism from $\Gamma^e\backslash\mathfrak{S}$ to V^e as in 5.3.5 of [Sh], and for brevity let θ be the analytic isomorphism from $\hat{\Gamma}^e\backslash\mathfrak{S}$ to $\hat{V}^e$ given by our map $\Theta(e\tau,0)$. Let μ be the natural surjection from $\hat{\Gamma}^e\backslash\mathfrak{S}$ to $\Gamma^e\backslash\mathfrak{S}$ induced by the inclusion of $\hat{\Gamma}^e$ in Γ^e.

Fix any τ in $\hat{\Gamma}^e\backslash\mathfrak{S}$ and put $x=\lambda(\theta(\tau))$, $y=\varphi(\mu(\tau))$. Then, as we have already said, the varieties A^e_x, $A^e_{\theta(\tau)}$ are e-isomorphic. Also 5.3.5 of [Sh] shows that A^e_y is e-isomorphic to a certain variety, which on following through the construction of section 3.3 of [Sh] we see also to be e-isomorphic to $A^e_{\theta(\tau)}$ $(=A(\tau)$ in the notation of our section 2). Therefore A^e_x, A^e_y are e-isomorphic. By 3.15.1 (p. 316) of [Sh] we conclude that x=y, or

$$\lambda(\theta(\tau)) = \varphi(\mu(\tau)) .$$

Since τ is arbitrary and φ, μ are surjective, this shows that λ is surjective. And since for any σ in $\Gamma^e\backslash\mathfrak{S}$ the cardinality of $\mu^{-1}(\sigma)$ is at most the index of $\hat{\Gamma}^e$ in Γ^e, we deduce that $\lambda^{-1}(v)$ is finite for all v. This establishes the claims asserted a few lines above.

We can now prove the Theorem for A^e over $k(V^e)$ with respect to D^e. Again it is crucial not to omit any closed subset of V^e. Fix $d\geq1$, $h\geq1$, an extension K of k of relative degree at most d, and v in $V^e(K)$ with height at most h with respect to some suitable projective embedding of V^e. There exists w in $\lambda^{-1}(v)$, and height estimates (see e.g. the Heights Lemma of [Ma 3]) show that w is in $\hat{V}^e(L)$ for some extension L of K of relative degree at most c, and has height at most ch, where c is independent of h and K. For any $P\neq0$ in $A^e_v(K)$ the point $Q=\Lambda^{-1}_w(P)$ lies in $\hat{A}^e_w(L)$, and since $Q\neq0$ the Theorem for $\hat{A}^e$ gives (with the obvious notation)

$$\max(\hat{q}^e_w(Q), \hat{r}^e_w(Q)) \geq C^{-1}(ch)^{-1} .$$

But by (9.1) and functoriality we have

$$s_w q_v^e(P) = t_w \hat{q}_w^e(Q)\,, \quad s_w r_v^e(P) = t_w \hat{r}_w^e(Q)\,,$$

and so we deduce the desired lower bound for $\max(q_v^e(P), r_v^e(P))$. This completes the proof of the Theorem for A^e over $k(V^e)$ with respect to D^e. As before, similar arguments work for Lemmas 4.1 and 4.2. Of course we have only considered the special field $k=\mathbb{Q}(\varepsilon(1/e))$; but it is easy to extend the results to any number field.

We next generalize to arbitrary families using the same sort of reasoning. At first we have to impose some conditions. Namely, suppose that A and D are defined over $k(V)$, that A is principally polarized, that D corresponds to the principal polarization, and that the points of order dividing e on A are also defined over $k(V)$. Then we are able to construct a corresponding fibre system, at least over some non-empty open subset V_0 of V, and a morphism λ from V_0 to V^e again using Theorem 5.5 of [Sh]. The Theorem for v in V_0 and points of $A_v(K)$ can then be deduced from the Theorem for points of $A^e_{\lambda(v)}(K)$ somewhat as above; we leave the details to the reader. Of course we now argue in the opposite (easier) direction, and there is no need to check that λ is surjective (it might not be); furthermore, we require only a trivial height estimate bounding the height of $\lambda(v)$ from above in terms of the height of v.

Again the same arguments work for Lemmas 4.1 and 4.2.

We can now remove most of the conditions imposed above. Let A be an abelian variety defined over $k(V)$, and let D be an ample divisor on A also defined over $k(V)$. Then A is isogenous to a principally polarized abelian variety A' (see [Mu 2] p. 234), and it is not difficult to show that A', as well as the dual isogenies σ from A to A' and σ' from A' to A, are defined over a finite extension F of $k(V)$. We can even suppose that the points of order dividing e on A' are defined over F. We can now identify F with a function field $k(V')$ for some variety V' by means of a generically finite surjective map φ from V' to V defined over k, and then we can consider A, A' as families over $k(V')$ related by σ. Let D' be an ample divisor on A', also defined over $k(V')$ and corresponding to the principal polarization on A'. Thus by what we proved in the previous paragraph, the Theorem and Lemmas 4.1, Lemma 4.2 hold for A' over $k(V')$ with respect to D'. Let V_0' be the corresponding non-empty open subset of V', and let $D''=\sigma^*D'$. We will shortly also need the pullback $D'''=\sigma'^*D''$. We may suppose that for all v in V_0' we obtain suitable specializations $A_v, A_v', D_v, D_v', D_v'', D_v'''$ and σ_v, σ_v'. Further the cardinality M of the kernel of σ_v is independent of v.

We proceed to deduce the Theorem and Lemmas 4.1, 4.2 for A over $k(V')$, at first with respect to the divisor D'' and then with respect to D. Now functoriality holds only in the weaker form (with the obvious notation)

$$r''_v(P) \geq r'_v(\sigma_v(P)) \tag{9.2}$$

for v in $V'_0(K)$ and P in $A_v(\bar{k})$. We first extend Lemma 4.1, using a trick to avoid the kernel of σ_v. Let Q be a point of $A_v(\bar{k})$ of order $t>1$. Going back to the tangent space, it is easy to see that for each positive integer m we can find P_m in $A_v(\bar{k})$ with

$$mP_m = Q\,,\ m^2 r''_v(P_m) = r''_v(Q)\,. \tag{9.3}$$

In particular the P_m are all different. So we can find m with $m \leq M+1$ such that P_m is not in the kernel of σ_v. So $P' = \sigma_v(P_m)$ is a non-zero point of $A'_v(\bar{k})$ of order at most mt. Therefore by Lemma 4.1 for A' with respect to D' we have

$$r'_v(P') \geq C_1^{-1}(mt)^{-2}h^{-1}$$

for $h=\max(1,h(v))$. So by (9.2) and (9.3)

$$r''_v(Q) \geq m^2 r'_v(P') \geq C_1^{-1}t^{-2}h^{-1}\,.$$

This proves Lemma 4.1 for A over $k(V')$ with respect to D''.

To prove the Theorem in the same situation we take $P \neq 0$ in $A_v(K)$. If $P' = \sigma_v(P) \neq 0$ then we apply the Theorem to P' in $A'_v(K)$ and use functoriality (9.2); otherwise if $\sigma_v(P)=0$ then P is a point of order at most M and we use the version of Lemma 4.1 just established. In either case we deduce the Theorem for A over $k(V')$ with respect to D''.

Finally to extend Lemma 4.2 we have to use in a similar way the dual isogeny σ'_v from A'_v to A_v. Given $P_0,\ldots,P_B$ in $A_v(\bar{k})$ we apply Lemma 4.2 to any set of inverse images $P'_0 = \sigma'^{-1}_v(P_0),\ldots,P'_B = \sigma'^{-1}_v(P_B)$ in $A'_v(\bar{k})$. The desired result then follows easily from noting first that since D'''_v is the pullback of D'_v by multiplication by the degree M of σ_v, we have (again with the obvious notation)

$$r'''_v(Q) = M r'_v(Q)$$

for all Q in $A'_v(\bar{k})$, and secondly that by (9.2) we have also

$$r''_v(\sigma_v(Q)) \leq r'''_v(Q)$$

for all such Q.

This completes the proof of the Theorem and Lemmas 4.1, 4.2 for A over $k(V')$ with respect to D''. An application of Lemma 9.1 (with V' instead of V) enables us to deduce the same results for A over $k(V')$ with respect to the original divisor D, taking possibly a smaller non-empty open subset V_0' of V'.

Finally all these calculations are in terms of some height function h' on $V'(\bar{k})$. To go back to V we recall that the map φ from V' to V is generically finite and surjective. It follows that there is a non-empty open subset V_0 of V and a constant c with the following property. For each number field K and each v in $V_0(K)$ there exists an extension K' of K of relative degree at most c and v' in $V_0'(K')$ with $\varphi(v')=v$; and further we have the height estimates

$$h'(v') \leq c \max(1,h(v))$$

(see again the Heights Lemma in [Ma 3]). From these remarks we see immediately that our Theorem, together with Lemmas 4.1 and 4.2, holds in full generality with the open set V_0.

10. Further Remarks

We have now established versions of Lemmas 4.1 and 4.2 for arbitrary families. For ease of later reference we state these results explicitly here. Note again the presence of the non-empty open set V_0.

Torsion Lemma. *For each* $d\geq 1$ *there exists* $C_3>0$, *depending only on* k, A, D, d *and the embedding of* V, *such that for any extension* K *of* k *of relative degree at most* d, *any* v *in* $V_0(K)$, *and any integer* $t>1$ *we have*

$$r_v(P) \geq C_3^{-1}t^{-2}(\max(1,h(v)))^{-1}$$

for all P *in* $A_v(\bar{k})$ *of order* t.

Box Principle. *For each* $d\geq 1$ *there exists* $C_4>0$, *depending only on* k, A, D, d *and the embedding of* V, *such that for any extension* K *of* k *of relative degree at most* d *and any* v *in* $V_0(K)$ *the following holds. If for some integer*

$$B \geq C_4(\max(1,h(v)))^n$$

we are given points $P_0,\ldots,P_B$ *on* $A_v(\bar{k})$ *then we can find* a, b, *with* $0\le a<b\le B$, *such that*

$$r_v(P_a - P_b) \le C_4 B^{-1/n}.$$

Next we shall deduce Corollaries 1 and 2 of section 1 from the Theorem together with the Box Principle. For $d\ge 1$ let K be an extension of k of relative degree at most d, and let v be in $V_0(K)$. Write $h=\max(1,h(v))$.

For Corollary 1 let P be any non-torsion point of $A_v(K)$. Choose B as a sufficiently large multiple of h^n, and apply the Box Principle to the points $0,P,\ldots,BP$. We find an integer $m=b-a$ with

$$0 < m \le B, \quad r_v(mP) \le C^{-1}h^{-1},$$

where C is the constant of the Theorem. Since $mP\neq 0$ the Theorem now implies that $q_v(mP) > C^{-1}h^{-1}$, and therefore

$$q_v(P) = m^{-2}q_v(mP) \ge B^{-2}q_v(mP) > C_1^{-1}h^{-2n-1}.$$

This proves Corollary 1.

A similar argument gives Corollary 2. Suppose on the contrary that $A_v(K)$ contains B+1 torsion points, where again B is a sufficiently large multiple of h^n. Applying the Box Principle to these points, we find a non-zero torsion point P with

$$r_v(P) \le C^{-1}h^{-1},$$

again for the constant C of the Theorem. As $q_v(P)=0$, this already contradicts the Theorem and thereby proves Corollary 2.

We will now consider to what extent our results are best possible with respect to their dependence on $h=\max(1,h(v))$. First we prove that our Theorem is indeed best possible in this respect, even for $d=1$, $k=\mathbb{Q}$. To get extremal examples it is convenient to consider the elliptic curves E_v given by

$$Y^2 = X(X-1)(X-v)$$

and embedded in $\mathbb{P}_2$ in the obvious way. Suppose first $n=1$. If the period lattice is written as $\mathbb{Z}+\mathbb{Z}\tau_v$ for τ_v in the fundamental region, and a point P on E_v corresponds to $\lambda+\mu\tau_v$ for real λ, μ, it is implicit in the proof of Lemma 4.2 that

$$r_v(P) \le c(y_v^{-1}\lambda^2 + y_v\mu^2)$$

for c absolute and $y_v=\mathrm{Im}\,\tau_v$ (and there is even a similar inequality in the opposite direction, provided λ, μ are interpreted modulo $\mathbb{Z}$). We choose $\lambda=\frac{1}{2}$, $\mu=0$; then $Y(P)=0$, and so P is defined over $\mathbb{Q}$ provided v is in $\mathbb{Q}$. Since $q_v(P)=0$ we get

$$\max(q_v(P),r_v(P)) \leq \frac{1}{4} c y_v^{-1} . \tag{10.1}$$

Suppose in fact v is a large positive integer, so that we can take $h=\log v$. Since

$$j(\tau_v) = 2^8(v^2 - v + 1)^3 v^{-2}(v - 1)^{-2}$$

we have

$$|j(\tau_v)| \geq 2v^2 .$$

On the other hand it is well-known that

$$|j(\tau_v)| \leq c' + |\varepsilon(-\tau_v)| = c' + \exp(2\pi y_v)$$

for c' absolute, and so we deduce the inequality

$$y_v \geq h/2\pi$$

for h sufficiently large (which incidentally shows that the Matrix Lemma is best possible). Putting this into (10.1) we obtain an extremal example to our Theorem for $n=1$.

For $n>1$ such examples can be found just by taking the product n times. Presumably examples also exist for generically simple families of abelian varieties.

In a similar way, by considering the points corresponding to $(b/B)\tau_v$ $(0 \leq b \leq B)$, it is not difficult to verify that, at least for $n=1$, the lower bound for B in the Box Principle is best possible. And the same remark about products serves to extend this (after a moment's thought) to any $n>1$.

By way of contrast, it seems doubtful if either Corollary 1 or Corollary 2 is best possible in a similar sense. A conjecture of Lang [La] (p. 92), when suitably generalized (see Silverman [Sil 2]), would suggest that $q_v(P)$ is bounded below independently of h (and even tends to infinity in some way). Similarly it is conjectured that the torsion of an abelian variety depends only on the dimension and the field of definition, and thus ought to be bounded above independently of h. However, both conjectures seem very deep.

We next observe that the arguments of this section yield slightly more about $q_v(P)$ than we have recorded. For example, by applying the Box Principle to the points $b_1P_1+b_2P_2$, it is easy to show that

$$\max(q_v(P_1),q_v(P_2)) \geq C''^{-1}(\max(1,h(v)))^{-n-1}$$

for any two independent points P_1, P_2 on $A_v(K)$; thus the bound for the "second minimum" of q_v is substantially better than the bound for the first minimum. There are of course similar inequalities for higher successive minima.

Our final remark concerns the open subset V_0 of V in our results. As we have seen, we can take $V_0=V$ for both the maximal families $\hat{A}^e$, A^e considered; and it seems likely that this is also true in general, provided the results are formulated with sufficient care. But the proof of this probably requires properties of abelian varieties more delicate than any used in this paper, and so we did not attempt a verification.

REFERENCES.

[Bai 1] W. L. Baily, Jr., On the theory of θ-functions, the moduli of abelian varieties, and the moduli of curves, Annals of Math. **75** (1962), 342-381.

[Bai 2] W. L. Baily, Jr., Automorphic forms with integral Fourier coefficients, Several complex variables I, Lecture notes in Math. Vol. **155**, Springer, Berlin-Heidelberg-New York 1970 (pp.1-8).

[Bak] A. Baker, The theory of linear forms in logarithms, Transcendence theory; advances and applications, Academic Press, London 1977 (pp.1-27).

[Be] D. Bertrand, Galois orbits on abelian varieties and zero estimates, to appear.

[Co] P. Cohen, Explicit calculation of some effective constants in transcendence proofs, Ph. D. Thesis, University of Nottingham 1985 (Chapter 3).

[Fr] E. Freitag, Siegelsche Modulfunktionen, Grundlehren d. math. Wiss., Vol. **254**, Springer, Berlin-Heidelberg-New York 1983.

[Frey] G. Frey, Some aspects of the theory of elliptic curves over number fields, Expositiones Math. **4** (1986), 35-66

[Ig] J.-I. Igusa, Theta functions, Grundlehren d. math. Wiss., Vol **194**, Springer, Berlin-Heidelberg-New York 1972.

[La] S. Lang, Elliptic curves, diophantine analysis, Grundlehren d. math. Wiss., Vol. **231**, Springer, Berlin-Heidelberg-New York 1978.

[LR] H. Lange and W. Ruppert, Complete systems of addition laws on abelian varieties, Inventiones Math. **79** (1985), 603-610.

[M] R. C. Mason, Diophantine equations over function fields, London Math. Soc. Lecture Notes, Vol. **96**, Cambridge 1984.

[Ma 1] D. W. Masser, Small values of the quadratic part of the Néron-Tate height, Progress in Math., Vol. **12**, Birkhäuser, Boston-Basel-Stuttgart 1981 (pp.213-222).

[Ma 2] D. W. Masser, Small values of the quadratic part of the Néron-Tate height on an abelian variety, Compositio Math. **53** (1984), 153-170.

[Ma 3] D. W. Masser, Specializations of Mordell-Weil groups, in preparation.

[MW] D. W. Masser and G. Wüstholz, Zero estimates on group varieties II, Inventiones Math. **80** (1985), 233-267.

[Mu 1] D. Mumford, On the equations defining abelian varieties II, Inventiones Math. **3** (1967), 75-135.

[Mu 2] D. Mumford, Abelian varieties, Oxford 1974.

[MF] D. Mumford and J. Fogarty, Geometric invariant theory, Ergebnisse Math., Vol. **34**, Springer, Berlin-Heidelberg-New York 1982.

[N] J. V. Nesterenko, Bounds for the characteristic function of a prime ideal, Math. USSR Sbornik **51** (1985), 9-32 (Mat. Sbornik **123** (1984), 11-34).

[Ph] P. Philippon, Lemmes de zéros dans les groupes algébriques commutatifs, to appear in Bull. Soc. Math. France.

[PW] P. Philippon and M. Waldschmidt, Formes linéaires de logarithmes sur les groupes algébriques, in preparation.

[Sh] G. Shimura, Moduli and fibre systems of abelian varieties, Annals of Math. **83** (1966), 294-338.

[Sie] C. L. Siegel, Moduln Abelscher Funktionen, Ges. Abh. Vol III, Springer, Berlin-Heidelberg-New York 1966 (pp.373-435) (Nach Akad. Wiss. Göttingen, Math.-phys. Kl. **25** (1960), 365-427).

[Sil 1] J. H. Silverman, Heights and the specialization map for families of abelian varieties, J. reine angew. Math. **342** (1983), 197-211.

[Sil 2] J. H. Silverman, Lower bounds for height functions, Duke Math. J. **51** (1984), 395-403.

[SD] M. F. Singer and J. H. Davenport, Elementary and Liouvillian solutions of linear differential equations, to appear in J. Symbolic Computation.

[ZM] J. G. Zarhin and J. I. Manin, Height on families of abelian varieties, Math. USSR Sbornik **18** (1972), 169-179 (Mat. Sbornik **89** (1972), 171-181).

LARGE TRANSCENDENCE DEGREE REVISITED
I. EXPONENTIAL AND NON-CM CASES

W. Dale Brownawell*
Department of Mathematics, Penn State University
University Park, PA 16802

I. Introduction.

The first results on the algebraic independence of more than one number using elimination theory were obtained by A.O. Gelfond (see [Br1] for an account of the method and [Wa2] for an update with complete bibliography). He employed elementary properties of resultants to establish a criterion for algebraicity involving the values of a sequence of integral polynomials with slowly growing size. In order to establish the algebraic independence of at least two numbers out of rather small sets of numbers related by the exponential function, he also proved a bound for the number of zeros an exponential polynomial can have in a disk of radius $R > 0$. G.V. Chudnovsky [Ch] was the first to give an extension of Gelfond's method, based on the successive use of his semi-resultants, for showing the algebraic independence of arbitrarily many numbers out of (exponentially larger) sets of numbers related by the exponential function. For this he also required the "small values theorem" of R. Tijdeman [Tij], which was a stronger quantitative form of Gelfond's zero estimate. Further work on this method was carried out by P. Warkentin [War], R. Endell [En], E. Reyssat [Re] and P. Philippon [Ph1], and Yu.V. Nesterenko [Ne3].

Through several fundamental advances, D.W. Masser and G. Wüstholz [Ma-WÜ2] were able to establish the analogues of Chudnovsky's statements for Weierstrass elliptic functions without complex multiplications. They gave explicit bounds for the degrees in the Hilbert Nullstellensatz to provide a replacement for the elimination techniques via (semi-) resultants. They also used commutative algebra to refine their basic algebraic zero estimate of [Ma-WÜ1]. They developed an effective elliptic version of a theorem of E. Kolchin on subgroups of products of algebraic groups and applied it to guarantee that the polynomials used to express the values generated by the auxiliary function satisfy the conditions of Hilbert's Nullstellensatz. It became clear that an optimal improvement of the bounds in the degrees appearing in the Nullstellensatz would reduce the cardinality of the sets of numbers containing at least k algebraically independent members to $O(k^2)$. Such bounds have now been established in [Br2] using the powerful tools of Nesterenko [Ne1]-[Ne3] and Philippon [Ph4] for elimination and a generous suggestion of C. Berenstein and A. Yger to employ deep results from the theory of several complex variables.

*Research supported in part by NSF grant DMS-8503324

In the meantime, Philippon has obtained a version of the zero estimates for general commutative algebraic groups [Ph5], which is best possible in several respects. In addition he has obtained a generalization [Ph2],[Ph4] of Gelfond's criterion, now involving a sequence I_N of polynomial ideals over an algebraic number field, whose generators are small at a fixed point ω of $\mathbb{C}^n$ or $\mathbb{C}_p{}^n$, where each I_N has only finitely many zeros inside a ball about ω of radius $\rho_N > 0$. These results are of sufficient strength to yield the algebraic independence of k numbers out of specific sets of $O(k)$ numbers related by either the exponential function or a Weierstrass elliptic function without complex multiplication. Thus the consequences of Philippon's result are even stronger than those provided in the standard way [Ma-Wü2] by the sharp Nullstellensatz.

Quite recently the author has used the tools developed by Nesterenko and Philippon to establish sharp lower bounds on the maximum absolute values of integral polynomials having no common zero within a ball of radius $\rho > 0$ centered at a fixed point in $\mathbb{C}^n$ [Br3]. This allows us to relax a technical hypothesis in [Ph4] from a measure of linear independence to an intermittent lower bound which must hold "only" for infinitely many values of the parameter. The corresponding remark also applies to the principal results of [Wa3]. Our proof resembles that of [Ma-Wü2] insofar as no sequence of ideals or criterion for algebraic independence is invoked. However it resembles that of [Ph4] insofar as we concern ourselves with the question of possible zeros of the ideal only near ω. Moreover we obtain in a natural manner extensions of many of the quantitative applications of [Ph3],[Ne4], [Ne5],[Ja].

It is the purpose of this paper to carry out these applications and to indicate briefly how the two methods compare. If all the numbers involved in any one of the theorems could be shown to be algebraically independent, then the corresponding quantitative result would be a lower bound for any non-zero polynomial over $\mathbb{Z}$ in these numbers. Present independence results using these methods are unfortunately not so strong. Therefore we recall the definitions of the appropriate quantitative analogues, first devised by Philippon [Ph3].

Following that we briefly sketch the two elimination techniques. Then we derive the properties which are actually provided by the comcombination of Philippon's notion of redundant variables in the auxiliary functions, the Masser-Wüstholz effective version of Kolchin's theorem for elliptic functions without complex multiplications, and Philippon's zero estimates for algebraic groups. This is the raw material for both of the elimination techniques. Finally we deduce the results from the two approaches for algebraic independence. In the second part of this paper, which is joint work with R. Tubbs, we treat the case of elliptic functions with complex multiplications.

I am indebted to D.W. Masser and R. Tubbs for many helpful conversations. Apparently Tubbs was the first to prove zero estimates which correspond to zero-free regions [Tu]. As far as I am aware,

Masser first explicitly noted that the usual technical hypothesis below gives a much larger zero-free (or rather zero-dimensional in his case) region than that required by Waldschmidt-Zhu [Wa-Zh], R. Endell [En], and Philippon [Ph4]. He did this in his unpublished proof of the algebraic independence of at least three of the numbers $\alpha^{\beta}, \alpha^{\beta^2}, \alpha^{\beta^3}, \alpha^{\beta^4}$, when β is a quintic number and α is non-zero algebraic with $\log \alpha \neq 0$.

II. Integral Chow Forms, Sizes, and Absolute Values of Ideals

A. Basic Properties

In [Ne2], [Ne3], Nesterenko introduced the notions of the degree, height and absolute value at a point for any unmixed homogeneous ideal $\mathfrak{H}$, of rank $m-d \leq m$ in a polynomial ring $R = \mathbb{Z}[z_0,\ldots,z_m]$, $\mathfrak{H} \cap \mathbb{Z} = 0$. We shall say that $\mathfrak{H}$ has dimension d. Philippon has given another development [Ph4] for polynomials over an arbitrary algebraic number field, using the Mahler measure instead of the naive height. However we adopt the original definitions of Nesterenko here because of their relative ease of statement. Although several of the essential features were first proven by Philippon in his fundamental paper [Ph4], we shall quote them in the versions given by Nesterenko [Ne4], which are adapted for the present definitions.

If

$$\mathfrak{H} = \mathfrak{Q}_1 \cap \cdots \cap \mathfrak{Q}_r \cap \mathfrak{Q}_{r+1} \cap \cdots \cap \mathfrak{Q}_s$$

is an irredundant primary decomposition of $\mathfrak{H}$ with $\mathfrak{Q}_i \cap \mathbb{Z} = 0$, $i = 1,\ldots,r$, and $\mathfrak{Q}_{r+1} \cap \cdots \cap \mathfrak{Q}_s = (b)$, $b \in \mathbb{N}$, then the (integral) Chow form of $\mathfrak{H}$ takes the shape

$$F = b \prod_{i=1}^{r} F_i^{e_i} \tag{2.1}$$

where e_i is the exponent of $\mathfrak{Q}_i$ and F_i is the Chow form of the corresponding prime ideal $\mathfrak{P}_i$, $i = 1,\ldots,r$. Moreover each F_i can be described in the following manner: If one intersects the variety of zeros of $\mathfrak{P}_i$ with the hyperplanes given by

$$H_j:\ u_{j0}z_0 + \ldots + u_{jm}z_m = 0,$$

$j = 0,\ldots,d$, in the new variables u_{jk}, then one obtains $\deg \mathfrak{P}_i$ points $(\alpha_{k0},\ldots,\alpha_{km})$, $k = 1,\ldots,\delta_i = \deg \mathfrak{P}_i$, where one can assume for each i that there is an ℓ with all $\alpha_{k\ell} = 1$. These points are in a sense generic zeros of $\mathfrak{P}_i$. Using them, one can express F_i as

$$F_i = a_i \prod_{k=1}^{\delta_i} (\alpha_{k0}u_{d0} + \cdots + \alpha_{km}u_{dm}), \tag{2.2}$$

where $a_i \in \mathbb{Z}[u_{00},\ldots,u_{0m},\ldots,u_{d-1,0},\ldots,u_{d-1,m}] = \mathbb{Z}[\underline{u}_0,\ldots,\underline{u}_d]$, and the F_i are invariant up to change of sign under permutation of the

$\underline{u}_j$. These results correspond to classically known propositions on the properties of Chow forms over fields [Sa],[VdW].

We say that $\mathfrak{P}_1,\ldots,\mathfrak{P}_r$ are the prime ideals underlying the form $F(\underline{u}_0,\ldots,\underline{u}_d)$. Then Nesterenko defined $\deg \mathfrak{H} := \deg_{\underline{u}_0} F$, height $\mathfrak{H}$:= height F, and for non-zero $\omega = (\omega_0,\ldots,\omega_m) \in \mathbb{C}^{m+1}$,

$$|\mathfrak{H}|_\omega := |F|_\omega := \text{ht } F(S^{(0)}\omega,\ldots,S^{(d)}\omega)/(\max\{|\omega_i|\})^{(d+1)\deg \mathfrak{H}},$$

where the $S^{(i)}$ are generic skew-symmetric matrices; the height on the right hand side is computed considering the expressions as polynomials over $\mathbb{C}$ in the variables which are, say, the upper triangular entries of the $S^{(i)}$. This definition may look puzzling at first, or even second, glance. We note however that, when $\mathfrak{H}$ is a prime ideal, then the coefficients of the monomials in the entries of the $S^{(i)}$ which appear in $F(S^{(0)}\underline{z},\ldots S^{(d)}\underline{z})$ generate $\mathfrak{H}$, perhaps intersected with some embedded components or components intersecting $\mathbb{Z}$ [Ne1]. Therefore for our purposes these coefficients from $F(S^{(0)}\underline{z},\ldots,S^{(d)}\underline{z})$ may be treated rather like a canonical basis for $\mathfrak{H}$. From Gelfond's fundamental inequality concerning the product of the heights of factors of a polynomial, one obtains the following remark, which is essential to Philippon's criterion:

Remark 1. If for any constant $C > 0$, $\log |\mathfrak{H}|_\omega < -C\cdot\log \text{ht } \mathfrak{H}$, then $\mathfrak{H}$ has a prime component $\mathfrak{P}$ such that $\log |\mathfrak{P}|_\omega < -C\cdot\log \text{ht } \mathfrak{P} + m^3\deg \mathfrak{P}$.

Remark 2. The Chow form does not distinguish $\mathfrak{P}$-primary ideals of the same exponent. Therefore, even if it is the Chow form of a known ideal $\mathfrak{H}$, a power product of Chow forms of prime ideals of the same rank not meeting $\mathbb{Z}$ is the Chow form of the corresponding intersection of symbolic powers as well. Since our main concern will be properties of the Chow forms themselves, we ordinarily will not make a choice of underlying ideal. Of course the underlying prime ideals of the Chow form are uniquely determined.

B. Quantitative Measures of Independence

In [Ph3] Philippon introduced the appropriate quantitative analogs of the statement that at least d of the numbers $\alpha_1,\ldots,\alpha_m$ are algebraically independent. In terms of our definitions, this amounts to giving lower bounds for the absolute value at $(1,\alpha_1,\ldots,\alpha_m)$ of all unmixed homogeneous ideals $\mathfrak{H}$ in $\mathbb{Z}[x_0,\ldots,x_m]$ of rank at least $m-d$. The lower bound will be given in terms of the degree, height, and rank of the ideal. Philippon shows that the two following possible alternative definitions are in effect equivalent to the one just given:

a) Measures of relative transcendence in dimension d given by lower bounds for $\max_i |P_i(\alpha_1,\ldots,\alpha_m)|$ whenever $P_i(x_1,\ldots,x_m) \in \mathbb{Z}[x_1,\ldots,x_m]$ generate an ideal of dimension d, where the lower bounds are expressed in terms of the maximal size $t(P_i)$.

b) Approximation measures in dimension d defined as lower bounds for $\max_j |\theta_j-\alpha_j|$, expressed in terms of the maximal size of the minimal polynomials for the θ_j, which are all algebraic over, say, $\alpha_1,\ldots,\alpha_d$.

III. Elimination

A. Resultants

The central tool for obtaining such results, which we shall dub the *resultant* of a Chow form and a homogeneous polynomial, was introduced by Nesterenko [Ne2], who also established several extremely useful properties.

If the Chow form F is given by (2.1) and (2.2) with $b = 1$ and if $Q \in R = \mathbb{Z}[x_0,\ldots,x_m]$ is homogeneous, then the *resultant* of F and Q is defined by

$$\operatorname{Res}(F,Q) := \prod_{i=1}^{r}(a_i^{\deg Q} \prod Q(\alpha_{k0},\ldots,\alpha_{km}))^{e_i},$$

where for fixed i the inside products run over $k = 1,\ldots,\deg \mathfrak{P}_i$. Nesterenko showed (Lemma 6 of [Ne3]) that if $\operatorname{Res}(F,Q) \neq 0$, then $\operatorname{Res}(F,Q)$ is in turn itself a Chow form whose non-constant irreducible factors are the Chow forms of those minimal primes p of $(\mathfrak{P}_i,Q)$, $i = 1,\ldots,r$, for which $p \cap \mathbb{Z} = (0)$. This resultant has properties which are analogous to those of the classical resultant. For example, since Q is being evaluated at the generic zeros of the $\mathfrak{P}_i$, $\operatorname{Res}(F,Q) = 0$ if and only if $Q \in \bigcup \mathfrak{P}_i$. If we let size = deg + log ht, then Lemma 5 of [Ne3] implies that

$$\deg_{\underline{u}_0} \operatorname{Res}(F,Q) \leq \deg_{\underline{u}_0} F \cdot \deg Q$$

and that with a constant depending only on m,

$$\text{size} \operatorname{Res}(F,Q) << \deg F \cdot \text{size}\, Q + \deg Q \cdot \text{size}\, F.$$

Moreover the resultant behaves very well with respect to absolute values. Proposition 3 of [Ne3] shows that in analogy with ordinary resultants, if F and Q are small at ω, then so is $\operatorname{Res}(F,Q)$:

Proposition A (Nesterenko). *If we let* $d_F = \deg_{\underline{u}_1} F$, $d_Q = \deg Q$ *and denote the heights of* F *and* Q *by* H_F *and* H_Q, *respectively, then*

$$|\operatorname{Res}(F,Q)|_\omega \leq H_Q^{d_F} H_F^{d_Q} e^{3m^2 d_F d_Q} \max\{|F|_\omega/H_F, |Q|_\omega/H_Q\},$$

where $|Q|_\omega = |Q(\omega)|\cdot|\omega|^{-d_Q}$.

However Philippon noticed (Proposition 2.5 of [Ph4]) that there is another case in which the resultant can be shown to be small: when $|Q|_\omega$ is small relative to the distance from ω to the zeros of $\mathfrak{H}$. This idea will be crucial in the proof of Philippon's criterion. According to our philosophy, we state the result as given in Lemma 4 of [Ne4], using the notion of the projective distance, $d_{proj}(\omega,\beta)$, from ω to β:

$$d_{proj}(\omega,\beta) \;:=\; \max_{i,j} |\omega_i\beta_j-\omega_j\beta_i|/(\max|\omega_i|)(\max|\beta_i|).$$

Proposition B (Philippon). *If* $\mathfrak{H}$ *is an unmixed homogeneous prime ideal of* R *with* $\mathfrak{H}\cap\mathbb{Z} = (0)$ *and if for every non-trivial zero* β *of* $\mathfrak{H}$, *we have that*

$$|Q|_\omega \;\le\; (d_{proj}(\omega,\beta))^\mu,$$

where $0 < \mu \le 1$, *then*

$$|\mathrm{Res}(F,Q)|_\omega \;<\; (|\mathfrak{H}|_\omega)^\mu \cdot H_F{}^{d_Q} \cdot H_Q{}^{d_F} \exp(8m^2 d_F d_Q).$$

To make effective use of the zero-free regions and to apply Philippon's Proposition B, we need another very important property relating the smallness of a homogeneous ideal $\mathfrak{H}$ at ω to the (projective) distance from ω to $Z(\mathfrak{H})$, the zeros of $\mathfrak{H}$. This property was was established by P. Philippon as Lemme 2.7 of [Ph4], and we give it in the version of Lemma 6 of [Ne4].

Proposition C (Philippon). *If* $\mathfrak{H}$ *is a homogeneous unmixed ideal of* R *of dimension* d *with* $\mathfrak{H}\cap\mathbb{Z} = (0)$, *then for every non-trivial point* $\omega \in \mathbb{C}^{m+1}$, *there is a non-trivial zero* $\beta \in \mathbb{C}^{m+1}$ *of the ideal* $\mathfrak{H}$ *such that*

$$(d_{proj}(\omega,\beta))^{(d+1)d_F} \;\le\; |\mathfrak{H}|_\omega e^{3m^3(d+1)d_F}.$$

B. Direct Elimination

The basic proposition of [Br3] is proved by an induction reminiscent of the ideal enlargement in, say, [Br-Ma]. However as noted above, working with Chow forms does not tie one to any particular ideal except in the radical case. That proposition deals with the following situation: Let $0 < \rho \le 1$ and let $\omega \in \mathbb{C}^{m+1}$. Let $\mathfrak{H}$ be an unmixed homogeneous ideal of $\mathbb{Z}[t_0,\ldots,t_q]$ of rank m-d, degree D_d, size σ_d, $\sigma_d \ge 1$, having a Chow form F_d whose every prime component has a zero within a projective distance of $\rho^{D_{d-1}\cdots D_{d'}}$ of ω. For each dimension $j = d-1,\ldots,d'$, where $d' \ge 0$, let $P_j \in R\backslash U_{j+1}$ be a homogeneous polynomial of size at most $\sigma_j \ge 1$ and degree at most D_j, where

$\mathfrak{u}_{j-1}$ is the union of the prime ideals of R underlying F_{j+1}, and define F_j inductively as $F_j := R^*(F_{j+1},P_j)$, where the asterisk denotes that we have deleted from $\mathrm{Res}(F_{j+1},P_j)$ factors from $\mathbb{Z}$ and any other factors arising from prime ideals having no zeros within a ball of radius ρ, $0 < \rho < 1$, centered at $\omega \in \mathbb{C}^{m+1}$. Finally we assume that $F_{d'} = \pm 1$.

Theorem 3.1([Br3]). *Under the above circumstances,*

$$\log \max \{|\mathfrak{H}|_\omega, |P_{d-1}|_\omega, \ldots, |P_{d'}|_\omega\} \geq \tag{3.1}$$
$$-c_1 \Sigma_j D_d \ldots D_{j+1}\sigma_j D_{j-1} \ldots D_{d'} + c_2 D_d \ldots D_{d'} \log \rho,$$

where c_1, c_2 *are explicit constants depending only on* m.

We sketch the proof. Of course each $\Delta_j := \deg_{\underline{u}_0} F_j \leq D_d \ldots D_j$. It is also easy to verify inductively, using the propositions A and C above and Gelfond's well-known inequality, that each $\tau_j :=$ size $F_j <<$ $D_j \tau_{j+1} + \Delta_{j+1}\sigma_j$ and

$$|F_j|_\omega \leq \max \{|F_{j+1}|_\omega, |P_j|_\omega\} e^{C(D_j\tau_{j+1}+\Delta_{j+1}\sigma_j)} \rho^{-C(j+1)D_d \ldots D_j},$$

where $C > 0$ is a constant depending only on m. (In case $\mathfrak{H} = (0)$, one needs Proposition 1 of [Ne2] to deal with the principal ideal (P_1).) However by construction $F_{d'} = \pm 1$. Using the displayed inequality, we are done by induction.□

Corollary([Br3]). *Let* $\mathfrak{H}$ *and* ω *be as above. Let* $Q_1, \ldots, Q_k$ *be homogeneous polynomials of degree and size at most* D *and* t, *respectively. Assume that the ideal* $(\mathfrak{H}, Q_1, \ldots, Q_k)$ *has no zeros inside the projective ball of radius* ρ, $0 < \rho \leq 1$, *centered at* ω. *Then*

$$\log \max \{|\mathfrak{H}|_\omega, |Q_i|_\omega\} \geq -c_1' D_d D^d t + c_2' \sigma_d D^{d+1} + c_3' D^d \log \rho,$$

where c_1', c_2', c_3' *are constants depending effectively upon* m.

Proof. Assume that we have constructed the polynomials $P_{d-1}, \ldots, P_{k+1}$ as $\mathbb{Z}$-linear combinations of the polynomials $Q_i x_\ell^{D-\deg Q_i} = H_{\ell i}$, satisfying the above conditions for $j = d-1, \ldots, k+1$. Then if $F_{k+1} = \pm 1$, we are finished. Otherwise all the prime ideals underlying F_{k+1} have zeros within ρ of ω. Since the $H_{\ell i}$ do not have zeros this close, a sufficiently general $\mathbb{Z}$-linear combination of them, called F_k, will not lie in any of the primes of F_{k+1}. We continue until $F_{d'} = \pm 1$.□

C. The Gelfond-Philippon Criterion

Philippon's criterion has a weaker hypothesis on the zeros near ω in two respects. First of all it does not ask for a region without zeros, but rather one with only isolated zeros. Secondly to obtain an inequality with dependence on ρ as in the preceding result, Philippon needs to consider only a ball centered at ω of radius roughly $\exp(-(|\log \rho|^{d+1}))$. On the other hand, his criterion requires a sufficiently long sequence of ideals. We state a form of his result which is sufficient for quantitative applications. Note that the formulation in [Ph4] allows one to conclude only that ω is a zero of infinitely many J_N.

Theorem 3.2(Philippon). *Let* $\omega \in \mathbb{C}^{m+1}\backslash\{0\}$. *Suppose that* $\mathfrak{H}$ *is an unmixed homogeneous ideal of dimension* d *and size at most* $\sigma_d \geq 1$, *which vanishes at* ω. *For* $a > 1$ *and* $N \geq N_0$, *let* $\{D_N\},\{S_N\}$ *denote monotonically increasing, unbounded sequences of positive integers such that* $D_{N+1} \leq aD_N$, $S_{N+1} \leq aS_N$. *Assume that* $C > 0$ *is sufficiently large and that for each* $N \geq N_0$ *there is an ideal* J_N *generated by homogeneous polynomials* $P_k \in \mathbb{Z}[x_0,\ldots,x_m] = R$ *such that*

i) J_N *has only finitely many zeros within the ball* $B(\omega,\exp(-CD_N{}^dS_N\sigma_d))$,

ii) size $P_k \leq S_N$, deg $P_k \leq D_N$,

iii) $\log |P_k|_\omega \leq -C^{d+1}(D_N{}^d)S_N\sigma_d$.

Then for all $N \geq N_1$, *the point* ω *is a zero of* J_N.

Sketch of Philippon's Proof([Ph4]). For simplicity of notation, we suppose that $D_N = S_N = N$. We also use the convention that c will denote any constant which can be bounded effectively from above, independently of N. Half of the proof consists of showing inductively for $i = d-1,\ldots,0$, the statement:

A_i: For $N \geq CN_0{}^i$, there is a prime ideal $\mathfrak{P}_{N,i} \supset \mathfrak{P}_{N,i+1}$ of dimension i in R such that

a) size $\mathfrak{P}_{N,i} \leq c\, N^{d-i}\sigma_d$,

b) $\log |\mathfrak{P}_{N,i}|_\omega \leq -C^{i+1}(\text{size } \mathfrak{P}_{N,i})N^{i+1}$.

Here we let $\mathfrak{P}_{N,d}$ be a prime component of $\mathfrak{H}$ vanishing at ω. If $\mathfrak{H} = (0)$, we begin the induction by taking $\mathfrak{P}_{N,m-1}$ to be the principal ideal generated by any irreducible factor P of any non-zero generator of J_N for which $(-\log |P|_\omega)/\text{size } P \geq C^m N^{m-1}$, which is possible by

Gelfond's lemma.

Now assume A_i, $i > 0$. To carry out the induction step producing A_{i-1}, we denote by $\rho_{N,i}$ the minimum distance from ω to a zero of $\mathfrak{P}_{N,i}$, $i > 0$. Then $\rho_{N,i} \le \exp(-(C^{i+1}N^{i+1}/(i+1))-c)$ by Philippon's Proposition C above. Now select M maximal in the interval $[N_0,N]$ such that $\exp(-C^{d+1}M^{d+1}\sigma_d) \ge \rho_{N,i}$. For $N > (CN_0)^{(d+1)/2}$, note that $N_0 < M$. By hypothesis not all generators of J_M can vanish on the positive dimensional manifold of zeros of $\mathfrak{P}_{N,i}$ lying within $\rho_{N,i}$ of ω. Let $F_{N,i}$ be the resultant of the Chow form of $\mathfrak{P}_{N,i}$ and any generator P of I_M not lying in $\mathfrak{P}_{N,i}$. If $M = N$, then according to Nesterenko's Proposition A, $F_{N,i}$ will satisfy a) and moreover

$$\log |F_{N,i}|_\omega \le -C^{i+1}(\text{size } \mathfrak{P}_{N,i})N^{i+1}/3.$$

If $M < N$, then $|P|_\omega \le \rho_{N,i}{}^{(M/(M+1))^{d+1}}$ by the maximality of M. So now Philippon's Proposition B implies that $F_{N,i}$ satisfies a) and

$$\log |F_{N,i}|_\omega \le -C^{i+1}(\text{size } \mathfrak{P}_{N,i})N^{i+1}(M/(M+1))^{d+1} + cM\text{size } \mathfrak{P}_{N,i}.$$

In either of these two cases, since

$$\text{size } F_{N,i} \le cN\text{size } \mathfrak{P}_{N,i},$$

one then chooses a single prime ideal $\mathfrak{P}_{N,i-1}$ as in the remark at the end of section II.A to satisfy a) and b). Thus a) and b) hold for all $i = d-1,\dots,0$.

To conclude the argument we begin with the proposition A_0 and the zero dimensional ideals $\mathfrak{P}_{N,0}$ it describes. We first show that the size of $\mathfrak{P}_{N,0}$ is bounded above by $cN_0{}^d\sigma_d$. For that, choose M minimal in the interval $N_0 \le M \le N$, such that size $\mathfrak{P}_{N,0} \le cM^d\sigma_d$. We want to show that in fact $M = N_0$. First of all note that if M' lies in $[N_0,N]$ and

$$\exp(-C(M'+1)^{d+1}\sigma_d) < \rho_{N,0},$$

then Philippon's Proposition B shows the absolute value of the resultant of the Chow form of $\mathfrak{P}_{N,0}$ and any generator of $J_{M'}$, to be zero. Thus $\mathfrak{P}_{N,0}$ is an isolated prime ideal of $(J_{M'},\mathfrak{G})$, and therefore (Theorem II of [Ma-Wü2], Proposition 3.3 of [Ph5]), size $\mathfrak{P}_{N,0} \le c(M')^d\sigma_d$. By the minimality of M, such $M' = M$. As a consequence,

$$\rho_{N,0} < \exp(-CM^{d+1}\sigma_d).$$

If $M > N_0$, then applying Nesterenko's Proposition A shows the resultant of the Chow form for $\mathfrak{P}_{N,0}$ and any generator of J_{M-1} to be zero. Hence $\mathfrak{P}_{N,0}$ is an isolated primed ideal of J_{M-1} and size $\mathfrak{P}_{N,0} \le$

$c(M-1)^{d+1}$, contradicting the minimality of M. Thus $M = N_0$ and

$$\text{size } \mathfrak{P}_{N,0} \le cN_0^{d+1}\sigma_d,$$

independently of N. Therefore $\mathfrak{P}_{N,0}$ is one of a finite list of prime ideals. Since however

$$\log |\mathfrak{P}_{N,0}|_\omega \le -C(\text{size } P_{N,0})N,$$

we see that for all large enough N, $|\mathfrak{P}_{N,0}|_\omega = 0$, which establishes our claim.□

In a very interesting paper [Ph3], Philippon gives quantitative analogues of his basic theorem. He has since pointed out that when the regions $B(\omega, \exp(-CD_N{}^d S_N \sigma_d))$ are zero-free, one can in effect establish A_{-1} and thereby avoid the arguments in the second half of the proof of his criterion. Independently Nesterenko has taken up this line of reasoning in [Ne5] and obtained completely effective measures of algebraic independence in terms of additional constants appearing in place of our c and C^{d+1} in i) and ii).

IV. Ideals of Values Arising from Auxiliary Functions

A. Summary

In this section we record the properties of the polynomial ideals associated with the auxiliary functions occuring in independence proofs. As usual, the auxiliary function is a polynomial in the coordinate functions of the appropriate exponential map of a commutative algebraic group. But now the group is the product of the classical one with itself p times. Although this produces no new numerical values when $p > 1$, Philippon introduced in [Ph1] the remarkable technique of using this redundancy to select the coefficients of the auxiliary function to be rational integers. It is this technical advantage which enables us to apply the results of paragraph III.

In our considerations, we shall assume that $m, n \ge 1$, that $u_1, \ldots, u_m$ and $v_1, \ldots, v_n \in \mathbb{C}$, and that for $\varepsilon > 0$, small enough (in a sense to be made precise later), there is an $N \in \mathbb{N}$ such that

$$cN < \exp(N^\varepsilon),$$

and

$$|s_1u_1 + \ldots + s_m u_m| \ge \exp(-N^\varepsilon), \quad |t_1v_1 + \ldots + t_n v_n| \ge \exp(-N^\varepsilon) \quad (4.1)$$

for all non-trivial choices of integers $s_1, \ldots, s_m$ and $t_1, \ldots, t_n$ of absolute value at most N. Here and below we denote by c a positive constant which can be bounded effectively from above, independently of N and the basic parameter S of the auxiliary construction. For measures of independence we must require that (4.1) hold for *all* $N \ge N_\varepsilon$. Both these requirements can actually be weakened somewhat, as one does not have to take ε less than a positive number depending on the case considered. In each case the relevant constant for Philippon's approach is a bit larger than for the alternate approach afforded by (3.1), by a factor of approximately Δ from the table below. (Note that in this and ensuing sections, the letter m no longer refers to the number of variables (less 1) in a polynomial ring.)

In the table we record the properties of the polynomials P_k, furnished by the auxiliary function as mentioned above. We find that for S lying in a certain interval, the ideal of polynomials P_k has a zero-free region of radius $\exp(-N^{\varepsilon})$ centered at the point ω whose coordinates are given by the numbers among which we seek algebraically independent values. The Weierstrass elliptic functions $\wp(z)$ appearing in the table will be assumed to have algebraic invariants and to be without complex multiplications. As mentioned above, the sequel with R. Tubbs will deal with the complex multiplication case. In Appendix A we carry out the verification of the properties in some detail for one case of the table, and in Appendix B we give the parameters for the corresponding auxiliary functions in the remaining cases.

In the following table, c_0, c_p', c_p'', c_p''' denote constants depending only on p, m, n and the type of coordinates appearing in ω (i.e. on p and the row under consideration). In the last column we record the limit

$$\Delta := \lim_{S\to\infty} \lim_{p\to\infty} \log(-\log|P_k(\omega)|/(\sigma(P_k)))/\log S,$$

where $\sigma(P_k)$ denotes the upper bound obtained for size P_k. The relevance of Δ is, or will become, clear from an application of the elimination results cited in the preceding section.

Coordinates of Point ω	$\dfrac{\text{Size } P_k}{c_p(\log S)^{c_p'}}$	$\dfrac{-\log\lvert P_k(\omega)\rvert}{c_p''(\log S)^{c_p'''}}$	Δ
$\exp(u_i v_j)$	$S^{\frac{(p+n)(m+n)}{(p-m)n}}$	S^m	$\frac{mn}{m+n}$
$u_i, \exp(u_i v_j)$	$S^{\frac{(p+n+1)(m+n)}{(p-m)(n+1)}}$	S^m	$\frac{mn+m}{m+n}$
$u_i, v_j, \exp(u_i v_j)$	$S^{\frac{(pm+p+mn+m+n)(m+n)}{(np-mn-m-n)(m+1)}}$	$S^{\frac{(p-m)(mn+m+n)}{np-mn-m-n}}$	$\frac{mn+m+n}{m+n}$
g_2, g_3, $\wp(u_i v_j), \wp'(u_i v_j)$, no CM	$S^{\frac{(p+2n)(m+2n)}{(p-m)n}}$	S^m	$\frac{mn}{m+2n}$

g_2, g_3, u_i $\wp(u_i v_j), \wp'(u_i v_j)$ no CM	$S^{\frac{(p+2n+2)(m+2n)}{(p-m)(n+1)}}$	S^m	$\frac{mn+m}{m+2n}$
g_2, g_3, v_j $\wp(u_i v_j), \wp'(u_i v_j)$ no CM	$S^{\frac{(pm+2p+2mn+2n)(m+2n)}{(np-p-mn-n)(m+2)}}$	$S^{\frac{(mnp+2np-m^2n-mn)}{np-p-mn-n}}$	$\frac{mn+2n}{m+2n}$
g_2, g_3, u_i, v_j $\wp(u_i v_j), \wp'(u_i v_j)$ no CM	$S^{\frac{(pm+2p+2(mn+m+n))(m+2n)}{(pn-mn-m-n)(m+2)}}$	$S^{\frac{p(mn+m+2n)-m(mn+m+n)}{pn-mn-m-n}}$	$\frac{mn+m+2n}{m+2n}$

In case some $u_i v_j$ falls into the lattice of periods of $\wp(z)$, then one can translate as in [Ma-Wü2] to produce the effect of replacing the offending infinite coordinates $\wp(u_i v_j), \wp'(u_i v_j)$ by algebraic numbers or, at least when $m \geq 3$, one can more simply omit these values from our list. In other words when applying the addition formula to express $\wp(v_j(s_1 u_1 + \ldots + s_m u_m))$ in terms of the constituent $\wp(u_i v_j)$, $\wp'(u_i v_j)$, one can neglect with impunity those $u_i v_j$ lying in the lattice. Finally on replacing all u's by some large enough multiple, we may assume that none of the $u_i v_j$ are half-lattice points. Since the half-lattice points are the only places where $\wp'(z)$ vanishes and lattice points are the only poles of $\wp'(z)$, we may apply the following remark: If $z, z_0 \in \mathbb{C}$ such that

$$\max \{|\wp(z_0) - \wp(z)|, |\wp'(z_0) - \wp'(z)|\} < \varepsilon,$$

where $0 < \varepsilon < |\wp'(z_0)|/2$, then the integral formula

$$z - z_0 = \int_{\wp(z_0)}^{\wp(z)} \frac{dx}{y} \tag{4.2}$$

shows that $\|z - z_0\| < 2\varepsilon/|\wp'(z_0)|$, where the double bars indicate the distance to the nearest lattice point. In applications we can choose z so that

$$|z - z_0| < 2\varepsilon/|\wp'(z_0)|.$$

We emphasize that although the numbers g_2, g_3 occur in a polynomial way in the values produced by our auxiliary functions and therefore appear in our first column, they are not candidates for algebraically independent or even transcendental numbers by the present method. M. Waldschmidt has obtained some results in [Wa3] where these numbers are not assumed to be algebraic, but an additional technical hypothesis is imposed on the approximability of the related modular invariant by imaginary quadratic numbers. In a later paper we plan to generalize both techniques appearing here to fields with finite transcendence type and thereby develop another approach to algebraic independence where g_2, g_3 may be transcendental numbers.

V. Algebraic Independence

Once the sizes and absolute values of the table of the preceding section are known, it is easy to apply Philippon's criterion above to deduce a lower bound on the transcendence degree of the field generated by the coordinates of ω. Using the results of [Ph3] or [Ne4],[Ne5], one can even deduce lower bounds on the absolute values of integral ideals over $\mathbb{Z}$ of dimension less than the predicted lower bound on the transcendence degree. However these same lower bounds also follow almost immediately from our basic inequality (3.1). It appears that a supplementary argument would be necessary to make the constants appearing in [Ph3] effective, whereas those of [Ne5] are explicit, as are those of [Br2], although we have not recalled them in (3.1). Thus it is possible in each of the above cases to obtain completely effective quantitative results. Let us indicate briefly how that can be done.

Given the ideal $\mathfrak{A}$, of dimension d and with size bounded by σ, one selects the underlying parameter S so that if our bound on size $P_{\underline{S}}$ is now denoted as $t(S)$ and our lower bound on $-\log |P_{\underline{S}}(\omega)|$ by $\lambda(S)$, then

$$\sigma t(S)^{d+1} \approx \lambda(S).$$

Recall from the table of Section IV that

$$\lambda(S) \approx t(S)^{\Delta+\delta(p)},$$

where $\delta(p)$ depends on m and n, but tends to zero (from below) as p tends to ∞. This enables us to write

$$\sigma \approx t(S)^{\Delta-d-1+\delta(p)}.$$

Then to apply the corollary of the theorem of III.B., we make two further assumptions:

i) First that $N^{\varepsilon} \ll \sigma t(S) \approx t(S)^{\Delta-d+\delta(p)}$, in order that the first term of the right hand side of the corollary should dominate.

ii) Secondly that $\max\{cD^{24}, ct(S)^4\} \le N$.

Finally we recall the auxiliary assumption of (4.1):

iii) $cN < \exp(N^{\varepsilon})$.

The latter two assumptions are made in order that (4.1) should enable Philippon's zero estimate to provide a zero-free region centered at ω of radius roughly $\exp(-N^{\varepsilon})$. Assumption ii) insures that the coefficients occuring in (A.7) and (A.9) below are at most N. Assumption iii) insures that these coefficients do not swamp the closeness of the approximations from (A.3) and (A.4).

From the corollary of III.B. we can read off the lower bound

$$-c\sigma t(S)^{d+1+\delta(p)} \le \log \ |\mathfrak{A}|_{\omega''}, \tag{5.1}$$

where ω'' is the point in projective space with coordinates given by (1,coords of ω). For convenience of notation we replace the condition in ii) by the stricter inequality $t(S)^{24} \le N$. Let $\mathfrak{A}$ be a homogeneous ideal of dimension $d < \Delta-1$, and choose p large enough so that $\delta(p) < \Delta-d-1$. Then if σ lies in the interval

$$N^{\varepsilon\left[\frac{\Delta-d-1-\delta}{\Delta-d-\delta}\right]} \ll \sigma \ll N^{\frac{\Delta-d-1}{24}}, \tag{5.2}$$

the above choice of S can be made, and inequality (5.1) holds, where we have written δ for $\delta(p)$.

These simple considerations applied to the table of the preceding section then give rise to the following result.

Theorem 6.1. *Let* $\varepsilon, \delta > 0$, $\varepsilon < (\Delta-d-\delta)/24$. *Let* (4.1) *hold for some* N *with* $cN < \exp(N^{\varepsilon})$. *If* $\mathfrak{A}$ *is a homogeneous ideal of dimension* $d < \Delta-1$ *and size at most* σ *satisfying* (5.2), *then*

$$\log |\mathfrak{A}|_{\omega''} \geq -c\sigma^{(\Delta-\delta)/(\Delta-d-1-\delta)}.$$

An ineffective version of this result for the first three (exponential) cases was first established by Philippon (Theorem 2 [Ph3] and [Ph4]) under the stronger hypothesis that (4.1) hold for all ε and for all sufficiently large N. This result was rendered effective by Nesterenko [Ne5]. After the completion of this manuscript, effective versions for commutative algebraic groups under the strong technical hypothesis of [Wa4] on subgroups have been obtained by E.M. Jabbouri [Ja]. We obtain the following result as a corollary of our theorem.

Theorem 6.2. *Let* $0 < \varepsilon < 1/24$. *In each of the cases of the table of the preceding section if inequality* (4.1) *holds for infinitely many* N, *then*

$$\mathrm{tr}\deg_{\mathbb{Q}} \mathbb{Q}(\omega) \geq \Delta - 1.$$

Philippon's fundamental breakthrough [Ph4] established this result for the three exponential cases under the hypothesis that (4.1) hold for all sufficiently large N and all $\varepsilon > 0$. Philippon's earlier result [Ph2] established the lower bound for the general elliptic Gelfond-Schneider case. Under these stronger hypotheses, the inequality of Theorem 6.2 follows from results of M. Waldschmidt [Wa3] or [Wa4]. We also obtain as a corollary the following result.

Theorem 6.1. *Let* $0 < \varepsilon \leq 1/24$ *and* $\delta' > 0$. *Assume that* (4.1) *holds for all* $N \geq N_{\varepsilon}$. *Then there is a constant* c' *depending effectively on* N_{ε} *such that for all homogeneous ideals* $\mathfrak{A}$ *of dimension* $d < \Delta-1$ *and size* $\leq \sigma$, *we have the lower bound*

$$\log |\mathfrak{A}|_{\omega} \geq -c'\sigma^{\frac{\Delta-d}{\Delta-d-1}+\delta'}.$$

VI. Conclusion

In each case above we have seen that, given certain mild technical hypotheses, current transcendence techniques produce, for all integers S in a certain interval, ideals I(S) which are generated by polynomials P over $\mathbb{Z}$ with

a) size P $\le$ cS^{e_ω},

b) log $|P(\omega)|$ $\le$ $-c_\omega S^{f_\omega}$,

such that I(S) has no zero inside the ball $B(\omega, \exp(-cS^{\varepsilon}))$.

Somewhat different ways for showing the algebraic independence of at least $\Delta_\omega := [f_\omega/e_\omega] - 1$ coordinates of ω arise from taking advantage of different aspects of this information:

i) Philippon's criterion uses I(S) for *all* sufficiently large S, but only requires that I(S) be zero-*dimensional* in a rather small ball for each S. This gives also an (ineffective) measure of algebraic independence. See also [Ne4] and more recent work of Jabbouri effectivizing this measure.

ii) Our inequality (3.1) uses I(S) only for a *single* sufficiently large S, but requires a somewhat larger zero-*free* region. It gives a natural effectivization of Philippon's measures under the usual technical hypotheses. Moreover every instance of (4.1) with $\varepsilon < 1/24$ and N large enough provides a lower bound for the absolute value of ideals with a bound for size lying in a corresponding range.

iii) The measure of [Ne5], which came to our attention late in the preparation of this manuscript, is somewhat intermediate between i) and ii). It uses I(S) for all S in a *sufficiently long interval*, which can apparently be specified effectively, but it uses a zero-*free* ball of about the same radius as i).

Thus we now have a variety of elimination techniques available for questions of algebraic independence. They are of roughly equal strength for almost all numbers u_i, v_j. However it seems likely from research now in progress that there are situations where each approach will be particularly advantageous.

VII. Appendix 1. The Case of $\wp(u_i v_j)$.

The details of the construction of the polynomial ideals arising from values of auxiliary functions and the verification of the zero-free region are very similar in each of the above cases. Therefore we propose to give a rather detailed sketch in only one case and leave the others to the interested reader. For simplicity we shall assume that no $\Omega u_i v_j$ lies in the lattice of periods of the Weierstrass elliptic function $\wp(z)$ with algebraic invariants g_2, g_3. We assume also that the corresponding elliptic curve E admits no complex multiplications, so that we may apply the Kolchin-Masser- Wüstholz Theorem III of [Ma-Wü2].

A. The Auxiliary Function in Redundant (Faux) Variables

Although none of the details offered below are meant to surprise the expert, we give a rather complete outline, since so far no direct full proof has yet appeared explaining the existence of a large zero-free region. For $p \in \mathbb{N}$, we consider the p-parameter subgroup of $(E^n)^p$ given by

$$\underline{z} \longmapsto (\sigma^3(v_j z_k)\wp(v_j z_k), \sigma^3(v_j z_k)\wp'(v_j z_k), \sigma^3(v_j z_k)),$$

$\underline{z} = (z_1,\dots,z_p)$, $j = 1,\dots,n$, $k = 1,\dots,p$, where σ is the corresponding Weierstrass sigma function.

We let S be a sufficiently large integral parameter and set $D := [Cp^{m/(p-m)}S^{m(p+2n)/n(p-m)}]$, where C is a sufficiently large constant. For the undetermined polynomial $P = \sum a_{\underline{\lambda}}\underline{x}^{\underline{\lambda}}$ of degree D in each of its mn variables x_{jk}, where $\underline{x}^{\underline{\lambda}} = \prod_{jk} x_{jk}^{\lambda_{jk}}$, consider the auxiliary function

$$F(\underline{z}) = P(\wp(v_1 z_1),\dots,\wp(v_n z_p)) = \sum a_{\underline{\lambda}} \prod_{jk} \wp(v_j z_k)^{\lambda_{jk}}, \tag{A.1}$$

where $\underline{\lambda}$ runs through all elements of

$$\Lambda(D) := \{\underline{\lambda} \in \mathbb{Z}^{pn}: 0 \le \lambda_{jk} < D\}.$$

Since we are working with the $p^{\underline{th}}$ power of the same 1-parameter subgroup evaluated on the same finitely generated group, we obtain no new values when we take p to be large. Therefore the repeated copies of the 1- parameter subgroup are redundant ("faux" in [Ph1]) in a sense. However as Philippon ingeniously remarked, if p is large enough, we may with little loss choose the $a_{\underline{\lambda}}$ to be rational integers, not all zero, such that $F(\underline{z}) = 0$ for all $\underline{z} \in \Sigma(S)$, i.e. for all $\underline{z} \in \mathbb{C}^p$ with

$$z_i = \underline{\sigma}_i \cdot \underline{u} = s_{i1}u_1 + \dots + s_{im}u_m,$$

$i = 1,\dots,p$, where

$$\underline{\sigma}_i = (s_{i1},\dots,s_{im}) \in \mathbb{Z}^m(S) := \{\underline{\sigma} = (s_1,\dots,s_m) \in \mathbb{Z}^m : 0 \le s_j < S\}.$$

For $\underline{\sigma} \in \mathbb{Z}^m(S)$, let

$$F_{\underline{\sigma}}(x_1,y_1,z_1;\dots;x_m,y_m,z_m),$$
$$G_{\underline{\sigma}}(x_1,y_1,z_1;\dots;x_m,y_m,z_m),$$
$$H_{\underline{\sigma}}(x_1,y_1,z_1;\dots;x_m,y_m,z_m)$$

denote multihomogeneous polynomials in m triples (x_i,y_i,z_i) with coefficients which are algebraic integers in $\mathbb{Q}(g_2,g_3)$ of size at most $c(1+|\underline{\sigma}|^2)$, where $|\sigma| := \max |s_j|$, such that for $j = 1,\dots,n$,

$$X_{j\underline{\sigma}} = F_{\underline{\sigma}}(\wp(v_j u_1),\wp'(v_j u_1),1;\ldots;\wp(v_j u_m),\wp'(v_j u_m),1),$$
$$Y_{j\underline{\sigma}} = G_{\underline{\sigma}}(\wp(v_j u_1), \quad \ldots \quad ,\wp'(v_j u_m),1),$$
$$Z_{j\underline{\sigma}} = H_{\underline{\sigma}}(\wp(v_j u_1), \quad \ldots \quad ,\wp'(v_j u_m),1)$$

are projective coordinates for the point

$$(\sigma^3(v_j(\underline{\sigma}\cdot\underline{u}))\wp(v_j(\underline{\sigma}\cdot\underline{u})),\sigma^3(v_j(\underline{\sigma}\cdot\underline{u}))\wp'(v_j(\underline{\sigma}\cdot\underline{u})),\sigma^3(v_j(\underline{\sigma}\cdot\underline{u})))$$

(e.g. Lemma 7, p.176 of [Wa1]). Then each of our equations can be written as

$$P_{\sigma_1,\ldots,\underline{\sigma}_p}(\wp(v_1u_1),\wp'(v_1u_1),\ldots,\wp(v_nu_m),\wp'(v_nu_m),g_2,g_3) =$$
$$\Sigma a_{\underline{\lambda}}\prod_{jk}(X_{j\underline{\sigma}_k}{}^{\lambda_{jk}}Z_{j\underline{\sigma}_k}{}^{D-\lambda_{jk}}) = 0. \quad (A.2)$$

Let us agree to write these equations more compactly as

$$P_{\underline{s}}(\omega) = 0,$$

where $\underline{s} = (\underline{\sigma}_1,\ldots,\underline{\sigma}_p)$ with $\underline{\sigma}_i \in \mathbb{Z}^m(S)$ and

$$\omega := (\wp(v_1u_1),\ldots,\wp'(v_nu_m),g_2,g_3) \in \mathbb{C}^{2mn+2}.$$

This gives cS^{mp} equations whose coefficients are polynomials in the $2mn$ values $\wp(u_iv_j)$, $\wp'(u_iv_j)$ of degree $\leq cpDS^2$ in the $\wp(u_iv_j)$ and of degree ≤ 1 in the derivatives. The coefficients can be taken to be algebraic integers in $\mathbb{Q}(g_2,g_3)$ of size $\leq cpDS^2$. Thus over $\mathbb{Z}$ we must solve

$$cS^{mp}(cpDS^2)^{mn} \leq c^{1+mn}C^{mn}p^{pmn/(p-m)}S^{pm(p+2n)/(p-m)}$$

equations in the

$$D^{np} \geq (C/2)^{pn}p^{pmn/(p-m)}S^{pm(p+2n)/(p-m)}$$

unknowns $a_{\underline{\lambda}}$.

This is possible by the standard Thue-Siegel application of the box principle, and we find the desired $a_{\underline{\lambda}} \in \mathbb{Z}$, not all zero, with

$$\log|a_{k,\underline{\lambda}}| \leq C^{mn}p^{mn}S^{pm(m+2n)/(p-m)}.$$

From now on, the parameter p will be held fixed.

B. The Upper Bound

One verifies that the entire function of p variables

$$F(\underline{z}) := P(\wp(v_1z_1),\ldots,\wp(v_nz_p))\prod_{i,j}(\sigma(v_jz_i))^{3D}$$

vanishes at $\Sigma(S) = \mathbb{Z}^m(S)\cdot\underline{u} \times\ldots\times \mathbb{Z}^m(S)\cdot\underline{u}$ in $\mathbb{C}^p$. Thus as usual on $\Sigma(cS_1)$, where $S_1 = 2p^{n/(p-m)}S^{(p+2n)/(p-m)}$, we find that

$$\log|F| \leq -CS^m.$$

Let, as in [Ma-Wü2],

$$M(cS_1) = \max_{i,\underline{\sigma}_i \in \mathbb{Z}^m(cS_1)}\{|X_{j\underline{\sigma}_i}(\omega)|,|Y_{j\underline{\sigma}_i}(\omega)|,|Z_{j\underline{\sigma}_i}(\omega)|\},$$

$$M_0(cS_1) = \max_{z_i=\underline{\sigma}_i\cdot\underline{u}}\{|\sigma(v_jz_i)^3\wp(v_jz_i)|,|\sigma(v_jz_i)^3\wp'(v_jz_i)|,|\sigma(v_jz_i)^3|\},$$

where $\underline{z} = (z_1,\ldots,z_p) \in \Sigma(cS_1)$. Then since $(X_{j\underline{\sigma}_i}(\omega),Y_{j\underline{\sigma}_i}(\omega),Z_{j\underline{\sigma}_i}(\omega))$ and $(\sigma(v_jz_i)^3\wp(z_iv_j),\sigma(v_jz_i)^3\wp'(z_iv_j),\sigma(v_jz_i)^3)$ represent the same projective point, ratios of corresponding non-zero coordinates equal

$$M(cS_1)/M_0(cS_1).$$

But it is well-known, e.g. [Ma-Wü2], that $\log M(cS_1) \le cS_1^{\,2}$ and $\log M_0(cS_1) \ge -cS_1^{\,2}$. Thus when we stretch notation in the obvious way to let

$$\underline{z} = \underline{s}\cdot\underline{u} = (\underline{\sigma}_1\cdot\underline{u},\ldots,\underline{\sigma}_p\cdot\underline{u}) \in \Sigma(cS_1),$$

we find that

$$|P_{\underline{s}}(\omega)|\prod_{i,j}|\sigma(u_iv_j)|^{D'} = |F(\underline{z})|(M(cS_1)/M_0(cS_1))^{3mnD} \le \exp(-CS^m),$$

where $0 \le D' \le cDS_1^{\,2}$.

C. The Ideal and Zero-free Region

For now we work in the ring $\mathbb{Z}[x_{11},y_{11};\ldots;x_{mn},y_{mn};w_1,w_2]$ of integral polynomials in $2mn+2$ variables. For large enough S we consider the ideal $I(S)$ generated by

a) the minimal polynomials over $\mathbb{Z}$, in the variables w_1,w_2, for the algebraic numbers g_2,g_3,

b) $y_{ij}^{\,2}-4x_{ij}^{\,3}-w_1x_{ij}-w_2$, $i=1,\ldots,m;j=1,\ldots,n$, the equations specifying that the points $(\wp(v_ju_i),\wp'(v_ju_i),1)$ lie on the curve E, and

c) the polynomials $P_{\underline{s}}$, $\underline{s} \in \mathbb{Z}^{pm}(cS_1)$, defined as in (A.2), but now for a larger range of $\underline{s}$ as in the preceding section. These polynomials have

$$\text{size } P_{\underline{s}} \ll t(S) = p^{(m+2n)/p-m)}S^{(p+2m)(m+2n)/n(p-m)}.$$

We want to prove that if (4.1) holds, then for S satisfying inequality ii) of section V, the ideal $I(S)$ has no zeros in the ball $B(\omega,\exp(-4N^{\varepsilon}))$ in the usual Euclidean norm on $\mathbb{C}^{2mn+2}$. Let $\omega' = (\xi_{11},\zeta_{11};\ldots;\xi_{mn},\zeta_{mn};\gamma_2,\gamma_3) \in \mathbb{C}^{2mn+2}$ such that

$$\max_{ijk}\{|\wp(u_iv_j)-\xi_{ij}|,|\wp'(u_iv_j)-\zeta_{ij}|,|\gamma_k-g_k|\} < \exp(-4N^{\varepsilon}). \quad (A.3)$$

If ω' is a zero of $I(S)$ for S sufficiently large, then $\gamma_2 = g_2$ and $\gamma_3 = g_3$, since the minimal polynomials for g_2,g_3 lie in $I(S)$. Moreover from b) we see for each i,j that the point

$$(\xi_{ij},\zeta_{ij},1) \in E.$$

For $1 \le i \le m$, $1 \le j \le n$, select $\kappa_{ij} \in \mathbb{C}$ according to (4.2) such that

$$|u_i v_j - \kappa_{ij}| < \exp(-3N^{\varepsilon}) \qquad (A.4)$$

and

$$(\wp(\kappa_{ij}),\wp'(\kappa_{ij}),1) = (\xi_{ij},\zeta_{ij},1).$$

Let $\omega'' = (\xi_{11},\zeta_{11};\ldots;\xi_{mn},\zeta_{mn}) \in \mathbb{C}^{2mn}$. Then for $\underline{\sigma} = (s_1,\ldots,s_m) \in \mathbb{Z}^m$ and $j = 1,\ldots,n$,

$$\eta_{j\underline{\sigma}} := [X_{j\underline{\sigma}}(\omega''),Y_{j\underline{\sigma}}(\omega''),Z_{j\underline{\sigma}}(\omega'')] =$$

$$[\sigma(\underline{\sigma}\cdot\underline{\kappa}_j)^3\wp(\underline{\sigma}\cdot\underline{\kappa}_j),\sigma(\underline{\sigma}\cdot\underline{\kappa}_j)^3\wp'(\underline{\sigma}\cdot\underline{\kappa}_j),\ \sigma(\underline{\sigma}\cdot\underline{\kappa}_j)^3)]$$

in $\mathbb{P}^2$, where $\underline{\sigma}\cdot\underline{\kappa}_j := s_1\kappa_{1j}+\ldots+s_m\kappa_{mj}$.

Lemma. *Assume that the* u_i, v_j *satisfy the technical hypothesis* (4.1) *and that* κ_{ij} *satisfies* (A.4). *If* $Q \in \mathbb{C}[X_1,\ldots X_n]$ *has degree at most* D *and its multihomogenization* ${}^mQ(X_1,Z_1;\ldots;X_n,Z_n)$ *satisfies*

$${}^mQ(X_{1\underline{\sigma}}(\omega''),Z_{1\underline{\sigma}}(\omega'');\ldots;X_{n\underline{\sigma}}(\omega''),Z_{n\underline{\sigma}}(\omega'')) = 0$$

for all $\underline{\sigma} \in \mathbb{Z}^m(cS_1)$, *then* $Q \equiv 0$.

For $\underline{\sigma} \in \mathbb{Z}^m$, let $\underline{n}_{\underline{\sigma}} := (\ldots,\eta_{j\underline{\sigma}},\ldots)_{j=1,\ldots,n} \in E^n$ and for $S \in \mathbb{N}$, let $\underline{N}(S) = \{\underline{n}_{\underline{\sigma}} : 0 \le s_i \le S,\ i = 1,\ldots,m\}$. According to Theorem 2.1 of [Ph5], if we can show for every proper connected algebraic subgroup G' of E^n of codimension r which is an irreducible component of the intersection of E^n with the zeros of an ideal generated by multihomogeneous polynomials of degrees at most cD and which is contained in the translate in E^n of the zeros of Q, or rather its multihomogenization ${}^mQ$ displayed in the lemma, that

$$\mathrm{card}\{(\underline{N}(S_1)+G')/G'\}\cdot\deg G' \ge cD^r, \qquad (A.5)$$

then we will know that ${}^mQ$ vanishes identically on E^n.

Before we consider the various inequalities arising from the possibilities for r in (A.5), we observe that with the above notation, when $r > 0$, there cannot be three $\mathbb{Z}$-linearly independent $\underline{\sigma}^{(1)},\underline{\sigma}^{(2)},\underline{\sigma}^{(3)} \in \mathbb{Z}^m(cS_1)$ with $\underline{n}_{\underline{\sigma}^{(k)}} \in G'$, $k = 1,2,3$. For according to the Kolchin-Masser-Wüstholz Theorem III of [Ma-Wü2], there is a non-zero $\underline{\tau} = (t_1,\ldots,t_n) \in \mathbb{Z}^n(cD^2)$ such that for all $g = (g_1,\ldots,g_n) \in G'$,

$$\underline{\tau}(g) \ := \ t_1g_1 + \ldots + t_ng_n \ = \ 0 \qquad \text{(A.6)}$$

in E. If we write $\underline{\sigma}^{(k)} = (s_1^{(k)},\ldots,s_m^{(k)})$, $k = 1,2,3$, then equality (A.6) implies that $\Sigma_{ij}t_js_i\kappa_{ij}$ lies in the lattice of periods of $\wp(z)$. Thus with respect to a fixed basis ω_1,ω_2,

$$\Sigma t_js_i^{(k)}\kappa_{ij} \ = \ a_1^{(k)}\omega_1+a_2^{(k)}\omega_2,$$

with integers $a_\ell^{(k)}$ satisfying $|a_\ell^{(k)}| \le cD^2S_1$, $k = 1,2,3$. On eliminating ω_1,ω_2, we find $\underline{\sigma} = (s_1,\ldots,s_m) \in \mathbb{Z}^m((cD^2S_1)^4)$, a non-zero linear combination of $\underline{\sigma}^{(1)},\underline{\sigma}^{(2)},\underline{\sigma}^{(3)}$, such that

$$\Sigma t_js_i\kappa_{ij} \ = \ 0.$$

However from (A.4) we conclude that

$$|(\Sigma t_jv_j)(\Sigma s_iu_i)| \ = \ |\Sigma t_js_iu_iv_j| \ < \exp(-2N^\varepsilon), \qquad \text{(A.7)}$$

which is contradicted by our technical hypothesis (4.1) when $cDS_1{}^4 < \exp(N^\varepsilon)$. Thus there are no three such $\underline{\sigma}^{(k)} \in \mathbb{Z}^m(cS_1)$.

The coset map $\underline{n}_\sigma \longmapsto \underline{n}_\sigma+G'$ induces an equivalence relation on $\underline{N}(cS_1)$. By Lemma 3 of [Ma], if there are fewer than $(cS_1)^{m-t+1}$ classes, then there are at least t $\mathbb{Z}$-linearly independent $\underline{\sigma}^{(k)} \in \mathbb{Z}(cS_1)$ with $\underline{n}_\sigma \in G'$. But then by the preceding paragraph, $\mathrm{card}((\underline{N}(cS_1)+G')/G') \ge S_1{}^{m-2}$.

Proof of Lemma.

$r = 0$. Unless Q vanishes on E^n, as we wish to show, $\dim Z(Q)\cap E^n = n-1$. Therefore since $\dim G' = n$, G' cannot lie in a translate of $Z(Q)\cap E^n$, and there is nothing remaining to show.

$r = 1,2$. In this case, according to the remark above, we must satisfy the inequality

$$S_1{}^{m-2} \ > \ cD^r. \qquad \text{(A.8)}$$

One verifies that (A.8) holds for large enough p and S, since $\Delta > 1$ when the assertion of the theorem is non-trivial.

$r \ge 3$. In this case we apply the Kolchin-Masser-Wüstholz Theorem III of [Ma-Wü2] to obtain three $\mathbb{Z}$-linearly independent $\underline{\tau}$'s, say $\underline{\tau}^{(1)},\underline{\tau}^{(2)},\underline{\tau}^{(3)}$ in $\mathbb{Z}^n(cD^6)$ such that for all $g = (g_1,\ldots,g_n) \in G'$,

$$t_1^{(\ell)}g_1+\ldots+t_n^{(\ell)}g_n = 0$$

in E, $\ell = 1,2,3$. Thus if $\mathrm{card}((\underline{N}(cS_1)+G')/G') < S_1{}^m$, then there is some non-zero $\underline{\sigma} = (s_1,\ldots,s_m) \in \mathbb{Z}^m(cS_1)$ with

$$\sum t_j^{(\ell)} s_i \kappa_{ij} \in \mathbb{Z}\omega_1 + \mathbb{Z}\omega_2,$$

$\ell = 1,2,3$. As before, we produce a non-trivial $\underline{\tau} = (t_1,\dots,t_n) \in \mathbb{Z}^n(cD^{24})$ such that

$$\sum t_j s_i \kappa_{ij} = 0. \tag{A.9}$$

Again as before, this is contradicted by our technical hypothesis, so that when $n \geq r \geq 3$, the elements $\underline{n}_\sigma$ entering into the definition of $N(cS_1)$ are distinct modulo G'. Thus the inequality of Philippon will be satisfied if

$$S_1{}^m > cD^n,$$

which the reader can verify, first choosing p large enough and then S large enough, using the tacit assumption that $\Delta > 1$. This completes the proof of the lemma.□

Corollary. *For* $S \geq S_\varepsilon$ *and for any non-zero polynomial* $P(z_{11},\dots,z_{pn})$ *of degree at most* D, *the ideal* $I(S) := \{P_{\underline{s}} : \underline{s} \in \mathbb{Z}^{pm}(cS_1)\}$, *with* $P_{\underline{s}}$ *as in* (A.2), *has no zeros inside the Euclidean ball* $B_\varepsilon := B(\omega, \exp(-4N^\varepsilon))$.

Proof. The proof is carried out by induction on p. Let $\omega' \in B_\varepsilon$. The lemma establishes the claim for $p = 1$. When $p > 1$, write

$$P(\wp(v_1z_1),\dots,\wp(v_nz_1),\dots,\wp(v_nz_{p-1})) =$$

$$\sum P_{\underline{\lambda}_p}(\wp(v_1z_1),\dots,\wp(v_nz_{p-1})) \prod_{j=1}^n (\wp(v_jz_p))^{\lambda_{jp}},$$

where $\underline{\lambda}_p = (\lambda_{1p},\dots,\lambda_{np})$. Then since $P \neq 0$, some $P_{\underline{\lambda}_p} \neq 0$. Thus by the induction hypothesis, for some $\underline{s}' = (\underline{\sigma}_1,\dots,\underline{\sigma}_{p-1}) \in \mathbb{Z}^{m(p-1)}(c_1S_1)$,

$$C_{\underline{\lambda}_p',\underline{s}'} =$$

$$P_{\underline{\lambda}',\underline{s}'}(X_{1\underline{\sigma}_1}(\omega')/Z_{1\underline{\sigma}_1}(\omega'),\dots,X_{p-1,\underline{\sigma}_{p-1}}(\omega')/Z_{p-1,\underline{\sigma}_{p-1}}(\omega')) \prod_{i=1}^n \prod_{j=1}^{p-1} Z_{i\underline{\sigma}_j}(\omega')^D$$

is non-zero. Thus applying the lemma once more, this time to

$$\phi(z_p) = \sum C_{\underline{\lambda}_p,\underline{s}'} \prod_{j=1}^n P(v_jz_p)^{\lambda_{jp}},$$

we obtain $\underline{\sigma}_p \in \mathbb{Z}^m(cS_1)$ such that

$$\prod_{j=1}^n Z_{j\underline{\sigma}_p}(\omega')^D \prod C_{\underline{\lambda}_p,\underline{s}'} \prod_{j=1}^n (X_{j\underline{\sigma}_p}(\omega')/Z_{j\underline{\sigma}_p}(\omega'))^{\lambda_{jp}} \neq 0.$$

But this quantity is $P_{\underline{s}}(\omega')$, for $\underline{s} = (\underline{\sigma}_1,\dots,\underline{\sigma}_p)$, and the corollary is established.□

D. Proof of Theorem

For σ in the interval (5.2), we select S as discussed in section V so that the upper bound $t(S)$ for the sizes of the generators of $I(S)$ will statisfy $t(S) \approx \sigma^{1/(\Delta-d-1-\delta(p))}$. We consider the homogenizations ${}^hP$ of the generators of $I(S)$. Note that $|{}^hP|_{\omega''} \leq |P(\omega)|$. It is an exercise to verify that if $I(S)$ has no zeros within a Euclidean distance of $\exp(-4N^\varepsilon)$ from ω, then the polynomials ${}^hP$ have no common zeros in $B_{\rho'}(\omega'')$, where $\rho' = \exp(-5N^\varepsilon)$. Now a straightforward application of the Corollary of III.B. establishes the result.

VIII. Appendix B. Parameters.

In this appendix, we give parameters for the auxiliary functions used to establish the properties of $I(S)$ claimed in the table of section IV.A. The letter D_0 will denote the degree in z, D the degree of the underlying polynomials in the remaining functions. T will denote the order of vanishing obtained on $\Sigma(S)$. To apply the zero estimate, one considers the auxiliary function on the somewhat larger set $\Sigma(cS_1)$ and, if $T > 1$, to the higher order of vanishing T_1. We include the dependence on p more precisely than is comfortable, in order that the reader may convince himself that p can be chosen large enough in an effective manner. Of course once this has been done, all terms of the form $p^a\cdot(\log S)^b$ can thenceforth be ignored. On the other hand, for simplicity we shall not bother to make the constants involved in bounding T_1, S_1 explicit.

$\exp(u_iv_j)$: $T = T_1 = 1$, $D_0 = 0$.

$D = [2p^{m/\lambda}S^{m(p+n)/\lambda n}]$, $\lambda := p-m$

$S_1 \approx p^{n/\lambda}S^{(p+n)/\lambda}$

$u_i, \exp(u_iv_j)$: $T = T_1 = 1$.

$D = [2p^{(mn+m)/\lambda}S^{(mp-p+mn+m)/\lambda}(\log S)^{(p-m)/\lambda}]$, $\lambda := (p-m)(n+1)$

$D_0 = [2DS/(\log S)] = [2p^{(mn+m)/\lambda}S^{p(m+n)/\lambda}(\log S)^{n(m-p)/\lambda}]$

$S_1 \approx (D_0D^n)^{1/m} \approx p^{(n+1)^2/\lambda}S^{(p+n)(n+1)/\lambda}$

$u_i, v_j, \exp(u_iv_j)$

$D = [2p^{(mn+m+n)/\lambda}S^{(mp+mn+m+n)/\lambda}(\log S)^{-(m+n)/\lambda}]$, $\lambda := np-mn-m-n$

$D_0 = T = [2(DS/\log S)] = [2p^{(mn+m+n)/\lambda}S^{(m+n)p/\lambda}(\log S)^{n(m-p)/\lambda}]$

$$S_1 \approx (D^n S)^{1/(m+1)} = p^{n(mn+m+n)/(m+1)\lambda} S^{(n(m+1)p+(n-1)(mn+m+n))/(m+1)\lambda} \times (\log S)^{-n(m+n)/\lambda(m+1)}$$

$$T_1 \approx DS_1/\log S \approx p^{(m+n+1)(mn+m+n)/(m+1)\lambda} \times S^{(pm^2+(p+m+n)(mn+m+n))/(m+1)\lambda} (\log S)^{-n(mp+p-m^2+n)/(m+1)/\lambda}$$

$u_i, \wp(u_i v_j)$: $\quad T = T_1 = 1.$

$$D = [2p^{(mn+m)/\lambda} S^{(mp-2p+2mn+2m)/\lambda} (\log S)^{(p-m)/\lambda}], \qquad \lambda := (n+1)(p-m)$$

$$D_0 = [2DS^2/\log S] = [2p^{(mn+m)/\lambda} S^{(m+2n)p/\lambda} (\log S)^{-n(p-m)/\lambda}]$$

$$S_1 \approx (D_0 D^n)^{1/m} \approx p^{(n+1)^2/\lambda} S^{(p+2n)(n+1)/\lambda}$$

$v_j, \wp(u_i v_j)$: $\quad D_0 = 0.$

$$D = [2p^{(mn+n)/\lambda} S^{(mp+2p+2mn+2n)/\lambda} (\log S)^{-(p+n)/\lambda}], \qquad \lambda := np-p-mn-n$$

$$T \approx DS^2/\log S \approx p^{(mn+n)/\lambda} S^{(m+2n)p/\lambda} (\log S)^{n(m-p)/\lambda}$$

$$S_1 \approx D^{(n-1)/(m+2)} (\log S)^{1/(m+2)} \approx p^{(mn+n)(n-1)/(m+2)\lambda} \times S^{(mp+2p+2mn+2n)(n-1)/(m+2)\lambda} (\log S)^{-n(m+n)/(m+2)\lambda}$$

$$T_1 \approx DS_1{}^2/(\log S) \approx p^{(mn+n)(m+2n)/(m+2)\lambda} S^{(mp+2p+2mn+2n)(m+2n)/(m+2)\lambda} \times (\log S)^{(m^2 n-np(m+2)-2n^2)/(m+2)\lambda}$$

$u_i, v_j, \wp(u_i v_j)$

$$D = [2p^{(mn+m+n)/\lambda} S^{(mp+2mn+2m+2n)/\lambda} (\log S)^{-(m+n)/\lambda}], \qquad \lambda := np-mn-m-n$$

$$D_0 = T = [2DS^2/\log S] = [2p^{(mn+m+n)/\lambda} S^{p(m+2n)/\lambda} (\log S)^{(mn-np)/\lambda}]$$

$$S_1 \approx (D^n S^2)^{1/(m+2)} \approx p^{n(mn+m+n)/(m+2)\lambda} S^{(np(m+2)+2(mn+m+n)(n-1))/(m+2)/\lambda} \times (\log S)^{-n(m+n)/(m+2)\lambda}$$

$$T_1 \approx DS_1{}^2/(\log S) \approx p^{(m+2n+2)(mn+m+n)/(m+2)\lambda} \times S^{(m+2n)(p(m+2n)+2(mn+m+n))/(m+2)\lambda} (\log S)^{(n(m^2-2n)-(m+2)np)/(m+2)\lambda}$$

Bibliography

[Br-Ma] Brownawell, W.D. and Masser, D.W. Multiplicity estimates for analytic functions II, Duke Math J. 47(1980), 273-295.

[Br1] Brownawell, W.D. On the development of Gelfond's method, pp. 16-44 in *Proc. Number Theory Carbondale* 1979, M.V. Nathanson, ed., Lecture Notes in Math. 751, Springer Verlag, Berlin-Heidelberg- New York, 1979.

[Br2] Brownawell, W.D. Bounds on the degree in the Nullstellensatz, Annals of Math., to appear.

[Br3] Brownawell, W.D. A local Diophantine Nullstellen inequality, manuscript.

[Ch] Chudnovsky, G.V. Some analytic methods in the theory of transcendental numbers, Inst. of Math., Ukr. SSR Acad. Sci., Preprint IM 74-8 and IM 74-9, Kiev, 1974 = Chapter 1 in *Contributions to the Theory of Transcendental Numbers.* Am. Math. Soc., Providence, R.I. 1984.

[En] Endell, R. Zur algebraischen Unabhängigkeit gewisser Werte der Exponentialfunktion, in *Number Theory Noordwijkerhout* 1983, H. Jager, ed., Lecture Notes in Math. 1068, Springer Verlag, Berlin- Heidelberg-New York, 1984.

[Ge] Gelfond, A.O. Transcendental and Algebraic Numbers, GITTL Moscow 1952 = Dover, New York, 1960.

[Ja] Jabbouri, E.M. Mesures d'indépendance algébriques sur les groupes algébriques commutatifs, manuscript.

[Ma] Masser, D.W. On polynomials and exponential polynomials in several complex variables, Invent. Math. 63(1981), 81-95.

[Ma-Wü1] Masser, D.W. and Wüstholz, G. Zero estimates on group varieties I, Invent. Math. 64(1981), 489-516.

[Ma-Wü2] Masser,D.W. and Wüstholz, G. Fields of large transcendence degree generated by values of elliptic functions, Invent. Math. 72(1983), 407-463.

[Ne1] Nesterenko, Yu.V. Estimates for the orders of zeros of functions of a certain class and applications in the theory of transcendental numbers, Izv. Akad. Nauk SSSR Ser. Mat. 41 (1977), 253-284 = Math. USSR Izv. 11(1977), 239-270.

[Ne2] Nesterenko, Yu.V. Bounds for the characteristic function of a prime ideal, Mat. Sbornik 123, No. 1(1984), 11-34 = Math. USSR Sbornik 51(1985), 9-32.

[Ne3] Nesterenko, Yu.V. On algebraic independence of algebraic powers of algebraic numbers, Mat. Sbornik 123, No. 4(1984), 435-459 = Math. USSR Sbornik 51(1985), 429-454, brief version in *Approximations Diophantiennes et Nombres Transcendants,* D. Bertrand and M. Waldschmidt, eds, Birkhäuser Verlag, Verlag, Boston-Basel-Stuttgart, 1983, pp.199-220.

[Ne4] Nesterenko, Yu.V. On a measure of the algebraic independence of the values of some functions, Mat. Sbornik 128, No.4(1985), 545-568 = Math USSR Sbornik 56(1986), in press.

[Ne5] Nesterenko, Yu.V. On bounds of measures of algebraically independent numbers, pp.65-76 in *Diophantine Approximations*, P.L. Ulnova, ed., Moscow Univ. Press, Moscow, 1985 (Russian).

[Ph1] Philippon, P. Indépendance algébrique de valeurs des fonctions exponentielles p-adiques, J. reine angew. Math. 329 (1981), 42-51.

[Ph2] Philippon, P. Pour une théorie de l'indépendance algébrique, Thèse, Université de Paris XI, 1983.

[Ph3] Philippon, P. Sur les mesures d'indépendance algébrique, pp.219-233 in *Seminaire de Theorie des Nombres*, Catherine Goldstein, ed, Birkhäuser, Boston-Basel-Stuttgart, 1985.

[Ph4] Philippon, P. Critères pour l'indépendance algébrique, Inst. Hautes Etudes Sci. Publ. Math. No. 64, 1986, 5-52.

[Ph5] Philippon, P. Lemmes de zéros dans les groupes algébriques commutatifs, Bull. Soc. Math. France.114(1986), 355-383.

[Re] Reyssat,E. Un critère d'indépendance algébrique, J. reine angew. Math. 329(1981), 66-81.

[Sa] Samuel, P. *Méthodes d'Algèbre Abstraite en Géometrie Algébrique*, 2nd ed., Ergebnisse d. Math. u. ihrer Grenzgebiete, Band 4, Springer Verlag, Berlin-Heidelberg-New York, 1967.

[Tij] Tijdeman, R. An auxiliary result in the theory of transcendental numbers, J. Number Theory 5(1973), 80-94.

[Tu] Tubbs, R. Lower bounds on linear forms and algebraic independence, Special Session on Transcendence Theory and Diophantine Problems, New York Regional AMS Meeting, April 15, 1983.

[VdW] Van der Waerden, B.L. Zur algebraischen Geometrie 19, Grundpolynom und zugeordnete Form, Math. Ann. 136(1958), 139-155.

[Wa1] Waldschmidt, M. *Nombres Transcendants et Groupes Algébriques*, Astérisque 69-70, Soc. Math. France, Paris, 1979.

[Wa2] Waldschmidt, M. Algebraic independence of transcendental numbers. Gel'fond's method and its developments, pp.551-571, in *Perspectives in Mathematics, Anniversary of Oberwolfach*, W. Jager, J. Moser, R. Remmert, eds, Birkhäuser Verlag, Boston-Basel-Stuttgart, 1984.

[Wa3] Waldschmidt, M. Groupes algébriques et grands degrés de transcendance, Acta Math. 156(1986), 253-302.

[Wa4] Waldschmidt, M. Algebraic independence of values of exponential and elliptic functions, J. Indian Math. Soc. 48 (1984), 215-228.

[Wa-Zh] Waldschmidt, M. and Zhu Yao Chen. Une généralization en plusieurs variables d'un critère de transcendance de Gel'fond, C.R. Acad. Sc. Paris, Ser. I, 297(1983), 229-232.

[War] Warkentin, P. Algebraische Unabhängigkeit gewisser p-adischer Zahlen, Diplomarbeit, Freiburg, 1978.

LARGE TRANSCENDENCE DEGREE REVISITED
II. THE CM CASE

W. Dale Brownawell*
Department of Mathematics
Penn State University
University Park, PA 16802

and

Robert Tubbs*
Department of Mathematics
University of Colorado
Boulder, CO 80309

In this note we extend to the case of complex multiplication the results obtained in Part I [Br2]. The main new ingredient is an analog for elliptic curves with complex multiplication of the Kolchin-Masser-Wüstholz theorem (Theorem III of [Ma-Wü]). The second author established in [Tu] the existence of the appropriate number of linear relations in the tangent space but did not determine their rank. D.W. Masser suggested the use of elimination theory to obtain linearly independent linear relations in the tangent space. To effect this elimination in general, we were led to the surprising extreme of developing the notion of the semi-resultant of a Chow form and an ordinary form (Appendix B).

Let $\wp(z)$ be a Weierstrass elliptic function with algebraic invariants g_2, g_3 and with lattice of periods $\omega_1\mathbb{Z} + \omega_2\mathbb{Z}$. Assume for simplicity that $\wp(z)$ has complex multiplication by $\mathcal{O} = \mathbb{Z} + \tau\mathbb{Z}$, with $\tau = \omega_2/\omega_1$, which we assume to be an algebraic integer. The general case can be reduced to this one by a simple argument involving ideals including the relations between $\wp(z)$ and $\wp^*(z)$, where $\wp^*(z)$ has period lattice $\mathbb{Z}+d\tau\mathbb{Z}$ for a non-zero algebraic integer $d\tau$. Let

$$\mathcal{O}(S) = \{n_1+n_2\tau : 0 \le n_i < S\}.$$

As in [Br2] we begin with complex numbers $u_1,\dots,u_m$ and $v_1,\dots,v_n$ and $\varepsilon > 0$, $N \in \mathbb{N}$ such that $cN < \exp(N^\varepsilon)$ and

$$\begin{aligned} |s_1u_1 + \dots + s_mu_m| &> \exp(-N^\varepsilon) \\ |t_1v_1 + \dots + t_nv_n| &> \exp(-N^\varepsilon) \end{aligned} \tag{1}$$

for all non-trivial choices of $s_1,\dots,s_m$ and $t_1,\dots,t_n$ in $\mathcal{O}(N)$ for some positive integer N. Here as below we use the letter c to denote any effectively computable positive constant which is independent of the main parameter S of the auxiliary function. We remark that for given $\varepsilon > 0$ and for almost all choices of u's and v's, (1) will hold for all $N > N_\varepsilon$. We wish to investigate the independence properties of the numbers $\wp(u_iv_j)$. As in [Br2] in order to avoid (easier) special cases, we assume that no product u_iv_j lies in $\mathcal{O}$.

We introduce a quantity Δ, which here, as in the exponential case of [Br2], represents the ratio of the number of values involved

*Research supported in part by NSF

in the statement of the theorem and $m+n$. In the following table, K will be generated over $\mathbb{Q}$ by the values of the type indicated as i,j run over all possibilities $1 \le i \le m$, $1 \le j \le n$.

a) $K := \mathbb{Q}(\wp(u_i v_j))_{\forall i,j}$ $\qquad \Delta = \frac{mn}{m+n}$

b) $K := \mathbb{Q}(u_i, \wp(u_i v_j))_{\forall i,j}$ $\qquad \Delta = \frac{mn+m}{m+n}$

c) $K := \mathbb{Q}(u_i, v_j, \wp(u_i v_j))_{\forall i,j}$ $\qquad \Delta = \frac{mn+m+n}{m+n}$

Theorem. *Let* (1) *hold and* $\delta > 0$. *In each of the cases* a),b),c), *there are effectively computable constants such that if* $cN < \exp(N^{\varepsilon})$ *and if* $\sigma \in \mathbb{N}$ *lies in the interval*

$$N^{\varepsilon\left[\frac{\Delta-d-1-\delta}{\Delta-d-\delta}\right]} << \sigma << N^{\frac{\Delta-d-1}{12}},$$

then the following holds:

For any homogeneous ideal $\mathfrak{A}$ *in* $\Delta(m+n)+1$ *variables over* $\mathbb{Z}$ *of dimension* $d < \Delta-1-\delta$ *having* size $\mathfrak{A} \le \sigma$, *we have the lower bound:*

$$\log |\mathfrak{A}|_{\underline{\omega}} \ge -c\sigma^{(\Delta-\delta)/(\Delta-d-1-\delta)}.$$

Here $\underline{\omega}$ *is the point in* $\mathbb{P}_{\Delta(m+n)}(\mathbb{C})$ *whose coordinates are given by* 1 *and the generators of* K *displayed in* a),b),c).

An analogue of this result for commutative algebraic groups has been established by E.M. Jabbouri [Ja] after this manuscript was complete under the strong technical hypothesis on algebraic subgroups given in [Wa2]. As in Part I, the theorem has the following corollaries.

Corollary 1. *If for some* ε, $0 < \varepsilon < 1/12$, *the inequality* (1) *holds for infinitely many values of* N, *then*

$$\operatorname{tr}\deg_{\mathbb{Q}} K \ge \Delta - 1.$$

This lower bound was established by Philippon [Ph1] in the general elliptic Gelfond-Schneider situation. When one combines the considerations of Section 3 (effective Kolchin-Masser-Wüstholz in the CM case) with the results of [Wa2] or [Wa3], one obtains this lower bound when inequality (1) holds for all $\varepsilon > 0$ and every $N \ge N(\varepsilon)$.

Corollary 2. *If for some* ε, $0 < \varepsilon < 1/12$, *the inequality* (1) *holds for all* $N > N_{\varepsilon}$, *then the inequality of the theorem gives a measure of algebraic independence of the values in* a),b),c) *in dimension* $d < \Delta$.

Except for the ring over which the u_i and v_j are assumed to be linearly independent, these results are analogous to those holding for the ordinary exponential function. In both cases the bounds are the same, and in both cases one can establish b) by either Gelfond's method or Schneider's method.

We give a "detailed sketch" of the proof of the theorem in case c). The proofs of the other results are similar and slightly less complicated. Let $\sigma(z)$ denote the Weierstrass sigma function associated to the lattice $\mathcal{O}$ and take

$$h(z) = \sigma^3(z), \quad f(z) = \sigma^3(z)\wp(z), \quad g(z) = \sigma^3(z)\wp'(z).$$

Then

$$\pi(z) := (f(z), g(z), h(z))$$

gives a parametrization $\pi : \mathbb{C} \longrightarrow E(\mathbb{C})$ of the complex points of the elliptic curve E.

As our first lemma we record the result mentioned in the introductory paragraph, which we supplement in Appendix B (=Paragraph G) to give the effective analogue of the Kolchin-Masser-Wüstholz theorem which we require. We use the notation $\underline{X}_i = (X_i, Y_i, Z_i)$.

Lemma 0 (Lemma 3 of [Tu]). *Suppose that* $P \in \mathbb{C}[\underline{X}_1, \ldots, \underline{X}_n]$ *is a multihomogeneous polynomial of multidegree at most* $(D, \ldots, D)$ *and that* P *does not vanish identically on* E^n. *Suppose further that there exists a finitely generated* $\mathcal{O}$*-module* $W \subset \mathbb{C}^n$ *with*

$$P(\pi(w_1), \ldots, \pi(w_n)) = 0$$

for all $(w_1, \ldots, w_n) \in W$. *Then there exists a constant* c, *depending effectively on* E^n, *and* $t_1, \ldots, t_n \in \mathcal{O}(cD^{1/2})$ *not all zero, such that*

$$t_1 w_1 + \ldots + t_n w_n \in \mathcal{O}\cdot\omega_1$$

for all $(w_1, \ldots, w_n) \in W$.

Let

$$\varphi_{\underline{v}}(z) = (1, z; \pi(v_1 z); \ldots; \pi(v_n z))$$

and for $p \in \mathbb{N}$ define a p-parameter subgroup

$$\Phi_{\underline{v},p} = \Phi_{v_1,\ldots,v_n,p} : \mathbb{C}^p \longrightarrow (\mathbb{G}_a \times E^n)^p \subseteq (\mathbb{P} \times (\mathbb{P}_2)^n)^p$$

by

$$\Phi_{\underline{v},p}(z_1, \ldots, z_p) := (\varphi_{\underline{v}}(z_1), \ldots, \varphi_{\underline{v}}(z_p)).$$

Take S to be a sufficiently large integer and define parameters D, D_0, T, and T_1 for the remainder of the paper in terms of S by

$$D = [S^{2(pm+mn+m+n)/\lambda} (\log S)^{-(m+n)/\lambda}],$$

$$\lambda = pn-(mn+m+n), \quad D_0 = T = [DS^2 (\log S)^{-1}],$$

$$S_1 = [cS^{n(pm+mn+m+n)/\lambda(m+1)} (\log S)^{(-np-n^2+m+n)/2\lambda(m+1)}],$$

$$T_1 = [cDS_1^2 (\log S)^{-1}].$$

As in Part I, the inequality (1) will provide a zero-free region for ideals $I(S)$ when S lies in an interval determined by N.

A. The Auxiliary Function

We use the notation $\underline{X}_i^{(j)} = (X_i^{(j)}, Y_i^{(j)}, Z_i^{(j)})$, $i = 1, \ldots, n$, $j = 1, \ldots, p$.

Lemma 1. *For every large* S *there exists a non-zero multihomogeneous polynomial*

$$P(\varkappa_1, \underline{X}_1^{(1)}, \ldots, \underline{X}_n^{(1)}; \ldots; \varkappa_p, \underline{X}_1^{(p)}, \ldots, \underline{X}_n^{(p)})$$

with rational integer coefficients of absolute values at most $\exp(cDS^2)$ *and with*

$$\deg_{\varkappa_k} P \le cD_0, \ \deg_{\underline{X}_j^{(k)}} P \le cD, \ j = 1, \ldots, n, \ k = 1, \ldots, p,$$

such that

$$F(z_1, \ldots, z_p) = P \circ \Phi_{\underline{v}, p}(z_1, \ldots, z_p)$$

satisfies

$$\frac{\partial^{t_1 + \ldots + t_p}}{(\partial z_1)^{t_1} \ldots (\partial z_p)^{t_p}} F(z_1, \ldots, z_p) \Big|_{\underline{z} = y} = 0 \tag{2}$$

for all $y \in (\mathcal{O}(S)u_1 + \ldots + \mathcal{O}(S)u_m)^p =: Y^p(S)$ *and all* $0 \le t_1, \ldots, t_p < T$.

Proof. Briefly, through an application of the addition and multiplication formulae for $\wp(z)$ and $\wp'(z)$, as in Part I, and the so-called Anderson- Baker-Coates trick there exist ordinary polynomials $P_{\underline{t}, y} = P_{t_1, \ldots, t_p, y}$ over $\mathbb{Z}$ in the $2+m+n+2mn$ variables

$$\Gamma_2, \Gamma_3, x_1, \ldots, x_m, y_1, \ldots, y_n, X_{11}, Y_{11}, \ldots, X_{mn}, Y_{mn},$$

such that (2) holds if and only if $P_{\underline{t}, y}(\omega') = 0$, where ω' is the point $\omega' =$

$$(g_2, g_3, u_1, \ldots, u_m, v_1, \ldots, v_n, \wp(u_1v_1), \wp'(u_1v_1), \ldots, \wp(u_mv_n), \wp'(u_mv_n)).$$

Moreover we know that the polynomials $P_{\underline{t}, y}$ have coefficients which are themselves integral linear combinations of the coefficients of P. The coefficients in these linear combinations have absolute value at most $\exp(cD_0 \log S)$. In addition, the polynomials $P_{\underline{t}, y}$ have

$$\deg_{u_i}, \deg_{v_j} \le cD_0, \quad \deg_{X_k}, \deg_{Y_k}, \deg_{\Gamma_\ell} \le cDS^2. \tag{3}$$

On taking p to be sufficiently large and treating the coefficients of P as our unknowns, the system of equations (2) may be solved so that P satisfies the statement of the lemma.□

B. The Upper Bound

Then by standard estimates (e.g. Part I and Proposition 7.2.1 of [Wa]), we deduce that

$$\log |P_{\underline{t},y}(\underline{\omega}')| < -cTS^{2m}\log S,$$

even after extending the range of the indices to $t_1,\dots,t_p < cT_1$ and $y \in Y^p(S_1)$.

C. The Non-Zero Value

For the remainder of the paper, we make the following assumptions:

i) N is large enough so that $cN < \exp(N^{\varepsilon})$

ii) $S > c$, yet small enough that $\max\{cD^{3/2}, cS_1\} < N$. (4)

In the following result we use the inner product notation that for $\underline{\mu} = (\mu_1,\dots,\mu_m)$, $\underline{\sigma} = (\sigma_1,\dots,\sigma_m)$, $\underline{\kappa}_j = (\kappa_{1j},\dots,\kappa_{mj}) \in \mathbb{C}^m$, $\underline{\sigma}\cdot\underline{\mu} = \sigma_1\mu_1+\dots+\sigma_m\mu_m$ and so on. We let

$$N(S_1) := N_{\underline{\mu},\underline{\kappa}}(S_1) = \{(1,\underline{\sigma}\cdot\underline{\mu};\pi(\underline{\sigma}\cdot\underline{\kappa}_1);\dots;\pi(\underline{\sigma}\cdot\underline{\kappa}_n)) : \underline{\sigma} \in \mathcal{O}(S_1)\}. \quad (5)$$

Lemma 2. *In addition to the standing assumptions* (1),(4), *suppose that* $\underline{\mu},\underline{\kappa}_j \in \mathbb{C}^m$, $j = 1,\dots,n$, *satisfy*

$$\max_{i,j}\{|u_i-\mu_i|,|u_iv_j-\kappa_{ij}|\} < \exp(-3N^{\varepsilon}) \quad (6)$$

and that $\underline{\nu} = (\nu_1,\dots,\nu_n) \in \mathbb{C}^n$ *is arbitrary with associated one parameter subgroup* $\Phi_{\underline{\nu},1}$ *of* $\mathbb{G}_a\times E^n$ *defined by*

$$\Phi_{\underline{\nu},1} := (1,z;\pi(\nu_1 z);\dots;\pi(\nu_n z)).$$

Let $Q \in \mathbb{C}[x_0,x_1;\underline{X}_1;\dots;\underline{X}_n]$ *be a multihomogeneous polynomial of multi-degree at most* $(cD_0,cD,\dots,cD)$ *which vanishes at each point of* $N(cS_1)$ *along* $\Phi_{\underline{\nu},1}(z)$ *to an order of at least* T_1. *Then* Q *must vanish identically along a translate of* $\Phi_{\underline{\nu},1}(\mathbb{C})$.

Proof. Suppose that the hypotheses of the lemma hold but that Q does not vanish on a translate of $\Phi_{\underline{\nu},1}(\mathbb{C})$. Then we may apply the fundamental theorem 2.1 of [Ph2] to conclude that there exists a connected algebraic subgroup G' of G, which is incompletely defined by multihomogeneous polynomials of multidegree at most $(cD_0,cD,\dots,cD)$, such that if $G/G' = \mathbb{G}_a^{r_0}\times G''$, where $G'' = E^n/H'$ for some connected algebraic subgroup H' of E^n, is of dimension r_1, then

$$T_1\cdot\text{card}\,((N(S_1)+G')/G') \le cD_0^{r_0}D^{r_1}. \quad (7)$$

We consider the various possibilities.

If $r_0 = 1$, then $G' = \{0\}\times H'$. If $g \in N(S_1) \cap G'$, we project onto the first factor of G' to conclude that for some non-trivial

$(\sigma_1,\ldots,\sigma_m) \in \mathcal{O}^m(S_1)$,

$$\sigma_1\mu_1 + \ldots + \sigma_m\mu_m = 0.$$

This violates condition (1) concerning $u_1,\ldots,u_m$ in light of (4) and (6). Therefore if $r_0 = 1$, then $N(S_1) \cap G' = \{0\}$, and

$$\mathrm{card}\,((N(S_1)+G')/G') \;\; \geq \;\; S_1{}^{2m}.$$

Then (7) implies that

$$T_1S_1{}^{2m} \;\; \leq \;\; cD_0D^{r_1},$$

which cannot hold by our choices of parameters T_1,S_1,D_0,D. Therefore $r_0 = 0$.

Then inequality (7) becomes

$$T_1\cdot\mathrm{card}((N(S_1)+G')/G') \;\; \leq \;\; cD^r.$$

This cannot hold when $r = 1$ by our choice of parameters T_1,D. Hence $r > 1$.

Suppose that

$$\underline{g} = (1,\underline{\sigma}\cdot\underline{\mu};\underline{g}') := (1,\underline{\sigma}\cdot\underline{\mu};\pi(\underline{\sigma}\cdot\underline{\kappa}_1);\ldots;\pi(\underline{\sigma}\cdot\underline{\kappa}_n)) \in N(S_1) \cap G',$$

for some non-trivial $\underline{\sigma} \in \mathcal{O}^m(S_1)$, so that $g' \in H'$. Then $\mathcal{O}\cdot\underline{g}' \subset H'$, since H' is defined in the tangent space $\mathbb{C}^n$ of the identity element of E^n by homogeneous linear equations (see e.g. [Hu], p.87, and keep in mind that a commutative algebraic group has Lie algebra with trivial multiplication) and each copy of E has complex multiplications by $\mathcal{O}$.

Since G' is incompletely defined by multihomogeneous polynomials of multidegrees at most $(cD_0,cD,\ldots,cD)$, we see that H' is incompletely defined by polynomials of multidegrees at most $(cD,\ldots,cD)$. In fact by the usual device of taking sufficiently general linear combinations, say as is the proof of Proposition 3.3 of [Ph2] (or [Br4]), one can obtain r polynomials, say $P_1,P_2,\ldots,P_r$ incompletely defining H' in E^n. Thus the P_k vanish on $\mathcal{O}\cdot\underline{g}'$. If we set

$$W := \mathcal{O}\sigma_1(\kappa_{11},\ldots,\kappa_{1n}) + \ldots + \mathcal{O}\sigma_m(\kappa_{m1},\ldots,\kappa_{mn}),$$

then each

$$P_k(\pi(w_1),\ldots,\pi(w_n)) \;\; = \;\; 0, \tag{8}$$

for all $\underline{w} := (w_1,\ldots,w_n) \in W$.

By Lemma 0, there is a non-zero vector $\underline{t}^{(k)} = (t_{k1},\ldots,t_{kn}) \in \mathcal{O}(cD^{1/2})$ associated to P_k, $1 \leq k \leq r$, such that

$$t_{k1}w_1 + \ldots + t_{kn}w_n \in \mathcal{O}\omega_1 \tag{9}$$

for all $(w_1,\ldots,w_n) \in W$. Renumber if necessary so that $t_{11} \neq 0$. We argue that we can use the polynomials $P_1,\ldots,P_r$ to obtain (at least) two vectors $\underline{t},\underline{t}' \in \mathcal{O}^n(cD)$ which are $\mathcal{O}$-linearly independent. In fact, the argument of Appendix B would deliver r $\mathcal{O}$-linearly independent vectors.

Let $E_j = Z_jY_j^2-4X_j^3+g_2X_jZ_j^2+g_3Z_j^3$, $j = 1,\dots,n$. Since the ideal $\mathfrak{A}_r := (P_1,\dots,P_r,E_1,\dots,E_n)$ has an isolated prime component of dimension $n-r$ defining H' in $(\mathbb{P}_2)^n$, we know that the ideal $\mathfrak{A}_2 = (P_1,P_2,E_1,\dots,E_n)$ has an associated prime ideal of dimension $n-2$ vanishing on H'. Now consider the polynomials

$$\bar{P}_1(X_1,Y_1,Z_1) = P_1(X_1,Y_1,Z_1;\wp(z_2),\wp'(z_2),1;\dots;\wp(z_n),\wp'(z_n),1),$$

$$\bar{P}_2(X_1,Y_1,Z_1) = P_2(X_1,Y_1,Z_1;\wp(z_2),\wp'(z_2),1;\dots;\wp(z_n),\wp'(z_n),1),$$

$$E_1(X_1,Y_1,Z_1) = Z_1Y_1^2-4X_1^3+g_2X_1Z_1^2+g_2Z_1^3$$

as homogeneous polynomials over the field

$$L = \mathbb{C}(\wp(z_2),\wp'(z_2),\dots,\wp(z_n),\wp'(z_n)).$$

If $\mathfrak{A}_2$ is unmixed, then it can be verified rather easily that the classical resultant R of the three forms $\bar{P}_1,\bar{P}_2,E_1$ is non-zero. (The fact that $\mathfrak{A}_2$ is unmixed of rank $n+2$ implies that $\bar{P}_1,\bar{P}_2,E_1$ have no common zero in $\mathbb{P}_2(L)$.) It follows from §81,82 of [VdW] that

$$X_1^{4D+1}R = A_1\bar{P}_1 + A_2\bar{P}_2 + A_3E_1$$

for polynomials A_1,A_2,A_3 whose coefficients are polynomials over $\mathbb{Z}$ of degrees at most cD in the coefficients of $\bar{P}_1,\bar{P}_2$ and at most cD^2 in the coefficients of E_1. Analogous representations hold for $Y_1^{4D+1}R$ and $Z_1^{4D+1}R$, with similar bounds on the degrees. Therefore ${}^mR(\pi(z_2);\dots;\pi(z_n))$ vanishes on the $\mathcal{O}$-module

$$W' := \mathcal{O}\sigma_2(\kappa_{21},\dots,\kappa_{2n}) + \dots + \mathcal{O}\sigma_m(\kappa_{m1},\dots,\kappa_{mn}),$$

where ${}^mR$ denotes the appropriate multihomogenization of R. By Lemma 0, there exists a non-zero vector $(t_2',\dots,t_n') \in \mathcal{O}^{n-1}(cD)$ such that

$$t_2'w_2 + \dots + t_n'w_n \in \mathcal{O}\cdot\omega_1 \tag{10}$$

for all $(w_1,\dots,w_n) \in W$. For general $\mathfrak{A}_2$, the situation is more complicated, but the outcome the same. We refer to Appendix B for the procedure which furnishes the polynomial R giving the additional relation (10) in the general case.

Either the element in (10) is zero already or else using (9) and (10), we can eliminate ω_1 on the right hand sides to obtain

$$\tau_1\sum_{i=1}^m\sigma_i\kappa_{i1} + \dots + \tau_n\sum_{i=1}^m\sigma_i\kappa_{in} = 0.$$

Thus in either case we obtain a non-trivial relation of the preceding form with $\tau_i \in \mathcal{O}(cD^{3/2})$. Then

$$\sum_{i=1}^{m} \sum_{j=1}^{n} \sigma_i \tau_j \kappa_{ij} = 0.$$

Replacing each κ_{ij} by the corresponding $u_i v_j$ and recalling (4) and (6), we have the inequality

$$\left|\left(\sum_{i=1}^{m} \sigma_i u_i\right)\left(\sum_{j=1}^{n} \tau_j v_j\right)\right| < \exp(-2N^{\varepsilon})$$

for S large enough, which violates the technical hypothesis (1).

Therefore no connected algebraic subgroup G' of G can satisfy (7). This establishes the lemma.□

D. The Ideal and Zero-free Region

In paragraphs A and B we worked with inhomogeneous polynomials $P_{\underline{t},y}$ which give the dehomogenized values of the multihomogeneous polynomial P (or derivatives *along* $\Phi_{\underline{v},p}$) *at* points $\Phi_{\underline{v},p}(Y^p(S_1))$ of the p-parameter sub- group. In paragraph C we have established that we still obtain non-zero "values" from all points near $\Phi_{\underline{v},p}(Y^p(S_1))$, if we consider multiplicities *along* $\Phi_{\underline{v},p}$. As in Part I, these non-zero values will lead to a zero-free ball for an ideal of polynomials which are generated from P.

Let $G_2(x)$ and $G_3(x)$ denote the minimal polynomials of g_2 and g_3 over $\mathbb{Z}$. Then let $I(S)$ denote the homogeneous ideal in

$$R = \mathbb{Z}[Z,\Gamma_2,\Gamma_3,x_1,\ldots,x_m,y_1,\ldots,y_n,X_{11},Y_{11},\ldots,X_{mn},Y_{mn}]$$

generated by the homogeneous polynomials

(a) $H_k(Z,\Gamma_k) = Z^{\deg G_k} G_k(\Gamma_k/Z)$, $k = 2,3$,

(b) $E_{ij}(Z,\Gamma_2,\Gamma_3,X_{ij},Y_{ij}) = Z^2 Y_{ij}{}^2 - 4ZX_{ij}{}^3 + \Gamma_2 Z^2 X_{ij} + \Gamma_3 Z^3$, $i = 1,\ldots,m$, $j = 1,\ldots,n$,

(c) ${}^hP_{\underline{t},y}(Z,\Gamma_2,\Gamma_3,x_1,\ldots,x_m,y_1,\ldots,y_n,X_{11},Y_{11},\ldots,X_{mn},Y_{mn})$, $y \in Y^p(S_1)$, $0 \le t_1,\ldots,t_p < T_1$, where the superscript denotes homogenization with respect to the new variable Z.

Our goal is to show that if the hypotheses of the theorem are satisfied, then the ideal $I(S)$ does not have any zero in the ball $B(\omega'',\exp(-4N^{\varepsilon}))$ centered at $\omega'' =$
$(1,g_2,g_3,u_1,\ldots,u_m,v_1,\ldots,v_n,\wp(u_1v_1),\wp'(u_1v_1),\ldots,\wp(u_mv_n),\wp'(u_mv_n))$, where for any ℓ, we measure the distance between any two points $[x_i]$, $[x_i']$ of $\mathbb{P}_\ell(\mathbb{C})$ by $\max_{i,j} |x_i x_j' - x_i' x_j| / (\max |x_i|)(\max |x_i'|)$. Suppose on the contrary that

$$\omega^* = (\zeta,\gamma_2,\gamma_3,\mu_1,\ldots,\mu_m,\nu_1,\ldots,\nu_n,\xi_{11},\eta_{11},\ldots,\xi_{mn},\eta_{mn}) \tag{11}$$

is a zero of $I(S)$ lying inside $B(\omega'',\exp(-4N^{\varepsilon}))$. By the auxiliary hypothesis (4), S is sufficiently large to conclude that $\zeta \neq 0$, and upon dividing by ζ, we may as well assume that $\zeta = 1$. Hence if ω^* is a zero of $I(S)$, then

$$H_2(1,\gamma_2) = H_3(1,\gamma_3) = 0;$$

since G_2, G_3 have only finitely many zeros, S is large enough to guarantee that $\gamma_2 = g_2$ and $\gamma_3 = g_3$.

Therefore for each pair i,j we see that

$$(\xi_{ij},\eta_{ij},1) \in E.$$

As in Part I we choose $\kappa_{ij} \in \mathbb{C}$ with

$$\pi(\kappa_{ij}) = (\xi_{ij},\eta_{ij},1)$$

and

$$|u_i v_j - \kappa_{ij}| < c\exp(-4N^{\varepsilon}).$$

Again we use the notation

$$\underline{\kappa}_j := (\kappa_{1j},\ldots,\kappa_{mj}),$$

$1 \le j \le n$, and for $\underline{\sigma} \in \mathcal{O}^m$ we let $\underline{\sigma}\cdot\underline{\kappa}_j$ denote the usual scalar product. Let $N(S_1)$ be defined as in (5) and for $\underline{\sigma} \in \mathcal{O}^m$ and $j = 1,\ldots,n$, let

$$\eta_{j\underline{\sigma}} = (f(\underline{\sigma}\cdot\underline{\kappa}_j),g(\underline{\sigma}\cdot\underline{\kappa}_j),h(\underline{\sigma}\cdot\underline{\kappa}_j)).$$

Lemma 3. *Assume* (1) and (4). *Let the non-zero multihomogeneous polynomial*

$$P(\varkappa_1,\underline{X}_1^{(1)},\ldots,\underline{X}_n^{(1)};\ldots;\varkappa_p;\underline{X}_1^{(p)},\ldots,\underline{X}_n^{(p)})$$

satisfy

$$\deg_{\varkappa_k} P \le cD_0, \quad \deg_{\underline{X}_j^{(k)}} P \le cD.$$

Then the ideal $I(S)$ *associated with* P *as above has no zero in the ball* $B(\omega'',\exp(-4N^{\varepsilon}))$.

Proof. The proof is carried out by induction on p. When $p = 1$, we observe that if $\omega^* \in B(\omega'',\exp(-4N^{\varepsilon}))$, with ω^* represented by coordinates as in (11), is a zero of $I(S)$, then the hypotheses of Lemma 2 are satisfied. Hence P must vanish along a translate of $\Phi_{\underline{\nu},1}(\mathbb{C})$, say $g + \Phi_{\underline{\nu},1}(\mathbb{C})$ for some $g \in G$. Then applying the addition formula for G to absorb the effects of g into the coefficients shows that there exists a polynomial $\bar{P}$ with $\deg_x \bar{P} \le cD_0$ and $\deg_{\underline{X}} \bar{P} \le cD$ such that

$$\bar{P}(z,\pi(\nu_1 z),\ldots,\pi(\nu_n z)) = 0$$

for all $z \in \mathbb{C}$. Hence by Lemma 0, there is a non-trivial $(t_1,\ldots,t_n)$

$\in \mathcal{O}^n(cD^{1/2})$ such that for all z,

$$t_1\nu_1 z + \ldots + t_n\nu_n z \in \mathcal{O}\cdot\omega_1.$$

Consequently

$$t_1\nu_1 + \ldots + t_n\nu_n = 0.$$

Since $cD^{1/2} \le N$, if $\omega^* \in B(\omega'',\exp(-4N^{\varepsilon}))$, this violates our technical hypothesis (1) concerning $v_1,\ldots,v_n$. Hence P does not vanish on a translate of $\Phi_{\underline{\nu},1}(\mathbb{C})$. Therefore for any $\omega^* \in B(\omega'',\exp(-4N^{\varepsilon}))$, there are t_1, y_1 with $0 \le t_1 < T_1$, $y_1 \in \mathcal{O}(S_1)$, such that, in the notation of the proof of Lemma 1, $P_{t_1,y_1}(\omega^*) \ne 0$. This establishes our lemma when $p = 1$.

In analogy to the argument in Part I, we show by induction that in the general case, for each $\omega^* \in B(\omega'',\exp(-4N^{\varepsilon}))$, there are $\underline{t}$, y, with $0 \le t_1,\ldots,t_p < T_1$, $y \in \mathcal{O}^p(S_1)$ such that $P_{\underline{t},y}(\omega^*) \ne 0$. To carry out the induction, write $P = \sum P_\lambda M_\lambda$, where the M_λ are distinct multihomogeneous polynomials in $\varkappa_p$, $\underline{X}_j^{(p)}$ and the P_λ are non-zero multihomogeneous polynomials in the $\varkappa_k, \underline{X}_j^{(k)}$, $k = 1,\ldots,p-1$. Then for $\underline{t} = (\underline{t}^*,t_p)$ and $y = (y^*,y_p)$, we see that

$$P_{\underline{t},y} = \sum (P_\lambda)_{\underline{t}^*,y^*}(M_\lambda)_{t_p,y_p}.$$

By induction hypothesis, given any λ, we can find indices $y^* \in \mathcal{O}^{(p-1)}(S_1)$ and $\underline{t}^*$ with $0 \le t_1,\ldots,t_{p-1} \le T_1$ such that $(P_\lambda)_{\underline{t}^*,y^*}(\omega^*) \ne 0$. Having made that selection for an arbitrary λ, then by the case $p = 1$, we can choose t_p, y_p in the desired range such that $P_{\underline{t},y}(\omega^*) \ne 0$. This completes the proof of the lemma.□

E. Conclusion

For every S satisfying our standing assumptions (1) and (4), we apply Lemma 3 to the multihomogeneous polynomial P constructed in Lemma 1. By §B above and Lemma 3, we have established the existence of an ideal I(S) which is generated over $\mathbb{Z}$ by polynomials Q having

size $Q \le cS^{2(m+n)(pm+p+mn+m+n)/\lambda(m+1)}(\log S)^{-((p+n)n+m(m+n))/\lambda(m+1)}$,

and

$$\log|Q(\omega)| \le -cS^{2(p+n)(mn+m+n)/\lambda}(\log S)^{pn-mn-(n+2)(m+n)/\lambda},$$

where λ is as before and the constants c depend on p. Moreover the ideal generated by the homogenizations of the generators of I(S) has a zero-free ball of radius at least $\exp(-4N^{\varepsilon})$ centered at ω'' for every S in our range. As explained in Part I, we immediately have the claim of our Theorem for the case at hand.

F. Appendix A. Parameters

As in Part I, we list parameters for the auxiliary functions for the other cases. The letter D_0 will denote the degree in z, cS_1 the bound on the coefficients of the arguments necessary for the zero-free regions. Having made our point on their irrelevance in Appendix B of Part I, we sup- press the powers of p and $\log S$.

$\wp(u_i v_j)$: $T = T_1 = 1$, $D_0 = 0$.

$$D = [S^{2m(p+n)/n(p-m)}],$$
$$S_1 \approx S^{(p+n)/(p-m)}.$$

$u_i, \wp(u_i v_j)$: $T = T_1 = 1$.

$$D_0 = [S^{2p(m+n)/(p-m)(n+1)}],$$
$$D = [S^{2(pm-p+mn+m)/(p-m)(n+1)}]$$
$$S_1 \approx [S^{(p+n)/(n+1)(p-m)}].$$

G. Appendix B. Semi-resultants

As mentioned in Part I, Yu.V. Nesterenko [Ne2] introduced the notion of the resultant (our terminology) of a Chow form and an ordinary form. This idea has proved very fruitful in investigations of algebraic independence, e.g. [Ne3],[Ne4],[Ph2],[Br2]. In this appendix we introduce a generalization of that concept which was suggested by considerations of G.V. Chudnovsky [Ch].

Definition. Let $F(u_0,\dots,u_n) = a\prod_{i=1}^{D}(\alpha_{i0}u_0+\dots+\alpha_{in}u_n)$ and $Q(x_0,\dots,x_n)$ be forms over a field. For $d = 0,\dots,D$, let

$$R_d(F,Q) := \Sigma_{M_d} a^{\deg Q} \prod_{i\in M_d} Q(\alpha_{i0},\dots,\alpha_{in}),$$

where M_d runs over all the subsets of $\{1,\dots,D\}$ of cardinality d.

The reason for introducing this notion is that it allows us to control the resultant of Q with the Chow form made up of all factors of F whose underlying prime ideals do not contain Q, even though no assumption is made on the structure of the coefficients of F and Q.

Proposition B.1. $R_d(F,Q)$ *is a polynomial over* $\mathbb{Z}$ *of height at most* $2^D(2(n+1))^{D\cdot\deg Q}$ *in the coefficients of* F *and* Q. *Moreover* $R_d(F,Q)$ *has*

$$\deg_{\text{coeff } F} \le \deg Q,$$
$$\deg_{\text{coeff } Q} \le D.$$

For a proof, consult Lemma 1 of [Br1]. Compare also Lemma 4 of [Ne2]. We shall call $R_d(F,Q)$ the dth order *semi-resultant* of F and Q. Of course if F is a Chow form, then $R_D(F,Q)$ is the resultant of F and Q defined in [Ne2]. We restrict ourselves here to a few remarks which are needed in the present note. We first note that the definition of $\|F\|_\omega$ given in Part I makes sense even if F is not defined over $\mathbb{Z}$. We shall not here take advantage of the further refinement of Philippon [Ph2] which introduces an extra factor in the denominator of that definition in order that $\|\mathfrak{A}\|_\omega$ make sense in this context.

Remark B.2. *If* F *is the Chow form above,* $F_1 = a\prod_{i=1}^{D'}(\alpha_{i0}u_0+\ldots+\alpha_{in}u_n)$ *and* $Q(\alpha_{i0},\ldots,\alpha_{in}) = 0$, $i = D'+1,\ldots,D$, *then* $R(F_1,Q) = R_{D'}(F,Q)$.

The remark follows immediately from the observation that all summands other than $R(F_1,Q)$ in the definition of $R_{D'}(F,Q)$ are certainly zero. For the applications we have in mind, we need the following further remark:

Proposition B.3. *If* F *is any Chow form over* $\mathbb{C}$, Q *is any form over* $\mathbb{C}$, *and* $\omega \in \mathbb{C}^{n+1}$ *is non-zero, then*

$$\log|R(F,Q)|_\omega \le \delta_F\chi_Q + \delta_Q\chi_F + 6m^2\delta_F\delta_Q + \log\max\{|F|_\omega/H_F, |Q|_\omega/H_Q\},$$

where δ_F = degree F, χ_F = log height F = $\log H_F$ *and similarly for* Q. *Moreover if* F *has underlying prime ideals* $\mathfrak{P}_1,\ldots,\mathfrak{P}_t$ *and* $R(F,Q) \ne 0$, *then the underlying prime ideals of* $R(F,Q)$ *are the minimal prime ideals of the various* $(\mathfrak{P}_i,Q)$.

The proof of this result is contained in the proofs of Lemma 6 and Proposition 3 of [Ne3], although those results are stated for forms over $\mathbb{Z}$ only.

Now we want to apply these remarks to the situation of Section C above to obtain two linearly independent $\underline{t}^{(i)} \in \mathcal{O}^n(cD)$ with $\underline{t}^{(i)}\cdot W \subset \mathcal{O}\cdot\omega_1$. We may renumber if necessary to obtain that, say, $t_{11} \ne 0$. Our aim is to use elimination to obtain a non-trivial multihomogeneous polynomial in $X_2,Y_2,Z_2;\ldots;X_n,Y_n,Z_n$ vanishing on H'. Since $\mathfrak{A}_r$ has an isolated prime ideal (corresponding to H') of codimension r in G, $\mathfrak{A}_2$ must have an isolated prime ideal of codimension 2 in G, i.e. of rank $n+2$. Therefore since $(E_1,\dots,E_n)$ is a prime ideal of rank n, $\mathfrak{A}_1$ is a complete intersection of rank $n+1$.

As in the complete intersection case, consider the polynomials

$$\bar{P}_1(X_1,Y_1,Z_1) = P_1(X_1,Y_1,Z_1;\wp(z_2),\wp'(z_2),1;\dots;\wp(z_n),\wp'(z_n),1),$$

$$\bar{P}_2(X_1,Y_1,Z_1) = P_2(X_1,Y_1,Z_1;\wp(z_2),\wp'(z_2),1;\dots;\wp(z_n),\wp'(z_n),1),$$

$$E_1(X_1,Y_1,Z_1) = Z_1Y_1{}^2-4X_1{}^3+g_2X_1Z_1{}^2+g_2Z_1{}^3,$$

mentioned above for classical elimination, as homogeneous polynomials over $L := \mathbb{C}(\wp(z_2),\wp'(z_2),\dots,\wp(z_n),\wp'(z_n))$. Lemma 2 of [Ne3] gives the Chow form F_1 of the ideal (E_1) explicitly.

The remark on the rank of $\mathfrak{A}_1$ implies that the resultant $F_2 = R(F_1,\bar{P}_1) \neq 0$. However from (8) and Proposition B.3 we can conclude that for all $w = (w_1,\dots,w_n) \in W$, $F_2\circ(w_2,\dots,w_n)|_{\pi(w_1)} = 0$, where the notation $F_2\circ(w_2,\dots,w_n)$ means that we have set $z_j = w_j$, $j = 2,\dots,n$. Indeed since F_2 is multihomogeneous in the coefficients of $\bar{P}_1$, if all of the coefficients of $\bar{P}_1$ vanish at $(w_2,\dots,w_n)$, then certainly $F_2\circ(w_2,\dots,w_n) = 0$. On the other hand if some coefficient of $\bar{P}_1$ does not vanish at $(w_2,\dots,w_n)$, then Proposition B.3 applies directly, since

$$P_1(\pi(w_1),\dots,\pi(w_n)) = E_1(\pi(w_1)) = 0.$$

Now F_2 is a Chow form over L. Its underlying prime ideals correspond bijectively to those of $\mathfrak{A}_1 = (P_1,E_1,\dots,E_n)$. The polynomial P_2 may lie in some of those prime ideals, but not in all of them as $\mathfrak{A}_2$ has components of rank $n+2$. Write $F_2 = \prod(\alpha_{i0}X_1 + \alpha_{i1}Y_1 + \alpha_{i2}Z_1)$ and renumber if necessary so that $(\alpha_{i0},\alpha_{i1},\alpha_{i2})$, $i = 1,\dots,D'$, are all the triples, counting multiplicities, such that $\bar{P}_2(\alpha_{i0},\alpha_{i1},\alpha_{i2}) \neq 0$. Then $F_3 := R_{D_1}(F_2,\bar{P}_2) \neq 0$, but as above $F_3\circ(w_2,\dots,w_n)|_{\pi(w_1)} = 0$ for all $(w_1,\dots,w_n) \in W$. However F_3 is a polynomial in $\wp(z_2),\wp'(z_2),\dots,\wp(z_n),\wp'(z_n)$. Multihomogenizing thus gives the polynomial R we seek. Proposition B.1 shows that the polynomial has multidegrees at most $(cD^2,\dots,cD^2)$. Now Lemma 0 produces the second linear expression (10) as before.

Bibliography

[Br1] W.D. Brownawell, Some remarks on semi-resultants, Chap. 14, *Transcendence Theory: Advances and Applications*, A. Baker and D.W. Masser, eds, Academic Press, New York, 1977.

[Br2] W.D. Brownawell, Large transcendence degree revisited I. Exnential and non-CM cases, in these Proceedings.

[Br3] W.D. Brownawell, A local Diophantine Nullstellen inequality, manuscript.

[Br4] W.D. Brownawell, A note on a paper of P. Philippon, Mich. J. Math., to appear.

[Ch] G.V. Chudnovsky, Some analytic methods in the theory of transcendental numbers, Inst. of Math., Ukr. SSR Acad. Sci., Preprint IM 74-8 and IM 74-9, Kiev, 1974 = Chapter 1 in *Contributions to the Theory of Transcendental Numbers*, Am. Math. Soc. Providence, R.I. 1984.

[Hu] J.E. Humphreys, *Linear Algebraic Groups*, Springer Verlag, New York-Heidelberg-Berlin, 1975.

[Ja] E.M. Jabbouri, Mesures d'indépendance algébriques de valeurs des fonctions elliptiques et abéliennes, manuscript.

[Ma-Wü] D.W. Masser and G. Wüstholz, Fields of large transcendence degree generated by values of elliptic functions, Invent. Math. 72(1983), 407-463.

[Ne1] Yu.V.Nesterenko, Estimates for the orders of zeros of functions of a certain class and applications in the theory of transcendental numbers, Izv. Akad. Nauk SSSR Ser. Mat. 41 (1977), 253-284 = Math. USSR Izv. 11(1977), 239-270.

[Ne2] Yu.V. Nesterenko, Bounds for the characteristic function of a prime ideal, Mat. Sbornik 123, No. 1(1984), 11-34 = Math. USSR Sbornik 51 (1985), 9-32.

[Ne3] Yu.V. Nesterenko, On algebraic independence of algebraic powers of algebraic numbers, Mat. Sbornik 123, No. 4(1984), 435-459 = Math. USSR Sbornik 51(1985), 429-454, brief version in *Approximations Diophantiennes et Nombres Transcendants*, D. Bertrand and M. Waldschmidt, eds, Birkhäuser Verlag, Boston-Basel-Stuttgart, 1983, pp. 199-220.

[Ph1] P. Philippon, Pour une théorie de l'indépendance algébrique, Thèse, Université de Paris XI, 1983.

[Ph2] P. Philippon, Critères pour l'indépendance algébrique, Inst. Hautes Etudes Sci. Publ. Math., No. 64, 1986, 5-52.

[Ph3] P. Philippon, Lemmes de zéros dans les groupes algébriques commutatifs, Bull. Soc. Math. France 114(1986), 355-383.

[Tu] R. Tubbs, A diophantine problem on elliptic curves, Trans. Am. Math. Soc., to appear.

[VdW] B.L. van der Waerden, *Algebra*, Vol. II, Ungar Publishing, New York, 1950.

[Wa1] M. Waldschmidt, *Nombres transcendents et groupes algébriques*, Astérisque 69-70 (1979), Société Math. France.

[Wa2] M. Waldschmidt, Groupes algébriques et grands degrés de transcendance, Acta Math. 156(1986), 253-302.

[Wa3] M. Waldschmidt, Algebraic independence of values of exponential and elliptic functions, J. Indian Math. Soc. 48 (1984), 215-228.

A NEW APPROACH TO BAKER'S THEOREM ON LINEAR FORMS IN LOGARITHMS I

by

G. WÜSTHOLZ

Max - Planck - Institut für Mathematik
Gottfried - Claren - Str. 26
5300 Bonn 3

und

Universität - Gesamthochschule Wuppertal
Fachbereich Mathematik
Gausstr. 20
5600 Wuppertal 1

1. INTRODUCTION

In this note we prove the multiplicity estimates that are necessary in order to obtain lower bounds for linear forms in logarithms in the style given in [Ba]. Besides its theoretical interest it will also have a practical consequence in improving the bound for the constants. This will be proved in a joint paper with A. Baker. The main ingredients are the techniques of multiplicity estimates which are developed in [Wü] and which have to be refined in the style of zero estimates proved in [Ma-Wü 1] the so called "zero - estimates with knobs on". Further in order to obtain the final version of the multiplicity estimates needed for lower bounds for linear forms in logarithms these techniques have to be enriched by some geometry of numbers, an idea which already appears in [Ma-Wü 1] and which is modified here. This will be exposed in part II of these notes.

Furthermore we remark that all this can be generalized to arbitrary commutative algebraic groups. This will be done in another series of papers concerning Siegel's theorem on integer points on curves.

We begin now with recalling some basic facts before we state the main result of this paper. The group which appears in Baker's result is the group defined over $\overline{\mathbb{Q}}$

$$G = \mathbb{G}_a \times \mathbb{G}_m^n$$

where $\mathbb{G}_a$ is the additive and $\mathbb{G}_m$ the multiplicative group. This group is nothing else than

$$G = \operatorname{spec} R$$

where

$$R = \overline{\mathbb{Q}}[X_0, X_1, X_1^{-1}, \ldots, X_n, X_n^{-1}]\,.$$

For an element $g \in G(\overline{\mathbb{Q}})$ we denote by T_g the morphism from G to G given by translation with g. This morphism can be described as follows. Denote by $(g_0, \ldots, g_n)$ the coordinates of g i.e. $g_i = X_i(g)$ $(0 \le i \le n)$ then the morphism T_g induces a homomorphism of R into R which we denote by $E(g)$ and which is given explicitly by the formulas

$$\begin{aligned} X_0 &\longrightarrow X_0 + g_0 \\ X_i &\longrightarrow g_i X_i \qquad (1 \le i \le n)\,. \end{aligned}$$

This homomorphism extends to a mapping on the set of ideals into itself. It preserves properties like being prime or primary and commutes with intersections and other typical properties and notions attached to ideals. This mapping is also denoted by $E(g)$. On R we have the following translations invariant derivations

$$(1.1) \qquad \frac{\partial}{\partial X_0}\,,\ X_1 \cdot \frac{\partial}{\partial X_1}\,,\ \ldots\,,\ X_n \cdot \frac{\partial}{\partial X_n}\,.$$

These derivations form a basis for the Lie algebra $\operatorname{Lie} G$ of G. By definition any $D \in \operatorname{Lie} G$ satisfies $T_g \circ D = D \circ T_g$ for arbitrary $g \in G(K)$. This relation can also be used to find the derivations above.

The algebraic group G is considered to be embedded in $\mathbb{A}^{n+1}$ and this space itself can be embedded into the multiprojective space $(\mathbb{P}^1)^{n+1}$ in the way described in [Fa-Wü]. So we get an embedding

$$i : G \hookrightarrow (\mathbb{P}^1)^{n+1}\,.$$

On $(\mathbb{P}^1)^{n+1}$ we have the notion of multidegrees given by the multi-Hilbert-polynomials. For details we refer to the appendix of [Ma-Wü 2]. So for an ideal $I \subseteq R$ we can associate multidegrees as follows. Let r be the rank of I and choose integers $i_1,\ldots,i_r$ such that $0 \leq i_1 < \ldots < i_r \leq n$. Then

$$\delta_{i_1,\ldots,i_r}$$

is defined as the multidegree of the image of $\mathrm{spec}(R/I)$ in $(\mathbb{P}^1)^{n+1}$ with respect to the directions $i_1,\ldots,i_r$. So $\delta_{i_1,\ldots,i_r}(I)$ is just the coefficient of the monomial $t_{j_1,\ldots,j_r}$, where

$$\{j_1,\ldots,j_{n+1-r}\} \cup \{i_1,\ldots,i_r\} = \{0,1,\ldots,n\},$$

in the multi-Hilbert-polynomial of the ideal J of $i(\mathrm{spec}(R/I))$ denoted by

$$H(J;t_0,\ldots,t_n)$$

of the monomial $t_{i_0,\ldots,i_n}$. We call this Hilbert-polynomial the Hilbert-polynomial of I and denote it by

$$H(I;t_0,\ldots,t_n).$$

In this way the degree theory on $(\mathbb{P}^1)^{n+1}$ induces a degree theory on G.

Let $\alpha_1,\ldots,\alpha_n,\beta_0,\ldots,\beta_n$ be algebraic numbers such that $\beta_n \neq 0$. Define derivations $\Delta_0,\ldots,\Delta_{n-1}$ as

$$\Delta_0 = \frac{\partial}{\partial X_0} - \frac{\beta_0}{\beta_n} X_n \frac{\partial}{\partial X_n}$$

$$\Delta_i = X_i \frac{\partial}{\partial X_i} - \frac{\beta_i}{\beta_n} X_n \frac{\partial}{\partial X_n} \qquad (1 \leq i \leq n-1).$$

Then the main result can be stated as follows.

Theorem 1.1. *Suppose that* $S>0$, $T>0$, $D_0,\ldots,D_n \geq 0$, $S_1 \geq \ldots \geq S_{n+1}$, $T_1 \geq \ldots \geq T_{n+1} \geq 0$ *are integers such that for* $0 \leq i_1 \leq \ldots \leq i_r \leq n$, $1 \leq r \leq n+1$

$$\binom{T+r-\delta_{r,n+1}}{r-\delta_{r,n+1}} S_r \geq r!\, D_{i_1}\cdots D_{i_r}$$

and

(1.3) $\qquad T \geq T_1 + \ldots + T_{n+1}\,,\quad S \geq S_1 + \ldots + S_{n+1}$

with at least one strict inequality. Then if $P(X_0,\ldots,X_n)$ *is a polynomial such that*

(1.4) $\qquad \deg_{X_i} P \leq D_i \qquad\qquad (0 \leq i \leq n)$

and

(1.5) $\qquad \Delta_0^{\tau_0}\cdots\Delta_{n-1}^{\tau_{n-1}}\; P(s,\alpha_1^s,\ldots,\alpha_n^s) = 0$

for $0 \leq s \leq S$ *and* $0 \leq \tau_0,\ldots,\tau_{n-1}$ *with* $\tau_0+\ldots+\tau_{n-1} < T$ *then either* $P=0$ *or there exists a connected algebraic subgroup* H *of* G *of rank* $r \leq n$ *such that one of the following properties is satisfied:*

(i) Lie $H \subseteq (\Delta_0,\ldots,\Delta_{n-1})$ *and for* $0 \leq i_1 < \ldots < i_r \leq n$ *we have*

$$\delta_{i_1,\ldots,i_r}(H) \leq r!\, \frac{D_{i_1}\cdots D_{i_r}}{S_r\binom{T_r+r-1}{r-1}},$$

(ii) $(1,\alpha_1,\ldots,\alpha_n) \in H(\overline{\mathbb{Q}})$ *and for* $0 \leq i_1 < \ldots < i_r \leq n$ *we have*

$$\delta_{i_1,\ldots,i_r}(H) \leq r!\, \frac{D_{i_1}\cdots D_{i_r}}{\binom{T_r+r-1}{r-1}}.$$

The theorem improves the main theorem in [Wü] in several directions. Firstly it takes into account different degrees in the style of [Ma-Wü 2] and secondly its conditions (1.2) are weakened compared with the main theorem mentioned above.

This leads to an improvement of the constant in the above mentioned theorem and this is the main point for any improvement of the constant appearing in Baker's work.

We end this section with some remarks on the condition (1.2). The most troublesome factor here is the $r!$ which appears explicitly on the right-hand side. We have no idea why this factor comes in. It appears in the multihomogeneous degree theory which we use. But it should not be there if one counts conditions as usual in transcendence. It seems to have something to do with infinity. In any case the illumination of this phenomena would be of great interest.

2. Bounding Degrees and Lengths of Ideals

In this section we shall state and prove several Lemmas which are of technical nature.

Lemma 2.1. *Suppose that* $I \subset R$ *is an ideal of rank* r *generated by polynomials* $P_1,\dots,P_k$ *whose degrees are bounded by*

$$\deg_{X_i} P_j \leq D_i \qquad (0 \leq i \leq n\ ,\ 1 \leq j \leq k).$$

Then for $0 \leq i_1 < \dots < i_r \leq n$ *we have the estimate*

$$\delta_{i_1,\dots,i_r}(I) \leq r!\, D_{i_1} \cdots D_{i_r}\,.$$

Proof. In the usual way one finds linear combinations $Q_1,\dots,Q_r$ of $P_1,\dots,P_k$ such that $\operatorname{rank}(Q_1,\dots,Q_r)=r$,

$$I \supseteq (Q_1,\dots,Q_r) =: J \tag{2.1}$$

and

$$\deg_{X_i} Q_j \leq D_i \qquad (0 \leq i \leq n\ ,\ 1 \leq j \leq r)\,.$$

Since for $0 \leq i_1 < \dots < i_r \leq n$ we have

$$\delta_{i_1,\dots,i_r}(I) \leq \delta_{i_1,\dots,i_r}(J)$$

because of (2.1) and the anti - monotony of δ for fixed rank (see [Ma-Wü 2]) it suffices to bound the partial degrees of J. This is done by induction and using Lemma A5 in [Ma-Wü 2]. This proves the Lemma 2.1.

In order to state the next Lemma we introduce the exponent ρ. Let $\mathcal{P} \subset R$ be a prime ideal of R generated by $P_1,\ldots,P_k$ of rank r. Then let

$$\hat{\pi} : R \longrightarrow S = R/\mathcal{P}$$

be the canonical projection. Let $J(\mathcal{P})$ be the matrix

$$\begin{pmatrix} \Delta_0 P_1 & \cdots & \Delta_0 P_k \\ \vdots & & \vdots \\ \Delta_{n-1} P_1 & \cdots & \Delta_{n-1} P_k \end{pmatrix}$$

in the matrix ring $M(n,k)(R)$ of $n \times k$ - matrices with coefficients in R and

$$J(\mathcal{P};\alpha) = \begin{pmatrix} \hat{\pi}(\Delta_0 P_1) & \cdots & \hat{\pi}(\Delta_0 P_k) \\ \vdots & & \vdots \\ \hat{\pi}(\Delta_{n-1} P_1) & \cdots & \hat{\pi}(\Delta_{n-1} P_k) \end{pmatrix}$$

the image of $J(\mathcal{P})$ in $M(n,k)(S)$. Then we put

(2.2) $\qquad \rho(\mathcal{P}) = \operatorname{rank} J(\mathcal{P};\alpha)$.

Then it is easy to see that $\rho(\mathcal{P})$ does indeed only depend on $\mathcal{P}$ and α and not of the choice of generators of either.

Let now $T>0$ be an integer and define for an ideal I of R

$$I^{(T)} = \{r \in R;\ \Delta_0^{\tau_0} \cdots \Delta_{n-1}^{\tau_{n-1}} r \in \sqrt{I} \ \text{ for } 0 \le \tau_0 + \ldots + \tau_{n-1} \le T\} .$$

Then this operation commutes with intersection. If $\mathcal{P}$ is a prime ideal then $\mathcal{P}^{(T)}$ is a primary ideal as can easily be verified by induction on the lexicographic order of $(\tau_0,\ldots,\tau_{n-1})$ and its associated prime ideal is $\mathcal{P}$.

Lemma 2.2. *Let* $\mathcal{P}$ *be a prime ideal and* $T>0$ *an integer. Then the length* $l(\mathcal{P}^{(T)})$ *is bounded by*

$$l(\mathcal{P}^{(T)}) \ge \binom{T + \rho(\mathcal{P})}{\rho(\mathcal{P})} .$$

Proof. This is Lemma 3 in [Wü].

Let next A be an arbitrary subset of G(K). Then the stabilizer of A is the set

$$\{g \in G(K);\ g+A \subseteq A\}\ .$$

It is an abstract group that is represented by a subgroup scheme S(A) of G. If I is an ideal then we put

$$S(I) = S(\mathcal{V}(I)(K))$$

where $\mathcal{V}(I)$ is the variety of I. Now let I be an ideal, $\mathcal{P}$ an isolated prime component and $V=\mathcal{V}(\mathcal{P})$ and $W=\mathcal{V}(I)$. Denote by S(V;W) the set

$$S(V;W) = \{g \in G(K); g+V \subseteq W\}\ .$$

Then S(V;W) is a union of cosets of the stabilizer of V.

Since

$$S(V;W) + V \subseteq W$$

we have

$$S(V;W) \subseteq \bigcap_{v \in V(\bar{K})} (W - v)$$

and therefore by an easy calculation

$$S(V;W) = \bigcap_{v \in V(\bar{K})} (W - v) = \bigcap_{i=1}^{N} (W - v_i)$$

for elements $v_1,\dots,v_N \in V(\bar{K})$.

Hence

$$J = (E(v_1)I,\dots,E(v_N)I)$$

is an ideal with

$$\mathcal{V}(J) = S(V;W)\ .$$

So we have proved the following Lemma.

Lemma 2.3. *Let* I *be an ideal,* $\mathfrak{P}$ *an isolated prime component of minimal rank ,* $V = \mathfrak{V}(\mathfrak{P})$, $W = \mathfrak{V}(I)$. *Then there exist elements* $v_1,\ldots,v_N \in V(\overline{K})$ *such that the ideal*

$$J = (E(v_1)I,\ldots,E(v_N)I)$$

satisfies

$$\mathfrak{V}(J) = S(V;W) .$$

Lemma 2.4. *Suppose that* I *is generated by polynomials of degree at most* D_i *in* X_i *for* $0 \le i \le n$. *Let* r *be the codimension of* S(V;W). *Then for* $0 \le i_1 < \ldots < i_r \le n$ *we have*

$$\delta_{i_1,\ldots,i_r}(J) \le r!\, D_{i_1} \cdots D_{i_r} .$$

Proof. We have

$$J = (E(v_1)I,\ldots,E(v_N)I)$$

and since I is generated by polynomials of degree at most D_i in X_i $(0 \le i \le n)$ so is J. The desired estimates now follow from Lemma 2.1.

3. Proof of Theorem 1.1

Let $\mathfrak{m}(s.\gamma)$ be the maximal ideal in R of $s.\gamma$. Then denote by $M = M(\Sigma)$ the multiplicative set of elements $r \in R$ such that

$$r \notin \bigcup_{s=0}^{S-1} \mathfrak{m}(s.\gamma) .$$

For any ideal $I \subseteq R$ we write

$$I^* = IRM^{-1} \cap R .$$

Further we put for $1 \le r \le n+1$

$$T^{(r)} = T - \sum_{i=1}^{r-1} T_i ,$$

$$S^{(r)} = S - \sum_{i=1}^{r-1} S_i ,$$

and

$$I_r = (E(s.\gamma)\Delta_0^{\tau_0}\cdots\Delta_{n-1}^{\tau_{n-1}} P;\ 0\le\tau_0+\ldots+\tau_{n-1}\le T-T^{(r)},\ 0\le s<S-S^{(r)})$$

where P is the polynomial in Theorem 1.1.

Proposition 3.1. *Either Theorem 1.1 holds or for* $1\le r\le n+1$ *we have*

(i) $\quad \operatorname{rank} I_r = r$,

(ii) $\quad \delta_{i_1,\ldots,i_r}(I_r) \le r!\, D_{i_1}\cdots D_{i_r} \qquad (0\le i_1<\ldots<i_r\le n)$,

(iii) $\quad I_r \subseteq \bigcap_{s=0}^{S^{(r)}} \mathfrak{m}(s.\gamma)(T^{(r)})$.

Corollary 3.2. *Proposition 3.1 implies Theorem 1.1.*

Proof. For $r=n+1$ we get from (iii)

$$\delta_{0,\ldots,n}(I_{n+1}^*) \ge (S^{(n+1)}+1)\binom{T^{(n+1)}+n}{n}$$

and from (ii)

$$\delta_{0,\ldots,n}(I_{n+1}) \le (n+1)!\, D_0\cdots D_n .$$

From (1.3) we deduce that

$$T^{(n+1)}\ge T_{n+1} \text{ and } S^{(n+1)}\ge S_{n+1} .$$

This implies that

$$S^{(n+1)}\binom{T^{(n+1)}+n}{n} > S_{n+1}\binom{T_{n+1}+n}{n}.$$

This contradicts (1.2) with $r=n+1$. Therefore Theorem 1.1 holds, as claimed.

4. PROOF OF PROPOSITION 3.1

In order to prove Proposition 3.1 we shall proceed as follows.

Proposition 4.1. *Suppose that (i) - (iii) in Proposition 3.1 are false and* $P\neq 0$. *Then there exists an algebraic subgroup* $H\subseteq G$ *such that one of the following properties is satisfied.*

(i) $\mathrm{Lie}\, H\subseteq(\Delta_0,\ldots,\Delta_{n-1})$ *and for* $0\leq i_1<\ldots<i_r\leq n$ *we have*

$$\delta_{i_1,\ldots,i_r}(H)\leq r!\,\frac{D_{i_1}\cdots D_{i_r}}{(S_r+1)\binom{T_r+r-1}{r-1}},$$

(ii) $(1,\alpha_1,\ldots,\alpha_n)\in H(\overline{\mathbb{Q}})$ *and for* $0\leq i_1<\ldots<i_r\leq n$ *we have*

$$\delta_{i_1,\ldots,i_r}(H)\leq r!\,\frac{D_{i_1}\cdots D_{i_r}}{\binom{T_r+r-1}{r-1}}.$$

In order to prove Proposition 4.1 we assume that $\mathcal{P}\neq 0$. Then let $r\geq 1$ be the largest integer such that (i)-(iii) in Proposition 3.1 are true up to r. This integer exists since by hypothesis of Theorem 1.1 (i)-(iii) in Proposition 3.1 are true for $r=1$. So (i)-(iii) in Proposition 3.1 are false for $r+1$ which we may assume to be less or equal to $n+1$. Since (ii) and (iii) are easy consequences of (i) and the definition of I_r and Lemma 3.1 we may assume that (i) is false for $r+1$. Hence

(4.1) $\qquad \mathrm{rank}\, I_r = \mathrm{rank}\, I_{r+1} = r$

It follows that there exists an isolated associated prime $\mathcal{P}$ of I_r of rank r which is also an associated prime of I_{r+1}. Hence

$$E(s.\gamma)\Delta_0^{\tau_0}\cdots\Delta_{n-1}^{\tau_{n-1}} I_r \subseteq I_{r+1} \subseteq \mathcal{P}$$

for $0\le\tau_0+\ldots+\tau_{n-1}\le T_r$ and $0\le s\le S_r$. From this we deduce

$$I_r \subseteq \bigcap_{s=0}^{S_r}(E(-s.\gamma)\mathcal{P})^{(T_r)}.$$

If $E(-s.\gamma)\mathcal{P}\ne E(-s'.\gamma)\mathcal{P}$ for $s\ne s'$ we get from (4.2) by degree theory and the Lemma 2.2

$$\text{(4.3)}\qquad D_{i_1}\cdots D_{i_r} \ge \delta_{i_1,\ldots,i_r}(I_r) \ge (S_r+1)\binom{T_r+\rho(\mathcal{P})}{\rho(\mathcal{P})}$$

for $0\le i_1<\ldots<i_r\le n$. Write $\rho=\rho(\mathcal{P})$. Then if $\rho(\mathcal{P})=r$ the inequalities (4.3) contradict (1.2). Therefore $\rho=r-1$. Let $V=\mathcal{V}(\mathcal{P})$ and $W=\mathcal{V}(I_r)$. Then the ideal J in Lemma 2.3 satisfies by Lemma 2.4 with $r'=\mathrm{rank}(J)$

$$\text{(4.4)}\qquad (r')!\, D_{i_1}\cdots D_{i_r} \ge \delta_{i_1,\ldots,i_{r'}}(J)\,.$$

Let $\mathcal{Y}=\mathfrak{I}(S(V))$ be the ideal of the stabilizer of V. Then we may assume that

$$\mathcal{Y} \supseteq (E(v_1)\mathcal{P},\ldots,E(v_N)\mathcal{P})$$

and $\mathcal{Y}$ is the radical of the right-hand side. We have by (4.2) for $0\le s\le S_r$

$$\begin{aligned} J &= (E(v_1)I_r,\ldots,E(v_N)I_r) \\ &\subseteq ((E(v_1)E(-s.\gamma)\mathcal{P})^{(T_r)},\ldots,(E(v_N)E(-s.\gamma)\mathcal{P})^{(T_r)}) \\ &\subseteq E(-s.\gamma)(E(v_1)\mathcal{P},\ldots,E(v_N)\mathcal{P})^{(T_r)} \\ &\subseteq (E(-s.\gamma)\mathcal{Y})^{(T_r)}\,. \end{aligned}$$

It follows then that

$$J \subseteq \bigcap_{s=0}^{S_r}(E(-s.\gamma)\mathcal{Y})^{(T_r)}.$$

If now $E(-s.\gamma)\mathcal{Y}\ne E(-s'.\gamma)\mathcal{Y}$ for $s\ne s'$ we get since $r'=\mathrm{rank}\,\mathcal{Y}$ by Lemma 3.2

$$\delta_{i_1,\ldots,i_{r'}} \geq (S_r+1)\binom{T_r+\rho(\mathfrak{Y})}{\rho(\mathfrak{Y})}$$

$$\geq (S_{r'}+1)\binom{T_{r'}+\rho(\mathfrak{Y})}{\rho(\mathfrak{Y})}$$

If now $\rho(\mathfrak{Y})=r'$ or $\rho(\mathfrak{Y})=n$ this inequalities together with the inequalities (4.4) contradict (1.2). Hence $\rho(\mathfrak{Y})=r'-1<n$. The variety of $\mathfrak{Y}$ is an algebraic subgroup H of G for which we get

$$(5.7) \qquad \delta_{i_1,\ldots,i_{r'}}(H) \leq r!\, \frac{D_{i_1}\cdots D_{i_{r'}}}{\binom{T_{r'}+r'-1}{r'-1}}.$$

for $0 \leq i_1 < \ldots < i_{r'} \leq n$. Furthermore since $\rho(\mathfrak{Y})=r'-1$ we obtain

$$\operatorname{Lie} H \subseteq (\Delta_0,\ldots,\Delta_{n-1}).$$

This is property (i) of Proposition 4.1.

If now $E(-s.\gamma)\mathfrak{Y}=E(-s'.\gamma)\mathfrak{Y}$ for some pair of integers s, s' with $s \neq s'$ then we get $E(-s.\gamma)=E(-s'.\gamma)$. This implies immediately that

$$\gamma \in H(\overline{\mathbb{Q}})$$

which means that there exists an algebraic subgroup H of G such that $\gamma \in H(\overline{\mathbb{Q}})$ and

$$(5.8) \qquad \delta_{i_1,\ldots,i_{r'}}(H) \leq (r')!\, \frac{D_{i_1}\cdots D_{i_{r'}}}{\binom{T_{r'}+r'-1}{r'-1}}.$$

This is property (ii) of Proposition 4.1. The proof of Proposition 4.1 is therefore completed.

5. SUBGROUPS OF G

Suppose that H is a subgroup of G. Then we can write

$$(5.1) \qquad H = H_a \times H_m$$

where $H_a = H \cap \mathbb{G}_a$, $H_m = H \cap \mathbb{G}_m^n$ and $\mathbb{G}_a$, $\mathbb{G}_m^n$ are considered as embedded in G in the obvious way. Let $T_0,\dots,T_n$ ne coordinates on Lie G. Then H is defined in Lie G by

$$L_1(T_0,\dots,T_n) = \dots = L_r(T_0,\dots,T_n) = 0\,,$$

where r=rank H and $L_i \in \mathbb{Z}[T_0,\dots,T_n]$ for $1 \leq i \leq r$. Furthermore one of the L_i is T_0 if $H_a=0$ and the other do not contain T_0 and none of the L_i contains L_i if $H_a=\mathbb{G}_a$. This follows ealsily from (5.1). Since H is assumed to be connected the determinants of the minors

$$\Lambda_{i_1,\dots,i_r} \qquad\qquad (0 \leq i_1 < \dots < i_r \leq n)$$

of the system of linear forms $L_1,\dots,L_r$ have no common factor. Then it is easily seen that

$$|\det(\Lambda_{i_1,\dots,i_r})| \leq \delta_{i_1,\dots,i_r}(H)$$

if $0 \leq i_1 < \dots < i_r \leq n$.

Furthermore if Lie $H \subseteq (\Delta_0,\dots,\Delta_{n-1})$ we have by duality

$$\beta_0 T_0 + \dots + \beta_n T_n \in (L_1,\dots,L_r) \otimes_{\mathbb{Z}} \overline{\mathbb{Q}}\,.$$

Therefore we have proved the following simple Lemma.

Lemma 5.1. *If* $H \subseteq G$ *is an algebraic subgroup of* G *of codimension* r *such that* Lie $H \subseteq (\Delta_0,\dots,\Delta_{n-1})$ *or* $(1,\alpha_1,\dots,\alpha_n) \in H(\overline{\mathbb{Q}})$ *then* H *is defined in* Lie G *by the linear forms* $L_1(T_0,\dots,T_n),\dots,L_r(T_0,\dots,T_n)$ *over* $\mathbb{Z}$ *with the following properties:*

(i) *Either* T_0 *does not appear in* $L_1,\dots,L_r$ *or it appears in exactly one, say* L_1*, and then* $L_1 = T_0$.

(ii) *For* $0 \le i_1 < \ldots < i_r \le n$ *the minors* $\Lambda_{i_1,\ldots,i_r}$ *of the systems* $L_1,\ldots,L_r$ *satisfy* $|\det(\Lambda_{i_1,\ldots,i_r})| \le \delta_{i_1,\ldots,i_r}(H)$.

(iii) $\beta_0 T_0 + \ldots + \beta_n T_n \in (L_1,\ldots,L_r)\otimes_{\mathbb{Z}}\overline{\mathbb{Q}}$ *if* $\mathrm{Lie}\, H \subseteq (\Delta_0,\ldots,\Delta_{n-1})\otimes_{\mathbb{Z}}\overline{\mathbb{Q}}$.

(iv) $L_1(1,\log\alpha_1,\ldots,\log\alpha_n)=\ldots=L_r(1,\log\alpha_1,\ldots,\log\alpha_n)$ *if* $(1,\alpha_1,\ldots,\alpha_n)\in H(\overline{\mathbb{Q}})$.

Remark. Lemma 5.1 implies immediately that the properties (i) and (ii) of Proposition 4.1 contradict the hypothesis of Theorem 1.1.

6. REFERENCES

[Ba] A. Baker, The theory of linear forms in logarithms, Transcendence theory: Advances and applications, Academic Press, London, 1-27 (1977).

[Fa-Wü] G. Faltings, G. Wüstholz, Einbettungen kommutativer algebraischer Gruppen und einige ihrer Eigenschaften, Journ. r. u. ang. Math. **354**, 175-205 (1984).

[Ma-Wü 1] D. W. Masser, G. Wüstholz, Fields of large transcendence degree generated by the values of elliptic functions, Inv. math. **72**, 407-464 (1983).

[Ma-Wü 2] D. W. Masser, G. Wüstholz, Zero estimates on group varieties II, Inv. math. **80**, 233-267 (1985).

[Wü] G. Wüstholz, Multiplicity estimates on group varieties, to appear in Ann. Math..

A NEW APPROACH TO BAKER'S THEOREM ON LINEAR FORMS IN LOGARITHMS II

by

G. WÜSTHOLZ

Max - Planck - Institut für Mathematik
Gottfried - Claren - Str. 26
5300 Bonn 3

und

Universität - Gesamthochschule Wuppertal
Fachbereich Mathematik
Gausstr. 20
5600 Wuppertal 1

1. INTRODUCTION

In the first part of these series of notes [Wü] we have proved a modified version of multiplicity estimates for the algebraic group $\mathbb{G}_a \times \mathbb{G}_m^n$ which appears in Baker's work. The main point was the existence of a certain subgroup which has certain properties and which forms an obstacle for the proof that the original polynomial that goes through all transcendence proofs vanishes identically. Now it turns out that algebraic subgroups of tori, i.e. subgroups of the form $\mathbb{G}_m^n$, are very easily described in their Lie algebras. Here one immediately is led to the study of lattices or more generally to the study of $\mathbb{Z}$-submodules of $\mathbb{Z}^n$. But for such objects one has a beautiful theory which is very classical: the geometry of numbers.

What we want to do in this part II in the series of three notes is to work out this connection between algebraic groups in the simplest possible case, namely for tori, and the geometry of numbers. This enables us to give an alternative proof of Baker's results. It should be mentioned that the idea to use geometry of numbers in the context of zero or multiplicity estimates can already be found in [Ma-Wü] in a slightly different context. We should also not keep in secret that things generalize to arbitrary commutative algebraic groups. Our present intention however is to work this out clearly in the most simplest case.

2. Some Geometry Of Numbers

Let $\mathcal{M}$ be a $\mathbb{Z}$-submodule of $\mathbb{Z}^n$ of rank r with $1 \le r \le n$. To $\mathcal{M}$ can be attached a volume in the following way. Take a basis $m_1, \ldots, m_r$ of $\mathcal{M}$ and write M for the $r \times n$-matrix having m_i in the i-th row. Then for $1 \le i_1 < \ldots < i_r \le n$ let $M_{i_1,\ldots,i_r}$ be the $r \times r$-minor of M corresponding to this set of indices. For later use we have to define weighted volumes. For this let $a_1, \ldots, a_n \ge 1$ be real numbers and put for $a = (a_1, \ldots, a_n)$

$$\mathrm{vol}_a(\mathcal{M}) = \max_{1 \le i_1 < \ldots < i_r \le n} (a_{i_1} \cdots a_{i_r} |\det(M_{i_1,\ldots,i_r})|) .$$

Next we define the distance function F on $\mathbb{R}^n$ by

$$F_a(x) = \max_{1 \le i \le n} (a_i |x_i|) \qquad \text{where } x = (x_1, \ldots, x_n) .$$

We now take integers $1 \le j_1 < \ldots < j_r < n$ such that

$$\mathrm{vol}_a(\mathcal{M}) = a_{j_1} \cdots a_{j_r} |\det(M_{j_1,\ldots,j_r})|$$

and consider the lattice $\mathcal{M}'$ generated in $\mathbb{Z}^r$ by the rows $m'_1, \ldots, m'_r$ of the matrix $M_{j_1,\ldots,j_r}$ above. F_a restricts to a distance function F'_a on $\mathbb{R}^r$ embedded in the obvious way in $\mathbb{R}^n$. Then the set $\mathcal{B}$ defined by

$$\mathcal{B} := \{F'_a ; F'_a < 1\}$$

is a convex body of volume $2^r (a_{j_1} \cdots a_{j_r})^{-1}$. Write $a'_k = a_{j_k}$ for $1 \le k \le r$. Then by Minkowski's Theorem on successive minima we obtain for the successive minima $\lambda_1, \ldots, \lambda_r$ of $\mathcal{M}'$ the inequalities (see [Ca])

$$\lambda_1 \cdots \lambda_r \le a_1' \cdots a_r' |\det(M')|$$

where M' is the matrix corresponding to $\mathcal{M}'$. By the definition of the successive minima there exist elements $y_1', \dots y_r' \in \mathcal{M}'$ such that

$$F_a'(y_i') = \lambda_j \qquad (1 \le j \le r)$$

and these elements form a set of linearly independent elements. We can write

$$y_j' = \sum_{i=1}^{r} s_{ij} m_i' \qquad (1 \le j \le r)$$

and define

$$y_j = \sum_{i=1}^{r} s_{ij} m_i \qquad (1 \le j \le r) .$$

The same calculations as in the proof of Lemma 4 in [Ma-Wü] yield

$$F_a(y_j) \le r.\lambda_j \qquad (1 \le j \le r) .$$

This finally leads to

(2.1) $$F_a(y_1) \cdots F_a(y_r) \le r^r \mathrm{vol}_a(\mathcal{M}) .$$

For later use we have also to introduce an other distance function. For this we write for $x=(x_1,\dots,x_n) \in \mathbb{R}^n$

$$G_a(x) = \sum_{i=1}^{n} a_i |x_i| .$$

Then $G_a(x)$ is a distance function related to $F_a(x)$ by

$$G_a(x) \le n F_a(x) .$$

We now fix a and simply write F and G and for $\mathrm{vol}_a(\mathcal{M})$ we write $\mathrm{vol}(\mathcal{M})$. Putting all together we have proved the following result.

Lemma 2.1. *Let* $\mathcal{M} \subseteq \mathbb{Z}^n$ *be a free submodule of rank* r. *Then there exist linearly independent elements* $m_1, \dots, m_r$ *of* $\mathcal{M}$ *such that*

$$G(m_1)\cdots G(m_r) \le (nr)^r \mathrm{vol}(\mathcal{M}) .$$

SUBGROUPS OF $\mathbb{G}_m^n$

Let $H \subseteq \mathbb{G}_m^n$ be a connected algebraic subgroup of $\mathbb{G}_m^n$. Then in the Lie algebra of $\mathbb{G}_m^n$ the Lie algebra of H is defined by linear forms $L_1(T_1,\ldots,T_n),\ldots,L_r(T_1,\ldots,T_n)$ with integer coefficients where r denotes the codimension of H and $T_1,\ldots,T_n$ are the canonical coordinates of $\mathrm{Lie}\,\mathbb{G}_m^n$. The vectors formed by the coefficients of $L_1,\ldots,L_r$ generate a $\mathbb{Z}$-submodule $\mathcal{M}(H)$ of $\mathbb{Z}^n$ of rank r.

By Lemma 1.1 there exist linearly independent elements $m_1,\ldots,m_r$ of $\mathcal{M}(H)$ such that

$$G(m_1)\cdots G(m_r) \le (nr)^r \mathrm{vol}(\mathcal{M}(H)) .$$

Lemma 3.1. *Let* $H \subseteq \mathbb{G}_m^n$ *be a connected algebraic subgroup of codimension* r. *Then there exists real numbers* $\lambda_1,\ldots,\lambda_r$ *with the following properties.* H *is defined in* $\mathrm{Lie}\,\mathbb{G}_m^n$ *by linear forms*

$$L_i = \sum_{j=1}^{n} m_{ij} T_j \qquad (1 \le i \le r)$$

with integer coefficients and

(i) $\sum_{j=1}^{n} |m_{ij}| a_j \le nr\lambda_i \qquad (1 \le i \le r)$,

(ii) the heights $H(L_i)$ *of* L_i *are bounded by* $nr\lambda_i$ *for* $1 \le i \le r$,

(iii) $\lambda_1 \cdots \lambda_r \le \mathrm{vol}(\mathcal{M}(H))$.

Proof. Put in Lemma 2.1 $\mathcal{M} = \mathcal{M}(H)$. Then for the vectors $m_i = (m_{i1},\ldots,m_{in})$ with $1 \le i \le r$ of Lemma 2.1 write

$$L_i(T_1,\ldots,T_n) = \sum_{j=1}^{n} m_{ij} T_j .$$

Then let $\lambda_1,\dots\lambda_r$ be the successive minima constructed in section 2. It follows that

$$|L_i(a_1,\dots,a_n)| \le \sum_{j=1}^{n} |m_{ij}| a_j = G(m_i)$$

and from Lemma 2.1 we get the desired inequalities. Note that $|m_{ij}| a_j \le n r \lambda_i$ and $a_i \ge 1$ for $1 \le i \le r$, $1 \le j \le n$.

4. BAKER'S THEOREM

Let $\beta_0,\dots,\beta_n$ be algebraic numbers not all zero and $\alpha_1,\dots,\alpha_n$ non-zero algebraic numbers. Suppose that the height of the β's is bounded by B which we suppose to be at least 4 and that the heights of $\alpha_1,\dots,\alpha_n$ are bounded by $A_1,\dots,A_n$, also at least equal to 4. Put

$$\Omega = \log A_1 \cdots \log A_n .$$

Then we have the following famous results.

Theorem 4.1 **(Baker [Ba]).** *If*

$$\Lambda = \beta_0 + \beta_1 \log \alpha_1 + \dots + \beta_n \log \alpha_n \ne 0$$

then

$$|\Lambda| > (B\Omega)^{-c\Omega\log\Omega}$$

for some effective constant $c>0$.

If $\beta_0=0$ and $\beta_1,\dots,\beta_n$ are rational integers, the so-called rational case, then one gets an improvement.

Theorem 4.2 **(Baker [Ba]).** *If, in the rational case* $\Lambda \ne 0$, *then*

$$|\Lambda| > B^{-c\Omega\log\Omega} .$$

Remark 4.3. Actually in Baker's results the terms $\log\Omega$ are replaced by $\log\Omega'$ for $\Omega'=\frac{\Omega}{\log A_n}$.

We shall now sketch the proof of Theorem 4.1 by using the result proved in part I of these notes. We shall not specify the constant c in Theorem 4.1. The proof of Theorem 4.2 is only some minor modification of the proof we shall give here. In part IV of these series of papers written jointly with A. Baker we shall work out the explicit constant c.

5. PROOF OF THEOREM 4.1

We fix $\varepsilon>0$ arbitrary and choose a sufficiently large constant k. Then we put

$$\begin{aligned}
T &= [k\Omega] \\
T_i &= \left[\frac{T}{(n+1)}\right] && (1\le i\le n+1) \\
D_0 &= [k^{-\varepsilon}\log B\cdot T] \\
D_i &= \left[\frac{k^{-\varepsilon}T}{\log A_i}\right] && (1\le i\le n) \\
S &= [\log B] \\
S_i &= \left[\frac{S}{(n+1)}\right] && (1\le i\le n+1)\,.
\end{aligned}$$

Then one constructs as in [Ba] with T replaced by 2T the usual auxiliary polynomial $P(X_0,\dots,X_n)$ and proves by extrapolation in the usual way that it satisfies the hypothesis of Theorem 1.1 with T and with S replaced by $c'\cdot S$ with some large integer $c'>0$. Theorem 1.1 yields an algebraic subgroup H of $\mathbb{G}_a\times\mathbb{G}_m^n$ whose partial degrees are bounded explicitly.

According to Theorem 1.1 we have to consider the cases (i) and (ii). We start with case (i). Accordingly we have for possible partial degrees of H the estimate

$$\delta_{i_1,\dots,i_r}(H) \ll \frac{D_{i_1}\cdots D_{i_r}}{S\,T^{r-1}}.$$

The same estimates hold for the multilicative part H_m of H where we write $H=H_a\times H_m$, $H_a\subseteq\mathbb{G}_a$, $H_m\subseteq\mathbb{G}_m^n$. By Lemma 3.1 we can find linear forms defining H_m having the properties stated in that Lemma. Now for simplicity we assume that

$H=\mathbb{G}_a\times H_m$. One then easily sees that this is equivalent to $\beta_0=0$. The remaining case is left to the reader. The subgroup H_m of $\mathbb{G}_m^n \hookrightarrow \mathbb{G}_a\times\mathbb{G}_m^n$ leads to the $\mathbb{Z}$-module $\mathcal{M}(H_m)$ and its volume with respect to the weights $a_0=1$, $a_i=\log A_i$ $(1\le i\le n)$ satisfies

$$\mathrm{vol}(\mathcal{M}(H_m)) \ll \max_{1\le i_1<\ldots<i_r\le n}\left(\frac{D_{i_1}\cdots D_{i_r}a_{i_1}\cdots a_{i_r}}{ST^{r-1}}\right).$$

Here we have used that the determinant of the minor $M_{i_1,\ldots,i_r}$ of the $r\times n$-matrix M corresponding to the module $\mathcal{M}(H_m)$ can be estimated by

$$|\det(M_{i_1,\ldots,i_r})| \le \delta_{i_1,\ldots,i_r}(H).$$

This leads to

$$\mathrm{vol}(\mathcal{M}(H_m)) \ll T.$$

Let $L_1,\ldots,L_r$ be the linear forms in Lemma 3.1. We put

$$\alpha_i' = \prod_{j=1}^{n}\alpha_j^{m_{ij}}, \qquad (1\le i\le r).$$

The heights of α_i' can be estimated by Lemma 3.1 as follows:

$$\begin{aligned}H(\alpha_i') &\le \prod_{j=1}^{n}H(\alpha_j)^{|m_{ij}|}\\ &\le \exp\Big(\sum_{j=1}^{n}|m_{ij}|\log A_j\Big)\\ &\le \exp(nr\lambda_j).\end{aligned}$$

Hence the logarithm satisfies

$$h(\alpha_i') \le nr\lambda_j$$

and therefore

$$h(\alpha_1')\cdots h(\alpha_r') \ll T. \tag{5.1}$$

Since by (i) of Theorem 1.1 we have

$$\beta_1T_1+\ldots+\beta_nT_n \in (L_1,\ldots,L_r)$$

we can write

$$\beta_1 T_1 + \dots + \beta_n T_n = \beta_1' L_1 + \dots + \beta_r' L_r .$$

One easily obtains a bound B' for the heights $\beta_1', \dots, \beta_r'$ namely

$$\log B' \ll \log B + \log T . \tag{5.2}$$

By induction we get now

$$\begin{aligned} |\Lambda| &= |\beta_1' L_1(\log \alpha_1, \dots, \log \alpha_n) + \dots + \beta_r' L_r(\log \alpha_1, \dots, \log \alpha_n)| \\ &> (B'T')^{-c'T'} \\ &> (BT)^{-cT} \end{aligned}$$

for an appropriate constant c using (5.1) and (5.2). This is exactly the stated inequality.

Now we come to the case (ii). Again we restrict ourselves to the case $\beta_0 = 0$ and leave the remaining case to the reader. Then by Lemma 3.1 we get linear forms $L_1, \dots, L_r$ with integer coefficients. The heights $h(L_i)$ can be estimated by

$$h(L_i) \ll T \log(BT) .$$

Write

$$L = \beta_1 T_1 + \dots + \beta_n T_n .$$

Since

$$L(\log \alpha_1, \dots, \log \alpha_n) \neq 0$$

and

$$L_i(\log \alpha_1, \dots, \log \alpha_n) = 0$$

the linear forms $L, L_1, \dots, L_r$ are linearly independent. Therefore one can express L in terms of $L_1, \dots, L_r$ and $n\text{-}r$ of the variables $T_1, \dots, T_n$. Call these $n\text{-}r$ variables $T_1', \dots, T_{n-r}'$. Then we find

$$L = \beta_1' T_1' + \dots + \beta_{n-r}' T_{n-r}' + \beta_1'' L_1 + \dots + \beta_r'' L_r .$$

The heights of the new coefficients $\beta_1', \dots, \beta_{n-r}', \beta_1'', \dots, \beta_r''$ which are obviously algebraic can be bounded by some B' satisfying the inequalities

$$\log B' \ll \log T + \log B .$$

If we replace the variables $T_1,\dots,T_n$ by the corresponding $\log\alpha_1,\dots,\log\alpha_n$ one gets

$$\begin{aligned}\Lambda &= L(\log\alpha_1,\dots,\log\alpha_n)\\ &= \beta_1'\log\alpha_1' + \dots + \beta_{n-r}'\log\alpha_{n-r}' .\end{aligned}$$

Again by induction

$$|\Lambda| > (B'\Omega')^{-c'\Omega'\log\Omega'}$$

where

$$\Omega' = \log A_1' \cdots \log A_{n-r}' .$$

Putting everything together we finally obtain

$$|\Lambda| > (B\Omega)^{-c\Omega'\log\Omega'}$$

which is even better than the stated inequality. Theorem 4.1 is therefore proved.

6. REFERENCES

[Ba] A. Baker, The theory of linear forms in logarithms, Trancendence theory: Advances and applications, Academic press, London, 1-27 (1977).

[Ca] J. W. S. Cassels, An introduction to the geometry of numbers, Springer, Berlin-Göttingen-Heidelberg (1959).

[Ma-Wü] D. W. Masser, G. Wüstholz, Fields of large transcendence degree generated by values of elliptic functions, Inv. math. **72**, 407-464 (1983).

[Wü] G. Wüstholz, A new approach to Baker's Theorem on Linear Forms in Logarithms I, this volume.

ON THE THUE-MAHLER EQUATION

E. BOMBIERI[1]

I. Introduction

The classic Thue-Mahler equation is the diophantine-exponential equation

$$F(x,y) = p_1^{a_1} \dots p_s^{a_s} \tag{1.1}$$

where $F(x,y)$ is a binary form of degree $r \geq 3$ without multiple factors and with rational integral coefficients, where $p_1,\dots,p_s$ are fixed primes, to be solved in integers $x,y,a_1,\dots,a_s$. Equivalently, if S is the set $\{p_1,\dots,p_s,\infty\}$ and I_S is the ring of rational S-integers, i.e. of rational numbers of type $m/(p_1^{b_1}\dots p_s^{b_s})$ with $b_i \geq 0$, we can write (1.1) as $F(x,y) =$ S-unit, and ask for S-integer solutions x,y.

More generally, we consider the Thue-Mahler equation over number fields, as follows.

Let k be a number field and let S be a finite set of places of k, including all places at ∞. We abbreviate $d = [k:\mathbb{Q}]$, $d_v = [k_v:\mathbb{Q}_v]$ and we normalize the absolute value $|\ |_v$ so that

(i) if $v|p$, then

$$|p|_v = p^{-d_v/d};$$

(ii) if $v|\infty$, then

$$|\ |_v = \|\ \|^{d_v/d}$$

with $\|\ \|$ the euclidean absolute value. Let S be a finite set of places of k which includes all places at ∞ and let I_S be the ring of S-integers and U_S be the group of S-units of k, thus

---

[1] Supported in part by NSF grant #DMS 812 0790.

$$I_S = \left\{x \in k;\ |x|_v \leqq 1 \quad \text{for} \quad v \notin S\right\}$$

$$U_S = \left\{x \in k;\ |x|_v = 1 \quad \text{for} \quad v \notin S\right\}.$$

We also define the S-absolute value

$$|x|_S = \prod_{v \in S} |x|_v$$

and the S-projective height

$$h_S(x,y) = \prod_{v \in S} \max(|x|_v, |y|_v).$$

By the product formula, we see that if $x \in I_S$, $x \neq 0$ then $|x|_S \geqq 1$ and similarly $h_S(x,y) \geqq 1$ whenever $(x,y) \in I_S^2$ is not $(0,0)$. An S-integer x is an S-unit if and only if $|x|_S = 1$. This leads us to consider, for a binary form $F(x,y) \in I_S[x,y]$ of degree $r \geqq 3$, without multiple factors, the associated Thue-Mahler equation

$$(1.2) \qquad |F(x,y)|_S = 1,$$

to be solved in S-integers x,y. Solutions to (1.2) are naturally distributed into equivalence classes $(\eta x, \eta y)$ with $\eta \in U_S$, and the aim of this paper is to obtain upper bounds for the number of equivalence classes of the Thue-Mahler equation (1.2).

This problem is not new, and bounds depending on F have been obtained by several authors, in particular Lewis and Mahler [L.-M.] (see also Evertse [E] for further information on the literature). Evertse was the first to obtain bounds independent of the height of F, and depending only on the cardinality of S, the degree of F and the degree of the field k. Evertse's bounds are worse than exponential in the degree of F, since his method goes through the reduction of the Thue-Mahler equation to the so-called unit equation and then to an equation of binomial type.

A different method was used by Bombieri and Schmidt in the case of the classical Thue equation $F(x,y) = 1$ over the rational integers, with a bound which is linear in the degree of F.

In this paper, we treat the general Thue-Mahler equation over number fields, obtaining a bound which is polynomial in the degree of F. We prove:

Main Theorem. *Let* k *be a number field and let* S *be a finite set of places of* k *containing all places at* ∞. *The number of equivalence classes of solutions in* S*-integers of the Thue-Mahler equation* $|F(x,y)|_S = 1$ *where* F *is a binary form with* S*-integer coefficients without multiple factors and of degree* $r \geq 6$, *does not exceed*

$$(4\ \mathrm{Card}(S))^{2[k:\mathbb{Q}]}\ (4r)^{26\ \mathrm{Card}(S)}.$$

This bound can certainly be improved and the factor $(4\ \mathrm{Card}(S))^{2[k:\mathbb{Q}]}$ can be replaced by $(2p)^{[k:\mathbb{Q}]}$, where p is any rational prime for which there is $v \notin S$ with $v|p$. It should also be noted that Silverman's bounds [S] in terms of the rank of Mordell-Weil groups on Jacobians provide an independent way of obtaining bounds for the number of solutions of the Thue-Mahler equation.

Our proof follows the diophantine approximation method of [B.-S.], but this time there is an added serious difficulty since we may have solutions (x,y) with $|y|_v < 1$ for some $v \in S$, and then the diophantine approximation technique fails completely at the place v. One way out of this difficulty is to notice that we can find solutions $(\eta x, \eta y)$ in the equivalence class of (x,y), such that $|\eta y|_v \geq \delta > 0$ for all $v \in S$ and a suitable $\delta = \delta(k,S)$. Unfortunately this idea brings estimates which involve the regulator and discriminant of k, which is undesirable if uniformity with respect to k is wanted. We solve this difficulty by introducing a new notion of *reduced form* (Section II), although even then the technical details are not always pleasant.

Our Main Theorem is stated to hold for $\deg F \geq 6$, but undoubtedly it holds also for $\deg F = 5,4,3$, perhaps with somewhat larger constants. An improved version of Lemma 2, along the lines in [L.-M.], would certainly suffice to treat the case $\deg F = 5,4$, and then an improved treatment in Section IX should also cover the case $\deg F = 3$. In view of the complicated proof, we felt that the effort in extending our result to the remaining cases $\deg F = 5,4,3$ was not justified, in particular because if deg F is kept fixed we may appeal to Evertse's results.

I wish to express here my gratitude to all mathematicians who shared my thoughts on this problem while this work was in progress, especially F. Catanese, who pointed out the difficulty with the case $|y|_v < 1$, and G. Wüstholz, for his patient encouragements in writing this paper. A special thank you goes also to the Max Planck Institute in Bonn and the Mathematical Sciences Research Institute in Berkeley, for providing partial financial support for this research.

II. The Thue-Mahler equation

Let $F(\mathbf{x}) = F(x,y) = a_0x^r + a_1x^{r-1}y + \dots + a_ry^r$ be a binary form of degree $r \geqq 3$ without multiple factors, with coefficients in I_S, the ring of S-integers in k. We are interested in the Thue-Mahler equation

$$|F(x,y)|_S = 1, \tag{2.1}$$

to be solved in S-integers x,y.

If (x,y) is a solution to (2.1) then $(\eta x, \eta y)$ is another solution whenever η is an S-unit. We identify such solutions as belonging to a same equivalence class and choose x/y in $k \cup \infty$ as a representative for this equivalence class; N(F) will denote the set of such representatives.

Let U_S be the group of S-units in k and let U_S/T be its quotient by the torsion subgroup of roots of unity in k; this is free abelian of rank Card(S) - 1 by the Dirichlet unit theorem, so that U_S/TU_S^r is finite of order $r^{\mathrm{Card}(S)-1}$. Thus every S-unit can be written as $\eta = \zeta\eta_i\vartheta^r$ where $\zeta \in T$ is a root of unity, η_i belongs to a fixed set of $r^{\mathrm{Card}(S)-1}$ representatives in U_S of U_S/TU_S^r and where $\vartheta \in U_S$. The equation $F(x,y) = \eta$ becomes $\eta_i^{-1}F(\vartheta^{-1}x, \vartheta^{-1}y) = \zeta$, hence we can write

$$N(F) = \cup\, N_0(F_i) \tag{2.2}$$

where $N_0(G)$, for a form G, denotes the set of equivalence classes of solutions in S-integers of

$$(2.3) \qquad |G(x,y)|_v = 1 \quad \text{for} \quad v \in S,$$

and where F_i runs over the forms $F_i = \eta_i^{-1}F$. This yields

$$(2.4) \qquad \mathrm{Card}(N(F)) \leqq r^{\mathrm{Card}(S)-1} \max \mathrm{Card}(N_0(G))$$

where the maximum is taken over all forms of degree r, without multiple factors, with coefficients in I_S.

For $A \in GL(2,I_S)$ and $\mathbf{x} = (x,y)$ we write $A\mathbf{x} = (ax+by, cx+dy)$ and say that two forms F and G are equivalent, and write $F \sim G$, if there are $A \in GL(2,I_S)$ with $\det A = ad-bc \in U_S$ and a root of unity $\varsigma \in T$ such that

$$G(\mathbf{x}) = \varsigma F(A\mathbf{x}).$$

It is clear that if $F \sim G$ then $\mathrm{Card}(N_0(F)) = \mathrm{Card}(N_0(G))$ and more precisely $N_0(F) = AN_0(G)$ whenever (2.5) holds, where we let A act on $N_0(G)$ by the obvious fractional linear transformation.

Let F be a form and suppose that $N_0(F)$ is not empty, thus $F(x_0,y_0) = \varsigma_0$ for some $x_0, y_0 \in I_S$ and some root of unity ς_0. If $A = \begin{bmatrix} a & b \\ c & d \end{bmatrix}$ has entries

$$\begin{aligned} a &= \varsigma_0^{-1}(a_0x_0^{r-1}+a_1x_0^{r-2}y_0+\ldots+a_{r-1}y_0^{r-1}), \\ b &= \varsigma_0^{-1}a_ry_0^{r-1}, \\ c &= -y_0, \\ d &= x_0 \end{aligned}$$

then $\det A = 1$, $A\mathbf{x}_0 = (1,0)$ and the form $G(\mathbf{x}) = \varsigma_0^{-1}F(A^{-1}\mathbf{x})$ is equivalent to F with $G(1,0) = 1$, that is G has leading coefficient 1. We then say that the form G is *normalized* and denote by $\mathfrak{F}$ the set of normalized forms equivalent to F. For a given normalized F let $f(z) = F(z,1) = \prod (z-\alpha_i)$ be the associated monic polynomial. We consider the α_i as elements of a fixed finite extension K of k and for each $v \in S$ we choose a place w of K over v and a corresponding extension of the absolute value $| \; |_v$. We define the v-diameter of F to be

(2.5) $$d_v(F) = \max |\alpha_i - \alpha_j|_v$$

where the maximum is over i, j; the S-diameter of F is

(2.6) $$d_S(F) = \prod_{v \in S} d_v(F).$$

The diameter function so defined is independent of the choice of K and of the extension w of v to K.

Let F be normalized and let $D(F) = \prod_{i \neq j} (\alpha_i - \alpha_j)$ be its discriminant. It is obvious that

$$|D(F)|_S \leqq d_S(F)^{r(r-1)}$$

On the other hand, let $G(x) = \eta F(Ax)$ where η is an S-unit. Then

$$|D(G)|_S = |\det A|_S^{r(r-1)} |D(F)|_S$$

and we conclude that

(2.7) $$d_S(G) \geqq |\det A|_S.$$

In a similar way, we define

(2.8) $$m_v(F;y) = \max_i \max\left[1, |\alpha_i - \frac{1}{r} \text{trace } (\alpha)|_v |y|_v\right]$$

(2.9) $$m_S(F;y) = \prod_{v \in S} m_v(F;y).$$

It is clear that

(2.10) $$|y|_v \, d_v(F) \leq 2^{s(v)} m_v(F;y)$$

(2.11) $$|y|_S \, d_S(F) \leq 2m_S(F;y).$$

We define

(2.12) $$m_S(\mathfrak{F}) = \inf m_S(F;y)$$

where the infimum runs over all pairs (F,y) with $F \in \mathfrak{F}$ and $y \neq 0$, such that there is $\mathbf{x} = (x,y)$ with $|F(\mathbf{x})|_v = 1$ for all $v \in S$. The infimum in (2.12) is attained, though it is not unique (changing F by $A = \begin{bmatrix} 1 & b \\ 0 & 1 \end{bmatrix}$ with $b \in I_S$ does not affect $m_S(F;y)$). We say that a pair (F_0,y_0) is *reduced* if

$$m_S(\mathfrak{F}) = m_S(F_0;y_0).$$

Now (2.11) yields

Lemma 1. *For every* F *and* $y \neq 0$ *we have*

$$1 \leq d_S(F) \leq 2m_S(F;y)/|y|_S.$$

In particular, if (F_0,y_0) *is reduced we have*

$$1 \leq d_S(F_0) \leq 2m_S(\mathfrak{F})/|y_0|_S.$$

We fix a reduced form (F_0,y_0) and call solutions $x/y \in N_0(F)$ small, large and very large according to the following definition.

Definition. *Let* (F_0,y_0) *be reduced. A solution* (x,y) *is*

(i) *small, if*

$$h_S(x - \tfrac{1}{r}\,\mathrm{trace}(\alpha)y, y/y_0) \leq (2m_S(\mathfrak{F}))^{9r+18};$$

(ii) *large, if*

$$(2m_S(\mathfrak{F}))^{9r+18} < h_S(x - \tfrac{1}{r}\,\mathrm{trace}(\alpha)y, y/y_0) \leq (2m_S(\mathfrak{F}))^{(9r+18)(17(r-1)/32)^{200}};$$

(iii) *very large otherwise.*

The counting of solutions to the Thue-Mahler equation is done by different methods, according to the above classification.

III. Diophantine approximation properties of solutions

Let F be a normalized form

$$F(x,y) = x^r + a_1x^{r-1}y + \dots + a_ry^r = \prod (x-\alpha_iy);$$

we abbreviate $L_i(\mathbf{x}) = x - \alpha_iy$ and $\det(\mathbf{x},\mathbf{x}') = xy' - x'y$. Let $\mathbf{x}$ be a solution to the Thue-Mahler equation

$$|F(\mathbf{x})|_v = 1 \quad \text{for} \quad v \in S, \tag{3.1}$$

and let $\mathbf{z}$ be a point with S-integer coordinates such that $\det(\mathbf{x},\mathbf{z}) = 1$. For any $\mathbf{x}'$ with S-integer coordinates we can write $\mathbf{x}' = a\mathbf{x} - b\mathbf{z}$ for some S-integers a,b; in fact, $b = \det(\mathbf{x}',\mathbf{x})$. By linearity we deduce

$$L_i(\mathbf{x}')/L_i(\mathbf{x}) = a - \beta_ib \tag{3.2}$$

where

$$\beta_i = \beta_i(\mathbf{z},\mathbf{x}) = L_i(\mathbf{z})/L_i(\mathbf{x}). \tag{3.3}$$

Let $F(\mathbf{x}) = \varsigma$ and let $G(u,w)$ be the form

$$G(u,w) = \varsigma^{-1}F(u\mathbf{x}-w\mathbf{z}). \tag{3.4}$$

We have, noting that $\prod L_i(\mathbf{x}) = \varsigma$:

$$G(u,w) = \prod L_i(u\mathbf{x}-w\mathbf{z}) = \varsigma^{-1} \prod (uL_i(\mathbf{x})-wL_i(\mathbf{z})) = \prod (u-\beta_iw).$$

By (3.2) we have

$$L_i(\mathbf{x}')/L_i(\mathbf{x}) - L_j(\mathbf{x}')/L_j(\mathbf{x}) = -(\beta_j-\beta_i)\det(\mathbf{x}',\mathbf{x}). \tag{3.5}$$

Lemma 2. *Let* $F(x) = a_0 \prod (x-\alpha_iy)$. *For every* $\mathbf{x} = (x,y)$ *with* $y \neq 0$ *we have*

$$|a_0|_S \prod_{v\in S} \min_i |x-\alpha_i y|_v \leqq \Big[2|a_0|_S \prod_{v\in S} \max_i \max(1,|\alpha_i|_v)\Big]^{r-1}$$

$$|D(F)|_S^{-1/2} \cdot d_S(F)^{(r-1)(r-2)/2} \cdot |F(x)|_S \, h_S(x)^{-r+1}.$$

Proof. Let us abbreviate $\xi = x/y$ and let $i(v)$ be such that

$$|\xi-\alpha_{i(v)}|_v = \min_j |\xi-\alpha_j|_v.$$

We have

$$a_0 \prod (\xi-\alpha_i) = F(x)/y^r,$$

thus

$$(3.6) \qquad |\xi-\alpha_{i(v)}|_v = |a_0|_v^{-1}\Big[\prod{}' \, |\xi-\alpha_j|_v\Big]^{-1} |F(x)|_v |y|_v^{-r},$$

where the $'$ in the product means that the value $j = i(v)$ is omitted. To obtain a lower bound in (3.6) we proceed as follows. Clearly

$$|\alpha_{i(v)}-\alpha_j|_v \leqq 2^{s(v)}\max\Big[|\xi-\alpha_{i(v)}|_v, |\xi-\alpha_j|_v\Big] = 2^{s(v)}|\xi-\alpha_j|_v$$

hence

$$(3.7) \qquad \prod{}' \, |\alpha_{i(v)}-\alpha_j|_v \leqq 2^{s(v)(r-1)} \prod{}' \, |\xi-\alpha_j|_v.$$

Also

$$|D(F)|_v^{1/2} = |a_0|_v^{r-1} \prod_{j<k} |\alpha_j-\alpha_k|_v$$

so that

$$(3.8) \qquad \prod{}' \, |\alpha_{i(v)}-\alpha_j|_v = |D(F)|_v^{1/2} \Big/ \Big[|a_0|_v^{r-1} \prod_{j<k,\, j,k\neq i(v)} |\alpha_j-\alpha_k|_v\Big]$$

$$\geqq |D(F)|_v^{1/2} |a_0|_v^{-r+1} d_v(F)^{-(r-1)(r-2)/2};$$

collecting together (3.6), (3.7) and (3.8) we get

(3.9) $\min_i |x-\alpha_i y|_v$

$$\leqq 2^{s(v)(r-1)} |D(F)|_v^{-1/2} |a_0|_v^{r-2} d_v(F)^{(r-1)(r-2)/2} |F(x)|_v |y|_v^{-r+1}.$$

Let $H(x,y) = \prod (x-y/\alpha_i)$ and let $i'(v)$ be such that

$$|1/\xi-1/\alpha_{i(v)}|_v = \min_j |1/\xi-1/\alpha_j|_v.$$

We have

$$\prod (1/\xi-1/\alpha_i) = (-1)^r(\alpha_1\ldots\alpha_r)^{-1} a_0^{-1} F(x) x^{-r},$$

therefore

(3.10)

$$|1/\xi-1/\alpha_{i'(v)}|_v = |\alpha_1\ldots\alpha_r|_v^{-1} |a_0|_v^{-1} |F(x)|_v \Big[\prod{}' |1/\xi-1/\alpha_j|_v \Big]^{-1} |x|_v^{-r}.$$

We proceed as before and note first that

$$|1/\alpha_{i'(v)}-1/\alpha_j|_v \leqq 2^{s(v)} |1/\xi-1/\alpha_j|_v$$

thus

$$\prod{}' |1/\alpha_{i'(v)}-1/\alpha_j|_v \leqq 2^{s(v)(r-1)} \prod{}' |1/\xi-1/\alpha_j|_v. \tag{3.11}$$

Moreover

$$\prod{}' |1/\alpha_{i'(v)}-1/\alpha_j|_v = \prod_{j<k} |1/\alpha_j-1/\alpha_j|_v \Big/ \prod_{j<k;\,j,k\neq i'(v)} |1/\alpha_k-1/\alpha_j|_v \tag{3.12}$$

$$= |\alpha_1\ldots\alpha_r|_v^{-1} |a_0|_v^{-r+1} |\alpha_{i'(v)}|_v^{-r+2} |D(F)|_v^{1/2} \Big/ \prod_{j<k;\,j,k\neq i'(v)} |\alpha_j-\alpha_k|_v$$

$$\geqq |\alpha_1\ldots\alpha_r|_v^{-1} |\alpha_{i'(v)}|_v^{-r+2} |a_0|_v^{-r+1} |D(F)|_v^{1/2} \, d_v(F)^{-(r-1)(r-2)/2}.$$

By (3.10), (3.11) and (3.12) we infer

$$|1/\xi - 1/\alpha_{i'(v)}|_v$$

$$\leqq 2^{s(v)(r-1)}|a_0|_v^{r-2}|\alpha_{i'(v)}|_v^{r-2}|D(F)|_v^{-1/2}\ d_v(F)^{(r-1)(r-2)/2}|F(\mathbf{x})|_v\ |\mathbf{x}|_v^{-r}$$

$$\text{(3.13)}\qquad \min_i|x-\alpha_i y|_v \leqq |x-\alpha_{i'(v)}y|_v$$

$$\leqq 2^{s(v)(r-1)}|a_0|_v^{r-2}|\alpha_{i'(v)}|_v^{r-1}|D(F)|_v^{-1/2}\ d_v(F)^{(r-1)(r-2)/2}\ |F(\mathbf{x})|_v\ |\mathbf{x}|_v^{-r+1}.$$

By (3.9) and (3.13) we get

$$\text{(3.14)}\qquad \min_i|x-\alpha_i y|_v \leqq 2^{s(v)(r-1)}|a_0|_v^{r-2}\Big[\max_i\max(1,|\alpha_i|_v)\Big]^{r-1}$$
$$\cdot\ |D(F)|_v^{-1/2}d_v(F)^{(r-1)(r-2)/2}|F(\mathbf{x})|_v\Big[\max(|x|_v,|y|_v)\Big]^{-r+1},$$

and Lemma 2 follows by taking the product of inequalities (3.14) for all $v \in S$.

In practice, we apply Lemma 2 as follows. Let $f(\mathbf{x}) = \prod (x-\alpha_i y)$ be normalized, let $\alpha_i^* = \alpha_i - \frac{1}{r}\,\text{trace}(\alpha)$, $x^* = x - \frac{1}{r}\,\text{trace}(\alpha)y$, let $y_0 \neq 0$ and let $G(X,Y)$ be defined by

$$G(X,Y) = \prod (X-(\alpha_i^* y_0)Y).$$

Then

$$G(x^*,y/y_0) = F(\mathbf{x}),$$
$$D(G) = y_0^{r(r-1)}D(F),$$
$$d_S(G) = |y_0|_S\ d_S(F)$$

and if we apply Lemma 2 with $G,(x^*,y/y_0)$ in place of $F,(x,y)$ we get:

$$\prod_{v\in S}\min_i|x-\alpha_i y|_v = \prod_{v\in S}\min_i|x^*-\alpha_i^* y_0(y/y_0)|_v$$

$$\leqq \Big[2\prod_{v\in S}\max_i\max(1,|\alpha_i^* y_0|_v)\Big]^{r-1}$$

$$\cdot|y_0|_S^{-r(r-1)/2}|D(F)|_S^{-1/2}|y_0|_S^{(r-1)(r-2)/2}d_S(F)^{(r-1)(r-2)/2}|F(\mathbf{x})|_S\ h_S(x^*,y/y_0)^{-r+1}$$

$$= \Big[\frac{2m_S(F;y_0)}{|y_0|_S}\Big]^{r-1}|D(F)|_S^{-1/2}d_S(F)^{(r-1)(r-2)/2}|F(\mathbf{x})|_S\ h_S(x^*,y/y_0)^{-r+1}.$$

By Lemma 1, $d_S(F) \leqq \dfrac{2m_S(F,y_0)}{|y_0|_S}$. We have shown

Corollary. *Let* (F_0, y_0) *be reduced. Then*

$$\prod_{v\in S} \min_i |x-\alpha_i y|_v \leqq \left[\frac{2m_S(\mathcal{F})}{|y_0|_S}\right]^{r(r-1)/2} |F_0(x)|_S \, h_S(x^*, y/y_0)^{-r+1}.$$

IV. Classification of solutions

In this section we show how solutions are divided in classes according to their diophantine approximation properties. The main tool is

Mahler's Lemma. *Let* B *with* $3/4 < B < 1$ *and let* q *be a positive integer. Let* $\kappa = 2(1-B)/q$. *There exists a set* Γ *of cardinality*

$$\mathrm{Card}(\Gamma) \leqq \left[\frac{4}{1-B}\right]^{q-1}$$

consisting of q-uples $\gamma = (\gamma(1),\ldots,\gamma(q))$ *with* $\gamma(i) \geqq 0$, $i=1,\ldots,q$ *and with*

$$\sum_i \gamma(i) = 1,$$

with the following property:

For any set of reals $E_1,\ldots,E_q,\Lambda$ *with* $0 < E_i \leqq 1$ *for* $i=1,\ldots,q$ *and with* $\Lambda = \prod E_i$, *there exists* $\gamma \in \Gamma$ *such that*

$$\Lambda^{(\gamma(i)+\kappa)/B} \leqq E_i \leqq \Lambda^{B\gamma(i)}. \tag{4.1}$$

Also there exists a set Ξ *of cardinality*

$$\mathrm{Card}(\Xi) \leqq \left[\frac{4}{1-B}\right]^{q-1}$$

consisting of q-uples $\vartheta = (\vartheta(1),\ldots,\vartheta(q))$ *with* $\vartheta(i) \geqq 0$, $i=1,\ldots,q$ *and with*

$$\sum_i \vartheta(i) = 1,$$

with the following property:

For any set of reals $E_1,\ldots,E_q,\Lambda$ *with* $1 \leqq E_i$ *for* $i=1,\ldots,q$ *and with* $\Lambda = \prod E_i$, *there exists* $\vartheta \in \Xi$ *such that*

$$\Lambda^{B(\vartheta(i)-\kappa)} \leqq E_i \leqq \Lambda^{\vartheta(i)/B}. \tag{4.2}$$

Proof. This is a refinement of Lemma 4 of Evertse [E]; we sketch here Evertse's argument for completeness.

We may suppose $q \geq 2$. Let $u(i)$ be determined by $E_i = \Lambda^{u(i)}$, so that $\Sigma\, u(i) = 1$, and let m be the smallest integer $m \geq \frac{q-1}{1-B}$. Let $g(i) = [mu(i)/B]$ and let $h(i) = [Bm\, u(i)]$. In the case in which $E_i \leq 1$ we have

$$\sum g(i) > \frac{m}{B}\sum u(i) - q = \frac{m}{B} - q \geq m-1,$$

$$\sum h(i) \leq mB \sum u(i) = mB < m,$$

$$h(i) \leq g(i),$$

thus there are integers $f(i)$, $h(i) \leq f(i) \leq g(i)$, such that $\Sigma\, f(i) = m$. Now

$$E_i = \Lambda^{u(i)} \leq \Lambda^{Bg(i)/m} \leq \Lambda^{Bf(i)/m},$$

$$\Lambda^{(f(i)+1)/(Bm)} \leq \Lambda^{(h(i)+1)/(Bm)} \leq \Lambda^{u(i)} = E_i$$

and we can take $\gamma(i) = f(i)/m$ because $\kappa \geq 1/m$. The cardinality of Γ is at most the number of possible $(f(1),\dots,f(q))$, which is the binomial coefficient $\binom{m+q-1}{q-1}$. Also $m < \frac{q-1}{1-B} + 1$ and the required bound follows by a majorization of the binomial coefficient.

The proof of the second statement is practically the same.

Let $\alpha = (\alpha_{i(v)})_{v\in S}$ be a choice of a root of $F(x,1) = 0$ for each $v \in S$ and let

$$(4.3)\qquad N_0(F,\alpha) = \left\{x/y \in N_0(F);\ |x-\alpha_{i(v)}y|_v = \text{minimum for every } v \in S\right\};$$

there are $r^{\mathrm{Card}(S)}$ possible choices for α, therefore

$$\mathrm{Card}(N_0(F)) \leq r^{\mathrm{Card}(S)} \max \mathrm{Card}\,(N_0(F,\alpha)).$$

We note here that since $\prod (x-\alpha_i y) = F(x)$ is a root of unity, we have

$$|x-\alpha_{i(v)}y|_v \leq 1 \quad \text{for} \quad v \in S,\ x/y \in N_0(F,\alpha).$$

For every partition $S = S' \cup S''$ we define

$$(4.4)\qquad N_0(F,\alpha,S') = \left\{x/y\in N_0(F,\alpha);\ |y/y_0|_v \geq 1 \text{ if } v\in S',\ |y/y_0|_v < 1 \text{ if } v\in S''\right\};$$

there are $2^{\mathrm{Card}(S)}$ possible choices for the partition, hence

$$\mathrm{Card}(N_0(F,\alpha)) \leqq 2^{\mathrm{Card}(S)} \max_{S'} \mathrm{Card}(N_0(F,\alpha,S')).$$

We apply Mahler's Lemma a first time choosing $|y/y_0|_v$, $v \in S'$ for the E_i and obtain a set of at most $\left[\frac{4}{1-B}\right]^{\mathrm{Card}(S')}$ vectors $(\vartheta(v))_{v\in S'}$ with $\Sigma_{v\in S'} \vartheta(v) = 1$, such that for $x/y \in N_0(F,\alpha,S')$ we have

$$(4.5) \qquad |y/y_0|_{S'}^{B(\vartheta(v)-\kappa')} \leqq |y/y_0|_v \leqq |y/y_0|_{S'}^{\vartheta(v)/B} \quad \text{for} \quad v \in S',$$

where $\kappa' = 2(1-B)/\mathrm{Card}(S')$, for some vector ϑ in the set. A second application of Mahler's Lemma yields a set of at most $\left[\frac{4}{1-B}\right]^{\mathrm{Card}(S'')}$ vectors $(\gamma(v))_{v\in S''}$ with $\Sigma_{v\in S''} \gamma(v) = 1$, such that for $x/y \in N_0(F,\alpha,S')$ we have

$$(4.6) \qquad |y/y_0|_{S''}^{(\gamma(v)+\kappa'')/B} \leqq |y/y_0|_{S''}^{B\gamma(v)} \quad \text{for} \quad v \in S'',$$

where $\kappa'' = 2(1-B)/\mathrm{Card}(S'')$, for some vector γ in the set. We define $N_0(F,\alpha,S',\vartheta,\gamma;B)$ the set of elements of $N_0(F,\alpha,S')$ such that (4.5) and (4.6) hold and obtain

$$(4.7) \qquad \mathrm{Card}\left[x/y\in N_0(F)\cap A\right] \leqq \left[\frac{8r}{1-B}\right]^{\mathrm{Card}(S)} \mathrm{Card}\left[x/y\in N_0(F,\alpha,S',\vartheta,\gamma;B)\cap A\right]$$

for every subset $A \subset N_0(F)$.

For every $x/y \in N_0(F,\alpha,S',\vartheta,\gamma;B)$ we have $|x-\alpha_{i(v)}y|_v \leqq 1$; we apply Mahler's Lemma once more and obtain a set of at most $\left[\frac{4}{1-B}\right]^{\mathrm{Card}(S)}$ vectors $(\Gamma(v))_{v\in S'}$, with $\Sigma_{v\in S} \Gamma(v) = 1$, such that

$$(4.8) \qquad |x-\alpha_{i(v)}y|_v \leqq \left[\prod_{v\in S} |x-\alpha_{i(v)}y|_v\right]^{B\Gamma(v)} \quad \text{for} \quad v \in S.$$

We define $N_0(F,\alpha,S',\vartheta,\gamma,\Gamma;B)$ the set of elements of $N_0(F,\alpha,S')$ such that (4.5), (4.6) and (4.8) hold. We have

$$\text{(4.9)}\quad \mathrm{Card}\left[N_0(F)\cap A\right] \leqq \left[\frac{32r}{(1-B)^2}\right]^{\mathrm{Card}(S)} \max \mathrm{Card}\left[N_0(F,\alpha,S',\vartheta,\gamma,\Gamma;B)\cap A\right]$$

for every subset $A \subset N_0(F)$.

V. The Gap Principle

Let x/y, x'/y' be distinct elements of $N_0(F,\alpha,S',\vartheta,\gamma;B)$. We have

$$\text{(5.1)}\qquad \begin{aligned} |\det(x,x')|_v &= |(x-\alpha_{i(v)}y)y'-(x'\alpha_{i(v)}y')y|_v \\ &\leqq 2^{s(v)}\max\left[|(x-\alpha_{i(v)}y)y'|_v,\ |(x'-\alpha_{i(v)}y')y|_v\right]. \end{aligned}$$

We choose z with S-integers coordinates with $\det(x,z) = 1$ and use (3.5). We have

$$|\beta_i-\beta_j|_v|y|_v = |\frac{1}{L_i(x)} - \frac{1}{L_j(x)}|_v,$$

therefore

$$\text{(5.2)}\qquad \begin{aligned} \max_i\ |\beta_i-\frac{1}{r}\ \mathrm{trace}(\beta)|_v\ |y|_v &= \max_i\ |\frac{1}{r}\sum_j(\beta_i-\beta_j)|_v\ |y|_v \\ &\leqq |r|_v^{-1}\ r^{s(v)}\max_{i,j}\ |\beta_i-\beta_j|_v\ |y|_v \\ &\leqq |r|_v^{-1}\ r^{s(v)}2^{s(v)}\ \max_i\ |\frac{1}{L_i(x)}|_v = |r|_v^{-1}(2r)^{s(v)}\ \frac{1}{|x-\alpha_{i(v)}y|_v}\ ; \end{aligned}$$

a similar inequality with x' in place of x also holds. By (5.1) and (5.2) we get

$$\text{(5.3)}\qquad \begin{aligned} &\max_i\ |\beta_i-\frac{1}{r}\ \mathrm{trace}\ (\beta)|_v\ |\det(x,x')|_v \\ &\leqq |r|_v^{-1}\ (4r)^{s(v)}\ \max\ (|y/y'|_v,|y'/y|_v). \end{aligned}$$

Now we can prove:

Gap Principle. *Let* x/y, x'/y' *be distinct elements of* $N_0(F,\alpha,S',\vartheta,\gamma;B)$. *Then*

$$m_S(\mathcal{F}) \leq 4r \prod_{v \in S} \max\left(|y/y'|_v, |y'/y|_v\right).$$

Proof. The form $G(u,w) = F(x)^{-1}F(ux-wz) = \prod (u-\beta_i w)$ is normalized and equivalent to F. If $u' = (u',w')$ is the point such that $x' = u'x - w'z$ then $|G(u')|_S = |F(x')/F(x)|_S = 1$ and $w' = \det(x,x')$. The right-hand side of (5.3) is at least 1, hence we can rewrite (5.3) as

$$m_v(G,w') \leq |r|_v^{-1}(4r)^{s(v)} \max\left(|y/y'|_v, |y'/y|_v\right).$$

The Gap Principle follows by taking the product of the last inequality over all $v \in S$.

VI. The Strong Gap Principle

In this section we show that large solutions to the Thue-Mahler equation have very large gaps.

Strong Gap Principle. *Let* (F_0,y_0) *be reduced and let* x/y, x'/y' *be distinct elements of* $N_0(F_0,\alpha,S',\vartheta,\gamma,\Gamma;B)$ *such that*

$$(2m_S(\mathcal{F}))^{9r+18} \leq h_S(x^*,y/y_0) \leq h_S(x'^*,y'/y_0).$$

Then

$$h_S(x'^*,y'/y_0) \geq h_S(x^*,y/y_0)^{17(r-1)/32}.$$

We begin with an auxiliary result.

Lemma 3. *Let* (F_0,y_0) *be reduced and let* x *be a solution to the Thue-Mahler equation* (3.1). *Let* T *be any finite set of places in* S. *Then*

$$h_T(x^*,y/y_0) \geq \frac{1}{2m_S(\mathcal{F})},$$

$$|y/y_0|_T \leq 2m_S(\mathcal{F})h_S(x^*,y/y_0).$$

Proof. Let $G(X,Y) = \prod (X-(\alpha_i^* y_0)Y)$, where $\alpha_i^* = \alpha_i - \frac{1}{r}\,\mathrm{trace}(\alpha)$. Then $G(x^*,y/y_0) = F_0(x)$ is a root of unity and we deduce that

$$\max_i |x^* - (\alpha_i^* y_0)(y/y_0)|_v \geq 1$$

and *a fortiori*

(6.1) $$1 \leq 2^{s(v)} \max_i \max(1, |\alpha_i^* y_0|_v) \cdot \max(|x^*|_v, |y/y_0|_v).$$

We take the product of inequalities (6.1) for all $v \in T$, note that $\prod_{v\in T} m_v(F_0, y_0) \leq m_S(\mathcal{F})$, and get the first conclusion of Lemma 3. To obtain the second conclusion, let $S = T \cup T'$. Then

$$|y/y_0|_T \leq h_T(x^*, y/y_0) = h_S(x^*, y/y_0)/h_{T'}(x^*, y/y_0)$$

and Lemma 3 follows.

To prove the Strong Gap Principle we proceed as follows. By (5.1), (4.8) and Lemma 2, Corollary we have

$$|\det(x,x')|_v/|y_0|_v$$
$$\leq 2^{s(v)}\left\{\left[\frac{2m_S(\mathcal{F})}{|y_0|_S}\right]^{r(r-1)/2} h_S(x^*, y/y_0)^{-r+1}\right\}^{B\Gamma(v)} \max\left[|y/y_0|_v, |y'/y_0|_v\right]$$

and by (4.5), (4.6) we also have

$$\max(|y/y_0|_v, |y'/y_0|_v) \leq \max(|y/y_0|_{S'}, |y'/y_0|_{S'})^{\vartheta(v)/B} \quad \text{if} \quad v \in S'$$
$$\max(|y/y_0|_v, |y'/y_0|_v) \leq \max(|y/y_0|_{S''}, |y'/y_0|_{S''})^{\gamma(v)B} \quad \text{if} \quad v \in S''.$$

The last three inequalities yield

(6.2) $$|\det(x,x')|_S \leq 2|y_0|_S \left\{\left[\frac{2m_S(\mathcal{F})}{|y_0|_S}\right]^{r(r-1)/2} h_S(x^*, y/y_0)^{-r+1}\right\}^B$$
$$\max(|y/y_0|_{S'}, |y'/y_0|_{S'})^{1/B} \max(|y/y_0|_{S''}, |y'/y_0|_{S''})^B.$$

We apply the second part of Lemma 3 with $T = S'$, note that $|y/y_0|_{S''} \leq 1$, $|y'/y_0|_{S''} \leq 1$ and $|\det(x,x')|_S \geq 1$, and infer from (6.2) that

(6.3) $$|y_0|_S^{Br(r-1)/2-1} h_S(x^*, y/y_0)^{(r-1)B}$$
$$\leq 2(2m_S(\mathcal{F}))^{Br(r-1)/2+1/B} h_S(x'^*, y'/y_0)^{1/B}.$$

We can simplify the last inequality. Suppose

$$h_S(x^*, y/y_0) \geq (2m_S(\mathcal{F}))^{r/2+1} h_S(x^*, y/y_0)^a;$$

then (6.3) becomes, after noting that $|y_0|_S \geq 1$: $h_S(x^*, y/y_0)^{(r-1)Ba} \leq 2(2m_S(\mathcal{F}))^{1/B-(r-1)Ba} h_S(x'^*, y/y_0)^{1/B}$ and the Strong Gap Principle follows by choosing $a = 17/18$.

VII. Small solutions

Let (F_0, y_0) be normalized and reduced. We want to count the number of elements in $N_0(F_0, \alpha, S', \vartheta, \gamma; B)$ with height $h_S(x^*, y/y_0) \leq H$, where $H = (2m_S(\mathfrak{F}))^{9r+18}$. We suppose here that $m_S(\mathfrak{F})$ is sufficiently large, and more precisely

$$(7.1) \qquad m_S(\mathfrak{F}) \geq 2^{29} r^{10}.$$

We begin by introducing a parameter $K > 1$ to be chosen later, and define

$$N_0(F_0, \alpha, S', \vartheta, \gamma; B, K, m, n)$$
$$= \left\{ x/y \in N_0(F_0, \alpha, S', \vartheta, \gamma; B);\ K^m \leq |y/y_0|_{S'} < K^{m+1},\ K^{-n-1} < |y/y_0|_{S''} \leq K^{-n} \right\}$$

where m, n are non-negative integers.

Lemma 4. *If* $K = (m_S(\mathfrak{F})/4r)^{1/9}$ *and* $m+n \leq \frac{1}{1-B}$ *then* $N_0(F_0, \alpha, S', \vartheta, \gamma; B, K, m, n)$ *has at most one element.*

Proof. We apply the Gap Principle to two distinct elements of $N_0(F_0, \alpha, S', \vartheta, \gamma; B, K, m, n)$, assuming they exist. It is clear that

$$\max(|y'/y|_v, |y/y'|_v) \leq K^{(m+1)\vartheta(v)/B - mB(\vartheta(v) - \kappa')} \quad \text{if } v \in S',$$
$$\max(|y'/y|_v, |y/y'|_v) \leq K^{-nB\gamma(v) + (n+1)(\gamma(v) + \kappa'')/B} \quad \text{if } v \in S'',$$

where $\kappa' = 2(1-B)/\mathrm{Card}(S')$, $\kappa'' = 2(1-B)/\mathrm{Card}(S'')$. Now the Gap Principle yields

$$(7.2) \qquad m_S(\mathfrak{F}) \leq (4r)K^{(m+n)(1/B-B)+2/B+2mB(1-B)+2(n+1)(1-B)/B}$$
$$< (4r)K^{(m+n)5(1-B)+4}.$$

We choose $K = \left[\frac{m_S(\mathfrak{F})}{4r}\right]^{1/9}$ and now (7.2) implies $m+n > \frac{1}{1-B}$. Since this result has been obtained on the assumption that $N_0(F_0,\alpha,S',\vartheta,\gamma;B,K,m,n)$ contained at least two distinct elements, Lemma 4 follows.

Lemma 5. *Suppose that* $m_S(\mathfrak{F}) > 2^{29}r^{10}$ *and that* (F_0,y_0) *is reduced. Then*

$$\mathrm{Card}\left\{x/y\in N_0(F_0);\ h_S(x^*,y/y_0)<(2m_S(\mathfrak{F}))^{9r+18}\right\}$$
$$\leq 470(3r+7)^2(80r(18r+39))^{\mathrm{Card}(S)}.$$

Proof. Let $x/y \in N_0(F_0,\alpha,S',\vartheta,\gamma;B,K,m,n)$, with K as in Lemma 4. By Lemma 3 we have

$$|y/y_0|_{S'} \leq 2m_S(\mathfrak{F})\, h_S(x^*,y/y_0) < (2m_S(\mathfrak{F}))^{9r+19},$$

and $K^m \leq |y/y_0|_{S'}$ by definition of $N_0(F_0,\alpha,S',\vartheta,\gamma;B,K,m,n)$; this implies

$$m \leq 9(9r+19)\,\frac{\log(2m_S(\mathfrak{F}))}{\log\left[\frac{m_S(\mathfrak{F})}{4r}\right]}. \tag{7.3}$$

Also by Lemma 1 we have $|y_0|_S \leq 2m_S(\mathfrak{F})$, thus $K^{-n} \geq |y/y_0|_{S''} = |y/y_0|_S/|y/y_0|_{S'} \geq |y_0|_S^{-1}|y/y_0|_{S'}^{-1} \geq (2m_S(\mathfrak{F}))^{-1}K^{-m-1}$, therefore

$$n \leq m+1+9\,\frac{\log(2m_S(\mathfrak{F}))}{\log\left[\frac{m_S(\mathfrak{F})}{4r}\right]}. \tag{7.4}$$

If we assume that

$$2m_S(\mathfrak{F}) < \left[\frac{m_S(\mathfrak{F})}{4r}\right]^{10/9} \tag{7.5}$$

we can simplify (7.3) and (7.4) to

$$m < 90r + 190, \quad n \leq m + 10; \tag{7.6}$$

the number of pairs (m,n) which satisfy (7.6) does not exceed $470\ (3r+7)^2$, while $m+n < 180r + 390$. If we choose $B = 1 - \frac{1}{180r+390}$ then $m+n < \frac{1}{1-B}$ and Lemma 4 applies, with the conclusion that

$$\mathrm{Card}\left[x/y \in N_0(F_0,\alpha,S',\vartheta,\gamma;B);\ h_S(x^*,y/y_0) < H\right] \leqq 470\ (3r+7)^2.$$

By (4.7) we have

$$\mathrm{Card}\left[x/y \in N_0(F_0);\ h_S(x^*,y/y_0) < H\right]$$
$$\leqq \left[\frac{8r}{1-B}\right]^{\mathrm{Card}(S)} \max \mathrm{Card}\left[x/y \in N_0(F_0,\alpha,S',\vartheta,\gamma;B);\ h_S(x^*,y/y_0) < H\right]$$

and Lemma 5 follows.

VIII. Large solutions

Let (F_0,y_0) be normalized and reduced.

Lemma 6. *Let* $H \geqq (2m_s(\mathcal{F}))^{9r+18}$. *Then the number of elements of* $N_0(F_0,\alpha,S',\vartheta,\gamma,\Gamma;B)$ *with*

$$H \leqq h_S(x^*,y/y_0) < H^{(17(r-1)/32)^N}$$

does not exceed N.

Proof. Let $\mathbf{x}_1,\mathbf{x}_2,\ldots,\mathbf{x}_M$ be solutions of the Thue-Mahler equation satisfying the conditions of Lemma 6, and arranged by increasing height $h_S(x_i^*,y_i/y_0)$. The Strong Gap Principle applies here and we get

$$h_S(x_{i+1}^*,y_{i+1}/y_0) \geqq h_S(x_i^*,y_i/y_0)^{17(r-1)/32}.$$

This yields easily

$$H^{(17(r-1)/32)^{M-1}} \leqq h_S(x_M^*,y_M/y_0) < H^{(17(r-1)/32)^N}$$

and Lemma 6 follows.

Corollary 1. *Let* $H \geqq (2m_S(\mathcal{F}))^{9r+18}$. *Then the number of elements of* $N_0(F_0)$ *with*

$$H \leqq h_S(x^*,y/y_0) < H^{(17(r-1)/32)^N}$$

does not exceed

$$N(512r)^{\mathrm{Card}(S)}.$$

Proof. Clear from Lemma 6 and (4.9) with $B = 3/4$.

Corollary 2. *The number of large solutions to the Thue-Mahler equation does not exceed*

$$200(512r)^{\mathrm{Card}(S)}.$$

Proof. Choose $H = (2m_S(\mathcal{F}))^{9r+18}$.

IX. Application of the Thue Principle

Let k be a number field, K an extension of k of degree r and let α_1, α_2 be elements of K of degree r over k. Let S be a finite set of places of k; for each $v \in S$ we choose an extension of the absolute value $|\ |_v$ to K, again denoted by $|\ |_v$. Let $\beta_1, \beta_2 \in k$ be approximations in k to α_1, α_2 in the sense that

$$|\alpha_i - \beta_i|_v < 1 \quad \text{for} \quad v \in S$$

and $i=1,2$. From [B] we recall the following result.

Thue Principle. *Let* $\sqrt{\frac{2}{r+1}} < t < \sqrt{\frac{2}{r}}$, $0 < \tau < t$, $t < \vartheta < t^{-1}$, $\delta_1 > 0$, $\delta_2 > 0$. *Then; either*

$$\prod_{v\in S} \max\left[|\alpha_1-\beta_1|_v^{\vartheta\delta_1}, |\alpha_2-\beta_2|_v^{\vartheta^{-1}\delta_2}\right]$$

$$> \left[(3h(\alpha_1))^C h(\beta_1)\right]^{-\frac{\delta_1}{t-\tau}} \left[(3h(\alpha_2))^C h(\beta_2)\right]^{-\frac{\delta_2}{t-\tau}}$$

with

$$C = \frac{1}{1-r^2/2},$$

or

$$\frac{r}{2}\frac{\delta_2}{\delta_1} > \frac{r}{2}t^2 + \frac{1}{2}\tau^2 - 1.$$

This is a rephrasing of [B], Theorem 2; note that the condition $|\alpha_i-\beta_i|_v < 1$ for $v \in S$ is superfluous, because otherwise a stronger inequality is obtained by removing from S all places at which $|\alpha_i-\beta_i|_v \geqq 1$ for some i.

We apply the Thue Principle as follows. We choose

$$\alpha_1 = \alpha_2 = \alpha^* y_0,$$
$$\beta_i = y_0\, x_i^*/y_i,$$

and we extend v to K so that $|x-\alpha y|_v = \text{minimum}$.

We have with w, v running over the absolute values of K, k respectively:

$$h(\alpha^* y_0) = \prod \max\left[1, |\alpha^* y_0|_w\right] = \prod \left[\prod_{i=1}^{r} \max(1, |\alpha_i^* y_0|_v)\right]^{1/r}$$
$$\leqq \prod_{v \in S} \max_i \max(1, |\alpha_i^* y_0|_v) \cdot \prod_{v \notin S} \max_i \max(1, |\alpha_i^* y_0|_v).$$

If $v \notin S$ then

$$|\alpha_i^* y_0|_v = \left|\left[\alpha_i - \frac{1}{r}\,\mathrm{trace}(\alpha)\right] y_0\right|_v \leqq \left|\frac{1}{r}\right|_v$$

and the product over $v \notin S$ does not exceed r. Also

$$\prod_{v \in S} \max_i \max(1, |\alpha_i^* y_0|_v) = m_S(\mathcal{F})$$

if we assume that we deal with a reduced pair (F_0, y_0), and we conclude

$$(9.1) \qquad h(\alpha^* y_0) \leqq r\, m_S(\mathcal{F}).$$

We also have

$$(9.2) \qquad h(\beta_i) = \prod_v \max(1, |y_0 x_i^*/y|_v) = \prod_v \max(|x_i^*|_v, |y/y_0|_v)$$
$$= h_S(x_i^*, y/y_0) \prod_{v \notin S} (|x_i^*|_v, |y/y_0|_v).$$

If $v \notin S$ then $|x_i^*|_v \leqq \left|\frac{1}{r}\right|_v$, $\left|\frac{y}{y_0}\right|_v \leqq \frac{1}{|y_0|_v}$, hence (9.2) yields

$$h(\beta_i) \leqq r \prod_{v \notin S} |y_0|_v^{-1}\, h_S(x_i^*, y/y_0) = r\,|y_0|_S\, h_S(x_i^*, y/y_0)$$

because of the product formula. Also $|y_0|_S \leqq 2m_S(\mathcal{F})$ by Lemma 1, and we get

$$(9.3) \qquad h(\beta_i) \leqq 2r\, m_S(\mathcal{F}) h_S(x_i^*, y/y_0).$$

Let $x/y \in N(F_0,\alpha,S',\vartheta,\gamma,\Gamma;B)$. Then by Lemma 2, Corollary we have:

(9.4) $$|\alpha^* y_0 - x^* y_0/y|_v = |y/y_0|_v^{-1} |x-\alpha_{i(v)}y|_v$$

$$\leqq \begin{cases} |y/y_0|_{S'}^{-\vartheta(v)/B} M^{B\Gamma(v)} & \text{if } v \in S' \\ |y/y_0|_{S''}^{-(\gamma(v)+\kappa'')/B} M^{B\Gamma(v)} & \text{if } v \in S'' \end{cases}$$

with $\kappa'' = 2(1-B)/\mathrm{Card}(S'')$ and

(9.5) $$M = \left[\frac{2m_S(\mathfrak{F})}{|y_0|_S}\right]^{r(r-1)/2} h_S(x^*,y/y_0)^{-r+1}.$$

We have $|y/y_0|_{S'} \geqq 1$ and by Lemma 3 we also have

$$|y/y_0|_{S''} = |y/y_0|_S/|y/y_0|_{S'} \geqq |y/y_0|_S(2m_S(\mathfrak{F})h_S(x^*,y/y_0))^{-1}$$
$$\geqq |y_0|_S^{-1}(2m_S(\mathfrak{F})h_S(x^*,y/y_0))^{-1} \geqq (4m_S(\mathfrak{F})^2 h_S(x^*,y/y_0))^{-1}.$$

In view of these inequalities, (9.4) becomes

(9.6)
$$|\alpha^* y_0 - x_* y_0/y|_v \leqq \begin{cases} M^{B\Gamma(v)} & \text{if } v \in S' \\ (4m_S(\mathfrak{F})^2 h_S(x^*,y/y_0))^{(\gamma(v)+\kappa'')/B} M^{B\Gamma(v)} & \text{if } v \in S'' \end{cases}$$

with M given by (9.5). Let us abbreviate

(9.7) $$\begin{cases} C_v = (2m_S(\mathfrak{F}))^{r(r-1)B\Gamma(v)/2} & \text{if } v \in S', \\ C_v = (4m_S^2(\mathfrak{F}))^{(\gamma(v)+\kappa'')/B}(2m_S(\mathfrak{F}))^{r(r-1)B\Gamma(v)/2} & \text{if } v \in S'', \end{cases}$$

and note that $|y_0|_S \geqq 1$, so that (9.6) implies

(9.8) $$|\alpha^* y_0 - x^* y_0/y|_v \leqq C_v\, h_S(x^*,y/y_0)^{\epsilon(v)-B(r-1)\Gamma(v)}$$

with

$$(9.9)\qquad \epsilon(v) = \begin{cases} 0 & \text{if} \quad v \in S' \\ (\gamma(v)+\kappa'')/B & \text{if} \quad v \in S'' \end{cases}$$

We apply the Thue Principle choosing $\alpha_1 = \alpha_2 = \alpha^* y_0$, $\beta_1 = x_1^* y_0/y_1$, $\beta_2 = x_2^* y_0/y_2$ where x_1/y_1 and x_2/y_2 are two very large solutions in $N_0(F_0, \alpha, S', \vartheta, \gamma, \Gamma; B)$, and choosing

$$\vartheta = 1, \quad \delta_1 = 1/\log h_S(x_1^*, y_1/y_0),$$
$$\delta_2 = 1/\log h_S(x_2^*, y_2/y_0).$$

We have

$$(9.10)\qquad |\alpha_i - \beta_i|_v^{\vartheta\delta_i} \leq C_v^{\vartheta\delta_i} h_S(x_i^*, y_i/y_0)^{(\epsilon(v)-B(r-1)\Gamma(v))\vartheta\delta_i} = C_v^{\delta_i} e^{\epsilon(v)-B(r-1)\Gamma(v)}$$

by definition of δ_i. Assuming, as we may, that $\delta_1 \geq \delta_2$ we have $C_v^{\delta_1} \geq C_v^{\delta_2}$ and (9.10) and the Thue Principle yield:

either

$$(9.11)\qquad \prod_{v\in S} C_v^{\delta_1} e^{\epsilon(v)-B(r-1)\Gamma(v)}$$
$$> (3h(\alpha^* y_0))^{-C\frac{\delta_1}{t-\tau} - C\frac{\delta_2}{t-\tau}} h(x_1^* y_0/y_1)^{-\frac{\delta_1}{t-\tau}} h(x_2^* y_0/y_2)^{-\frac{\delta_2}{t-\tau}}$$

with $C = \dfrac{1}{1-rt^2/2}$, *or*

$$\frac{r}{2}\frac{\delta_2}{\delta_1} > \frac{r}{2} t^2 + \frac{1}{2}\tau^2 - 1.$$

By (9.3),

$$h(\beta_i)^{\delta_i} \leq (2rm_S(\mathcal{F}))^{\delta_i} e,$$

and by (9.1) we see that

$$(9.12)\qquad (3h(\alpha^* y_0))^{-C\frac{\delta_1}{t-\tau}-C\frac{\delta_2}{t-\tau}}\, h(x_1^* y_0/y_1)^{-\frac{\delta_1}{t-\tau}}\, h(x_2^* y_0/y_2)^{-\frac{\delta_2}{t-\tau}}$$

$$\geqq (3rm_S(\mathfrak{F}))^{-2(C+1)\frac{\delta_1}{t-\tau}}\, e^{-\frac{2}{t-\tau}}.$$

Also

$$\prod_{v\in S} C_v^{\delta_1} e^{\epsilon(v)-B(r-1)\Gamma(v)}$$

$$= \left[2m_S(\mathfrak{F})\right]^{(2(1+\kappa''\mathrm{Card}(S''))/B+r(r-1)B/2)\delta_1}\, e^{(1+\kappa''\mathrm{Card}(S''))/B-B(r-1)}$$

$$< \left[2m_S(\mathfrak{F})\right]^{((6-4B)/B+r(r-1)B/2)\delta_1}\, e^{(3-2B)/B-B(r-1)}$$

and (9.12), (9.11) imply

either

$$(9.13)\qquad (2m_S(\mathfrak{F}))^{((6-4B)/B+r(r-1)B/2)\delta_1}\, e^{(3-2B)/B-B(r-1)}$$

$$> (3m_S(\mathfrak{F}))^{-2(C+1)\frac{\delta_1}{t-\tau}}\, e^{-\frac{2}{t-\tau}}$$

with $C = \frac{1}{1-rt^2/2}$,

or

$$(9.14)\qquad \frac{r}{2}\frac{\delta_2}{\delta_1} > \frac{r}{2}t^2 + \frac{1}{2}\tau^2 - 1.$$

We claim that if $r \geqq 6$ and $m_S(\mathfrak{F}) \geqq \frac{3r}{4}$ we can choose t, τ, B such that $\tau > \sqrt{2-rt^2}$ and (9.13) does not hold. In fact, since x_1/y_1 is a very large solution, we have

$$\frac{1}{\delta_1} = \log h_S(x_1^*, y_1/y_0) \geqq (9r+18)\left(\frac{17(r-1)}{32}\right)^{200}\log(2m_S(\mathfrak{F}))$$

and thus the left-hand side of (9.13) satisfies the inequality

(9.15) $\quad$ l.h.s.(9.13) <

$$\exp\left[\left[\frac{6-4B}{B} + r(r-1)\frac{B}{2}\right](9r+18)^{-1}\left[\frac{17(r-1)}{32}\right]^{-200} + \frac{3-2B}{B} - B(r-1)\right].$$

Similarly, $3rm_S(\mathfrak{F}) \leqq (2m_S(\mathfrak{F}))^2$ and the right-hand side of (9.13) satisfies the inequality

(9.16) $\quad$ r.h.s. (9.13) >

$$\exp\left[-\left[\frac{1}{1-rt^2/2} +1\right] \frac{4}{t-\tau} (9r++18)^{-1}\left[\frac{17(r-1)}{32}\right]^{-200} - \frac{2}{t-\tau}\right]$$

and we conclude from (9.15) and (9.16) that if (9.13) holds then

$$\left[\frac{6-4B}{B} +r(r-1) \frac{B}{2}\right](9r+18)^{-1}\left[\frac{17(r-1)}{32}\right]^{-200} + \frac{3-2B}{B} - B(r-1)$$
$$> - \left[\frac{1}{1-rt^2/2} +1\right] \frac{4}{t-\tau} (9r+18)^{-1}\left[\frac{17(r-1)}{32}\right]^{-200} - \frac{2}{t-\tau}$$

or in other words

(9.17)

$$B(r-1) - \frac{2}{t-\tau} - \frac{3-2B}{B} < \left[\frac{6-4B}{B} +r(r-1) \frac{B}{2} + \left[\frac{1}{1-rt^2/2} +1\right]\frac{4}{t-\tau}\right](9r+18)^{-1}\cdot\left[\frac{17(r-1)}{32}\right]^{-200}.$$

We choose

(9.18)
$$t = \sqrt{\frac{2}{r+a^2}}\,, \quad \tau = bt$$

where $a < b < 1$, so that the left-hand side of (9.17) becomes

(9.19)
$$\text{l.h..s. } (9.17) = B(r-1) - \frac{\sqrt{2(r+a^2)}}{1-b} - \frac{3}{B} + 2.$$

If $b = a = \frac{8-\sqrt{50}}{7}$, $B = 1$, $r = 6$ the expression in (9.19) is equal to 0. If we choose

(9.20) $$a = \frac{1}{20}, \quad b = \frac{1}{10}, \quad B = \frac{99}{100}$$

the expression in (9.19) is an increasing function of r, and we get

(9.21) $$\text{l.h.s. } (9.17) > 0.069$$

as soon as $r \geqq 6$. On the other hand, the crudest estimation of the right-hand side of (9.17) suffices to contradict (9.21), and we conclude that with the choices (9.18), (9.20) the first alternative (9.13) in the Thue Principle does not hold. Thus (9.14) must hold and hence

(9.22) $$\log h_S(x_2^*, y_2/y_0) < \frac{r}{rt^2 + \tau^2 - 2} \log h_S(x_1^*, y_1/y_0)$$
$$< 100\, r^2 \log h_S(x_1^*, y_1/y_0).$$

We may interpret (9.22) as saying that the set of very large solutions in $N_0(F_0, \alpha, S', \vartheta, \gamma, \Gamma; B)$ with $B = \frac{99}{100}$ lies in some interval

$$(H, H^{100r^2}).$$

Since $100r^2 < (\frac{17(r-1)}{32})^N$ as soon as $r \geqq 6$ and $N = 9$, Lemma 6 shows that there are at most 9 large solutions in $N_0(F_0, \alpha, S', \vartheta, \gamma, \Gamma; B)$, with $B = \frac{99}{100}$. In view of (4.9), we have shown:

Lemma 7. *The number of very large solutions in* $N_0(F_0)$ *does not exceed*

$$9(320000r)^{\mathrm{Card}(S)}$$

provided $m_S(\mathcal{F}) \geqq 3r/4$ *and* $r \geqq 6$.

X. A bound for $N_0(F_0)$

Theorem 1. *Suppose that* $r \geqq 6$ *and* $m_S(\mathcal{F}) \geqq 2^{29} r^{10}$. *Then*

$$\mathrm{Card}(N_0(F)) < 300r^2\, (213r)^{2\mathrm{Card}(S)}.$$

Proof. By Lemma 5, Corollary 2 to Lemma 6 and Lemma 7 we have

$$\mathrm{Card}(N_0(F_0)) \leqq 470(3r+7)^2(80r(18r+39)))^{\mathrm{Card}(S)} + 9(320000r)^{\mathrm{Card}(S)} + 200(512r)^{\mathrm{Card}(S)}.$$

If $r \geqq 6$, the quantity on the right does not exceed $300r^2(213r)^{2\mathrm{Card}(S)}$ and Theorem 1 follows.

XI. The general bound

In this section we show how to remove the restriction $m_S(\mathfrak{F}) \geqq 2^{29}r^{10}$ which appears in Theorem 1, at the cost of increasing a little the final bound. The idea occurs already in [B.-S.].

Let $A = \begin{bmatrix} a & b \\ c & d \end{bmatrix} \in GL(2, I_S)$ and let $G(\mathbf{x}) = F(A\mathbf{x})$. Then (2.7) and (2.11) with G in place of F, together with the fact that $|y|_S \geqq 1$ for $y \in I_S$, $y \neq 0$, yield:

Lemma 7. *If* $G = F \circ A$ *then*

$$m_S(\mathfrak{G}) \geqq \frac{1}{2}\,|\det A|_S.$$

Let $\mathfrak{p}$ be the smallest prime ideal not in S. By the Tchebotarev density theorem, there is a prime ideal $\underline{q}$ such that $\mathfrak{p}\underline{q}$ is a principal ideal (π). Let $\tilde{v}$ be the place determined by $\underline{q}$; by replacing S by $\tilde{S} = S \cup \tilde{v}$, we may assume that $\tilde{v} \in S$, which may increase $\mathrm{Card}(S)$ by 1. Now π is a unit outside $\mathfrak{p}$, hence the product formula yields

$$|\pi|_S = |\pi|_{\mathfrak{p}}^{-1} = p^{f_{\mathfrak{p}}/[k:Q]}$$

where p is the rational prime with $\mathfrak{p}|p$ and $f_{\mathfrak{p}}$ is the residue class degree of $\mathfrak{p}$. Let $\mathbf{x} = (x,y)$ be a solution to the Thue-Mahler equation (3.1). Then $|\Sigma\, a_i x^i y^{r-i}|_v = 1$ for every v, and $|a_i|_v \leqq 1$, $|x|_v \leqq 1$, $|y|_v \leqq 1$ if $v \notin S$; it follows that

$$\max\left[|x|_v, |y|_v\right] = 1 \qquad (11.1)$$

whenever $v \notin S$. We apply (11.1) to $v = \mathfrak{p}$, so that either $|x|_{\mathfrak{p}} = 1$ or $|y|_{\mathfrak{p}} = 1$. In what follows, we assume that $|x|_{\mathfrak{p}} = 1$, the argument for the case $|y|_{\mathfrak{p}} = 1$ being the same.

Let m be any positive integer. Since $|x|_{\mathfrak{p}} = 1$, the image of x in $I_S/\pi^m I_S$ is a unit. We choose a set J of representatives of $I_S/\pi^m I_S$ in I_S and note that there is $j \in J$ such that $y \equiv jx (\mathrm{mod}\ \pi^m)$. This shows that there is $\tilde{\mathbf{x}} = (\tilde{x}, \tilde{y})$ with S-integer coordinates such that

$$\mathbf{x} = \begin{bmatrix} 1 & 0 \\ j & \pi^m \end{bmatrix} \tilde{\mathbf{x}}\ . \tag{11.2}$$

Let $G_j = F \circ \begin{bmatrix} 1 & 0 \\ j & \pi^m \end{bmatrix}$, $G_j' = F \circ \begin{bmatrix} \pi^m & j \\ 0 & 1 \end{bmatrix}$. Then (11.2) and its analogue for the case $|y|_{\mathfrak{p}} = 1$ show that

$$\mathrm{Card}(N_0(F)) \leqq \sum (\mathrm{Card}(N_0(G_j)) + \mathrm{Card}(N_0(G_j')))$$

where $j \in J$. Since $\mathrm{Card}(J) = p^{mf_{\mathfrak{p}}}$ we obtain

$$\mathrm{Card}(N_0(F)) \leqq 2p^{mf_{\mathfrak{p}}}\ \mathrm{Card}(N_0(G)) \tag{11.3}$$

where G runs over the forms G_j, G_j'. By Lemma 7,

$$m_S(G) \geqq \frac{1}{2}\ |\pi|_S^m = \frac{1}{2}\ p^{mf_{\mathfrak{p}}/[k:\mathbb{Q}]}\ . \tag{11.4}$$

It is trivial that the smallest prime $\mathfrak{p}$ not in S divides a rational prime $p \leqq (2\mathrm{Card}(S))^2$. Thus if $K \geqq (2\mathrm{Card}(S))^2$ we can always choose the integer m so that

$$K < p^{mf_{\mathfrak{p}}/[k:\mathbb{Q}]} \leqq K^2. \tag{11.5}$$

We do this with

$$K = \max\left[(2\mathrm{Card}(S))^2, 2^{30}r^{10}\right]$$

and note that (11.4) and (11.5) imply that $m_S(\mathfrak{q}) \geqq 2^{29}r^{10}$. From (11.5) we get $p^{mf_{\mathfrak{P}}} \leqq K^{2[k:\mathbb{Q}]}$; now Theorem 1 applies to the forms G and from (11.3) we infer

$$(11.6) \qquad N_0(F) \leqq 2 \max\left[(2\mathrm{Card}(S))^2, 2^{30}r^{10}\right]^{[k:\mathbb{Q}]} 300\, r^2 (213r)^{2\mathrm{Card}(S)},$$

where however we must keep in mind that S may have been enlarged so as to include the prime $\underline{q}$. It is also clear that $[k:\mathbb{Q}] \leqq 2\mathrm{Card}(S)$. We replace $\mathrm{Card}(S)$ in (11.6) by $\mathrm{Card}(S) + 1$, so that the condition that S contains $\underline{q}$ can be omitted, use (2.2) and conclude easily

$$\begin{aligned} N(F) &\leqq r^{\mathrm{Card}(S)-1} N_0(F) \\ &\leqq 600\, (4\mathrm{Card}(S))^{2[k:\mathbb{Q}]}\, 2^{30[k:\mathbb{Q}]}\, 213^{2\mathrm{Card}(S)+2}\, r^{3\mathrm{Card}(S)+10[k:\mathbb{Q}]+3} \\ &< (4\mathrm{Card}(S))^{2[k:\mathbb{Q}]}\, (4r)^{26\mathrm{Card}(S)}, \end{aligned}$$

which proves our Main Theorem.

The Institute for Advanced Study

Princeton, New Jersey 08540

REFERENCES

[B] Bombieri, E.: On the Thue-Siegel-Dyson theorem. Acta Math. 148 (1982), 255-296.

[B.-S.] Bombieri, E., Schmidt, W. M.: On Thue's equation. Inventiones Math. 88 (1987), 69-81.

[E] Evertse, J.-H.: Upper bounds for the number of solutions of diophantine equations. Math. Centrum, Amsterdam, (1983), 1-127.

[L.-M.] Lewis, D., Mahler, K.: Representation of integers by binary forms. Acta Arith. 6 (1961), 333-363.

[S] Silverman, J.-H.: Representation of integers by binary forms and the rank of the Mordell-Weil group. Inventiones Math. 74 (1983), 218-292.

LECTURE NOTES IN MATHEMATICS
Edited by A. Dold and B. Eckmann

Some general remarks on the publication of monographs and seminars

n what follows all references to monographs, are applicable also o multiauthorship volumes such as seminar notes.

. Lecture Notes aim to report new developments - quickly, informally, and at a high level. Monograph manuscripts should be reasonably self-contained and rounded off. Thus they may, and often will, present not only results of the author but also related work by other people. Furthermore, the manuscripts should provide sufficient motivation, examples and applications. This clearly distinguishes Lecture Notes manuscripts from journal articles which normally are very concise. Articles intended for a journal but too long to be accepted by most journals, usually do not have this "lecture notes" character. For similar reasons it is unusual for Ph.D. theses to be accepted for the Lecture Notes series.

Experience has shown that English language manuscripts achieve a much wider distribution.

. Manuscripts or plans for Lecture Notes volumes should be submitted either to one of the series editors or to Springer-Verlag, Heidelberg. These proposals are then refereed. A final decision concerning publication can only be made on the basis of the complete manuscripts, but a preliminary decision can usually be based on partial information: a fairly detailed outline describing the planned contents of each chapter, and an indication of the estimated length, a bibliography, and one or two sample chapters - or a first draft of the manuscript. The editors will try to make the preliminary decision as definite as they can on the basis of the available information.

. Lecture Notes are printed by photo-offset from typed copy delivered in camera-ready form by the authors. Springer-Verlag provides technical instructions for the preparation of manuscripts, and will also, on request, supply special staionery on which the prescribed typing area is outlined. Careful preparation of the manuscripts will help keep production time short and ensure satisfactory appearance of the finished book. Running titles are not required; if however they are considered necessary, they should be uniform in appearance. We generally advise authors not to start having their final manuscripts specially tpyed beforehand. For professionally typed manuscripts, prepared on the special stationery according to our instructions, Springer-Verlag will, if necessary, contribute towards the typing costs at a fixed rate.

The actual production of a Lecture Notes volume takes 6-8 weeks.

.../...

4. Final manuscripts should contain at least 100 pages of mathema-tical text and should include

   - a table of contents
   - an informative introduction, perhaps with some historical re-marks. It should be accessible to a reader not particularl familiar with the topic treated.
   - subject index; this is almost always genuinely helpful for th reader.

5. Authors receive a total of 50 free copies of their volume, bu no royalties. They are entitled to purchase further copies o their book for their personal use at a discount of 33 1/3 % other Springer mathematics books at a discount of 20 % directl from Springer-Verlag.

   Commitment to publish is made by letter of intent rather than b signing a formal contract. Springer-Verlag secures the copyrigh for each volume.

ECTURE NOTES

SSENTIALS FOR THE PREPARATION
F CAMERA-READY MANUSCRIPTS

he preparation of manuscripts which are to be reproduced by photo-ffset requires special care. Manuscripts which are submitted in echnically unsuitable form will be returned to the author for retyping. here is normally no possibility of carrying out further corrections fter a manuscript is given to production. Hence it is crucial that the ollowing instructions be adhered to closely. If in doubt, please send s 1 - 2 sample pages for examination.

yping area. On request, Springer-Verlag will supply special paper with he typing area outlined.
he CORRECT TYPING AREA is 18 x 26 1/2 cm (7,5 x 11 inches).

ake sure the TYPING AREA IS COMPLETELY FILLED. Set the margins so that hey precisely match the outline and type right from the top to the ottom line. (Note that the page-number will lie outside this area). ines of text should not end more than three spaces inside or outside he right margin (see example on page 4).

ype on one side of the paper only.

ype. Use an electric typewriter if at all possible. CLEAN THE TYPE be-ore use and always use a BLACK ribbon (a carbon ribbon is best).

hoose a type size large enough to stand reduction to 75%.

ord Processors. Authors using word-processing or computer-typesetting acilities should follow these instructions with obvious modifications. lease note with respect to your printout that
) the characters should be sharp and sufficiently black;
i) if the size of your characters is significantly larger or smaller han normal typescript characters, you should adapt the length and readth of the text area proportionally keeping the proportions 1:0.68.
ii) it is not necessary to use Springer's special typing paper. Any hite paper of reasonable quality is acceptable.
F IN DOUBT, PLEASE SEND US 1-2 SAMPLE PAGES FOR EXAMINATION. We will e glad to give advice.

pacing and Headings (Monographs). Use ONE-AND-A-HALF line spacing in he text. Please leave sufficient space for the title to stand out clear-y and do NOT use a new page for the beginning of subdivisions of chap-ers. Leave THREE LINES blank above and TWO below headings of such sub-ivisions.

Spacing and Headings (Proceedings). Use ONE-AND-A-HALF line spacing in he text. Start each paper on a NEW PAGE and leave sufficient space for he title to stand out clearly. However, do NOT use a new page for the eginning of subdivisions of a paper. Leave THREE LINES blank above and WO below headings of such subdivisions. Make sure headings of equal mportance are in the same form.

he first page of each contribution should be prepared in the same way. herefore, we recommend that the editor prepares a sample page and asses it on to the authors together with these ESSENTIALS. Please take

.../...

the following as an example.

MATHEMATICAL STRUCTURE IN QUANTUM FIELD THEORY

John E. Robert
Fachbereich Physik, Universität Osnabrück
Postfach 44 69, D-4500 Osnabrück

Please leave THREE LINES blank below heading and address of the author. THEN START THE ACTUAL TEXT OF YOUR CONTRIBUTION.

Footnotes. These should be avoided. If they cannot be avoided, place them at the foot of the page, separated from the text by a line 4 cm long, and type them in SINGLE LINE SPACING to finish exactly on the outline.

Symbols. Anything which cannot be typed may be entered by hand in BLACK AND ONLY BLACK ink. (A fine-tipped rapidograph is suitable for this purpose; a good black ball-point will do, but a pencil will not). Do not draw straight lines by hand without a ruler (not even in fractions).

Equations and Computer Programs. Equations and computer programs should begin four spaces inside the left margin. Should the equations be numbered, then each number should be in brackets at the right-hand edge of the typing area.

Pagination. Number pages in the upper right-hand corner in LIGHT BLUE OR GREEN PENCIL ONLY. The final page numbers will be inserted by the printer.

There should normally be NO BLANK PAGES in the manuscript (between chapters or between contributions) unless the book is divided into Part A, Part B for example, which should then begin on a right-hand page.

It is much safer to number pages AFTER the text has been typed and corrected. Page 1 (Arabic) should be THE FIRST PAGE OF THE ACTUAL TEXT. The Roman pagination (table of contents, preface, abstract, acknowledgements, brief introductions, etc.) will be done by Springer-Verlag.

Corrections. When corrections have to be made, cut the new text to fit and PASTE it over the old. White correction fluid may also be used.

Never make corrections or insertions in the text by hand.

If the typescript has to be marked for any reason, e.g. for TEMPORARY page numbers or to mark corrections for the typist, this can be done VERY FAINTLY with BLUE or GREEN PENCIL but NO OTHER COLOR: these colors do not appear after reproduction.

Table of Contents. It is advisable to type the table of contents later, copying the titles from the text and inserting page numbers.

Literature References. These should be placed at the end of each paper or chapter, or at the end of the work, as desired. Type them with single line spacing and start each reference on a new line.
Please ensure that all references are COMPLETE and PRECISE.

Vol. 1145: G. Winkler, Choquet Order and Simplices. VI, 143 pages. 1985.

Vol. 1146: Séminaire d'Algèbre Paul Dubreil et Marie-Paule Malliavin. Proceedings, 1983–1984. Edité par M.-P. Malliavin. IV, 420 pages. 1985.

Vol. 1147: M. Wschebor, Surfaces Aléatoires. VII, 111 pages. 1985.

Vol. 1148: Mark A. Kon, Probability Distributions in Quantum Statistical Mechanics. V, 121 pages. 1985.

Vol. 1149: Universal Algebra and Lattice Theory. Proceedings, 1984. Edited by S. D. Comer. VI, 282 pages. 1985.

Vol. 1150: B. Kawohl, Rearrangements and Convexity of Level Sets in PDE. V, 136 pages. 1985.

Vol 1151: Ordinary and Partial Differential Equations. Proceedings, 1984. Edited by B.D. Sleeman and R.J. Jarvis. XIV, 357 pages. 1985.

Vol. 1152: H. Widom, Asymptotic Expansions for Pseudodifferential Operators on Bounded Domains. V, 150 pages. 1985.

Vol. 1153: Probability in Banach Spaces V. Proceedings, 1984. Edited by A. Beck, R. Dudley, M. Hahn, J. Kuelbs and M. Marcus. VI, 457 pages. 1985.

Vol. 1154: D.S. Naidu, A.K. Rao, Singular Pertubation Analysis of Discrete Control Systems. IX, 195 pages. 1985.

Vol. 1155: Stability Problems for Stochastic Models. Proceedings, 1984. Edited by V.V. Kalashnikov and V.M. Zolotarev. VI, 447 pages. 1985.

Vol. 1156: Global Differential Geometry and Global Analysis 1984. Proceedings, 1984. Edited by D. Ferus, R.B. Gardner, S. Helgason and U. Simon. V, 339 pages. 1985.

Vol. 1157: H. Levine, Classifying Immersions into $\mathbb{R}^4$ over Stable Maps of 3-Manifolds into $\mathbb{R}^2$. V, 163 pages. 1985.

Vol. 1158: Stochastic Processes – Mathematics and Physics. Proceedings, 1984. Edited by S. Albeverio, Ph. Blanchard and L. Streit. VI, 230 pages. 1986.

Vol. 1159: Schrödinger Operators, Como 1984. Seminar. Edited by S. Graffi. VIII, 272 pages. 1986.

Vol. 1160: J.-C. van der Meer, The Hamiltonian Hopf Bifurcation. VI, 115 pages. 1985.

Vol. 1161: Harmonic Mappings and Minimal Immersions, Montecatini 1984. Seminar. Edited by E. Giusti. VII, 285 pages. 1985.

Vol. 1162: S.J.L. van Eijndhoven, J. de Graaf, Trajectory Spaces, Generalized Functions and Unbounded Operators. IV, 272 pages. 1985.

Vol. 1163: Iteration Theory and its Functional Equations. Proceedings, 1984. Edited by R. Liedl, L. Reich and Gy. Targonski. VIII, 231 pages. 1985.

Vol. 1164: M. Meschiari, J.H. Rawnsley, S. Salamon, Geometry Seminar "Luigi Bianchi" II – 1984. Edited by E. Vesentini. VI, 224 pages. 1985.

Vol. 1165: Seminar on Deformations. Proceedings, 1982/84. Edited by J. Ławrynowicz. IX, 331 pages. 1985.

Vol. 1166: Banach Spaces. Proceedings, 1984. Edited by N. Kalton and E. Saab. VI, 199 pages. 1985.

Vol. 1167: Geometry and Topology. Proceedings, 1983–84. Edited by J. Alexander and J. Harer. VI, [illegible] pages. 1985.

Vol. 1168: S.S. Agaian, Hadamard Matrices and their Applications. III, 227 pages. 1985.

Vol. 1169: W.A. Light, E.W. Cheney, Approximation Theory in Tensor Product Spaces. VII, 157 pages. 1985.

Vol. 1170: B.S. Thomson, Real Functions. VII, 229 pages. 1985.

Vol. 1171: Polynômes Orthogonaux et Applications. Proceedings, 1984. Edité par C. Brezinski, A. Draux, A.P. Magnus, P. Maroni et A. Ronveaux. XXXVII, 584 pages. 1985.

Vol. 1172: Algebraic Topology, Göttingen 1984. Proceedings. Edited by L. Smith. VI, 209 pages. 1985.

Vol. 1173: H. Delfs, M. Knebusch, Locally Semialgebraic Spaces. XVI, 329 pages. 1985.

Vol. 1174: Categories in Continuum Physics, Buffalo 1982. Seminar. Edited by F.W. Lawvere and S.H. Schanuel. V, 126 pages. 1986.

Vol. 1175: K. Mathiak, Valuations of Skew Fields and Projective Hjelmslev Spaces. VII, 116 pages. 1986.

Vol. 1176: R.R. Bruner, J.P. May, J.E. McClure, M. Steinberger, H_∞ Ring Spectra and their Applications. VII, 388 pages. 1986.

Vol. 1177: Representation Theory I. Finite Dimensional Algebras. Proceedings, 1984. Edited by V. Dlab, P. Gabriel and G. Michler. XV, 340 pages. 1986.

Vol. 1178: Representation Theory II. Groups and Orders. Proceedings, 1984. Edited by V. Dlab, P. Gabriel and G. Michler. XV, 370 pages. 1986.

Vol. 1179: Shi J.-Y. The Kazhdan-Lusztig Cells in Certain Affine Weyl Groups. X, 307 pages. 1986.

Vol. 1180: R. Carmona, H. Kesten, J.B. Walsh, École d'Été de Probabilités de Saint-Flour XIV – 1984. Édité par P.L. Hennequin. X, 438 pages. 1986.

Vol. 1181: Buildings and the Geometry of Diagrams, Como 1984. Seminar. Edited by L. Rosati. VII, 277 pages. 1986.

Vol. 1182: S. Shelah, Around Classification Theory of Models. VII, 279 pages. 1986.

Vol. 1183: Algebra, Algebraic Topology and their Interactions. Proceedings, 1983. Edited by J.-E. Roos. XI, 396 pages. 1986.

Vol. 1184: W. Arendt, A. Grabosch, G. Greiner, U. Groh, H.P. Lotz, U. Moustakas, R. Nagel, F. Neubrander, U. Schlotterbeck, One-parameter Semigroups of Positive Operators. Edited by R. Nagel. X, 460 pages. 1986.

Vol. 1185: Group Theory, Beijing 1984. Proceedings. Edited by Tuan H.F. V, 403 pages. 1986.

Vol. 1186: Lyapunov Exponents. Proceedings, 1984. Edited by L. Arnold and V. Wihstutz. VI, 374 pages. 1986.

Vol. 1187: Y. Diers, Categories of Boolean Sheaves of Simple Algebras. VI, 168 pages. 1986.

Vol. 1188: Fonctions de Plusieurs Variables Complexes V. Séminaire, 1979–85. Edité par François Norguet. VI, 306 pages. 1986.

Vol. 1189: J. Lukeš, J. Malý, L. Zajíček, Fine Topology Methods in Real Analysis and Potential Theory. X, 472 pages. 1986.

Vol. 1190: Optimization and Related Fields. Proceedings, 1984. Edited by R. Conti, E. De Giorgi and F. Giannessi. VIII, 419 pages. 1986.

Vol. 1191: A.R. Its, V.Yu. Novokshenov, The Isomonodromic Deformation Method in the Theory of Painlevé Equations. IV, 313 pages. 1986.

Vol. 1192: Equadiff 6. Proceedings, 1985. Edited by J. Vosmansky and M. Zlámal. XXIII, 404 pages. 1986.

Vol. 1193: Geometrical and Statistical Aspects of Probability in Banach Spaces. Proceedings, 1985. Edited by X. Fernique, B. Heinkel, M.B. Marcus and P.A. Meyer. IV, 128 pages. 1986.

Vol. 1194: Complex Analysis and Algebraic Geometry. Proceedings, 1985. Edited by H. Grauert. VI, 235 pages. 1986.

Vol. 1195: J.M. Barbosa, A.G. Colares, Minimal Surfaces in $\mathbb{R}^3$. X, 124 pages. 1986.

Vol. 1196: E. Casas-Alvero, S. Xambó-Descamps, The Enumerative Theory of Conics after Halphen. IX, 130 pages. 1986.

Vol. 1197: Ring Theory. Proceedings, 1985. Edited by F.M.J. van Oystaeyen. V, 231 pages. 1986.

Vol. 1198: Séminaire d'Analyse, P. Lelong – P. Dolbeault – H. Skoda. Seminar 1983/84. X, 260 pages. 1986.

Vol. 1199: Analytic Theory of Continued Fractions II. Proceedings, 1985. Edited by W.J. Thron. VI, 299 pages. 1986.

Vol. 1200: V.D. Milman, G. Schechtman, Asymptotic Theory of Finite Dimensional Normed Spaces. With an Appendix by M. Gromov. VIII, 156 pages. 1986.

Vol. 1201: Curvature and Topology of Riemannian Manifolds. Proceedings, 1985. Edited by K. Shiohama, T. Sakai and T. Sunada. VII, 336 pages. 1986.

Vol. 1202: A. Dür, Möbius Functions, Incidence Algebras and Power Series Representations. XI, 134 pages. 1986.

Vol. 1203: Stochastic Processes and Their Applications. Proceedings, 1985. Edited by K. Itô and T. Hida. VI, 222 pages. 1986.

Vol. 1204: Séminaire de Probabilités XX, 1984/85. Proceedings. Edité par J. Azéma et M. Yor. V, 639 pages. 1986.

Vol. 1205: B.Z. Moroz, Analytic Arithmetic in Algebraic Number Fields. VII, 177 pages. 1986.

Vol. 1206: Probability and Analysis, Varenna (Como) 1985. Seminar. Edited by G. Letta and M. Pratelli. VIII, 280 pages. 1986.

Vol. 1207: P.H. Bérard, Spectral Geometry: Direct and Inverse Problems. With an Appendix by G. Besson. XIII, 272 pages. 1986.

Vol. 1208: S. Kaijser, J.W. Pelletier, Interpolation Functors and Duality. IV, 167 pages. 1986.

Vol. 1209: Differential Geometry, Peñíscola 1985. Proceedings. Edited by A.M. Naveira, A. Ferrández and F. Mascaró. VIII, 306 pages. 1986.

Vol. 1210: Probability Measures on Groups VIII. Proceedings, 1985. Edited by H. Heyer. X, 386 pages. 1986.

Vol. 1211: M.B. Sevryuk, Reversible Systems. V, 319 pages. 1986.

Vol. 1212: Stochastic Spatial Processes. Proceedings, 1984. Edited by P. Tautu. VIII, 311 pages. 1986.

Vol. 1213: L.G. Lewis, Jr., J.P. May, M. Steinberger, Equivariant Stable Homotopy Theory. IX, 538 pages. 1986.

Vol. 1214: Global Analysis – Studies and Applications II. Edited by Yu.G. Borisovich and Yu.E. Gliklikh. V, 275 pages. 1986.

Vol. 1215: Lectures in Probability and Statistics. Edited by G. del Pino and R. Rebolledo. V, 491 pages. 1986.

Vol. 1216: J. Kogan, Bifurcation of Extremals in Optimal Control. VIII, 106 pages. 1986.

Vol. 1217: Transformation Groups. Proceedings, 1985. Edited by S. Jackowski and K. Pawalowski. X, 396 pages. 1986.

Vol. 1218: Schrödinger Operators, Aarhus 1985. Seminar. Edited by E. Balslev. V, 222 pages. 1986.

Vol. 1219: R. Weissauer, Stabile Modulformen und Eisensteinreihen. III, 147 Seiten. 1986.

Vol. 1220: Séminaire d'Algèbre Paul Dubreil et Marie-Paule Malliavin. Proceedings, 1985. Edité par M.-P. Malliavin. IV, 200 pages. 1986.

Vol. 1221: Probability and Banach Spaces. Proceedings, 1985. Edited by J. Bastero and M. San Miguel. XI, 222 pages. 1986.

Vol. 1222: A. Katok, J.-M. Strelcyn, with the collaboration of F. Ledrappier and F. Przytycki, Invariant Manifolds, Entropy and Billiards; Smooth Maps with Singularities. VIII, 283 pages. 1986.

Vol. 1223: Differential Equations in Banach Spaces. Proceedings, 1985. Edited by A. Favini and E. Obrecht. VIII, 299 pages. 1986.

Vol. 1224: Nonlinear Diffusion Problems, Montecatini Terme 1985. Seminar. Edited by A. Fasano and M. Primicerio. VIII, 188 pages. 1986.

Vol. 1225: Inverse Problems, Montecatini Terme 1986. Seminar. Edited by G. Talenti. VIII, 204 pages. 1986.

Vol. 1226: A. Buium, Differential Function Fields and Moduli of Algebraic Varieties. IX, 146 pages. 1986.

Vol. 1227: H. Helson, The Spectral Theorem. VI, 104 pages. 1986.

Vol. 1228: Multigrid Methods II. Proceedings, 1985. Edited by W. Hackbusch and U. Trottenberg. VI, 336 pages. 1986.

Vol. 1229: O. Bratteli, Derivations, Dissipations and Group Actions on C*-algebras. IV, 277 pages. 1986.

Vol. 1230: Numerical Analysis. Proceedings, 1984. Edited by J.-P. Hennart. X, 234 pages. 1986.

Vol. 1231: E.-U. Gekeler, Drinfeld Modular Curves. XIV, 107 pages. 1986.

Vol. 1232: P.C. Schuur, Asymptotic Analysis of Soliton Problems. VIII, 180 pages. 1986.

Vol. 1233: Stability Problems for Stochastic Models. Proceedings, 1985. Edited by V.V. Kalashnikov, B. Penkov and V.M. Zolotarev. VI, 223 pages. 1986.

Vol. 1234: Combinatoire énumérative. Proceedings, 1985. Edité par G. Labelle et P. Leroux. XIV, 387 pages. 1986.

Vol. 1235: Séminaire de Théorie du Potentiel, Paris, No. 8. Directeurs: M. Brelot, G. Choquet et J. Deny. Rédacteurs: F. Hirsch et G. Mokobodzki. III, 209 pages. 1987.

Vol. 1236: Stochastic Partial Differential Equations and Applications. Proceedings, 1985. Edited by G. Da Prato and L. Tubaro. V, 257 pages. 1987.

Vol. 1237: Rational Approximation and its Applications in Mathematics and Physics. Proceedings, 1985. Edited by J. Gilewicz, M. Pindor and W. Siemaszko. XII, 350 pages. 1987.

Vol. 1238: M. Holz, K.-P. Podewski and K. Steffens, Injective Choice Functions. VI, 183 pages. 1987.

Vol. 1239: P. Vojta, Diophantine Approximations and Value Distribution Theory. X, 132 pages. 1987.

Vol. 1240: Number Theory, New York 1984–85. Seminar. Edited by D.V. Chudnovsky, G.V. Chudnovsky, H. Cohn and M.B. Nathanson. V, 324 pages. 1987.

Vol. 1241: L. Gårding, Singularities in Linear Wave Propagation. III, 125 pages. 1987.

Vol. 1242: Functional Analysis II, with Contributions by J. Hoffmann-Jørgensen et al. Edited by S. Kurepa, H. Kraljević and D. Butković. VII, 432 pages. 1987.

Vol. 1243: Non Commutative Harmonic Analysis and Lie Groups. Proceedings, 1985. Edited by J. Carmona, P. Delorme and M. Vergne. V, 309 pages. 1987.

Vol. 1244: W. Müller, Manifolds with Cusps of Rank One. XI, 158 pages. 1987.

Vol. 1245: S. Rallis, L-Functions and the Oscillator Representation. XVI, 239 pages. 1987.

Vol. 1246: Hodge Theory. Proceedings, 1985. Edited by E. Cattani, F. Guillén, A. Kaplan and F. Puerta. VII, 175 pages. 1987.

Vol. 1247: Séminaire de Probabilités XXI. Proceedings. Edité par J. Azéma, P.A. Meyer et M. Yor. IV, 579 pages. 1987.

Vol. 1248: Nonlinear Semigroups, Partial Differential Equations and Attractors. Proceedings, 1985. Edited by T.L. Gill and W.W. Zachary. IX, 185 pages. 1987.

Vol. 1249: I. van den Berg, Nonstandard Asymptotic Analysis. IX, 187 pages. 1987.

Vol. 1250: Stochastic Processes – Mathematics and Physics II. Proceedings 1985. Edited by S. Albeverio, Ph. Blanchard and L. Streit. VI, 359 pages. 1987.

Vol. 1251: Differential Geometric Methods in Mathematical Physics. Proceedings, 1985. Edited by P.L. García and A. Pérez-Rendón. VII, 300 pages. 1987.

Vol. 1252: T. Kaise, Représentations de Weil et GL_2 Algèbres de division et GL_n. VII, 203 pages. 1987.

Vol. 1253: J. Fischer, An Approach to the Selberg Trace Formula via the Selberg Zeta-Function. III, 184 pages. 1987.

Vol. 1254: S. Gelbart, I. Piatetski-Shapiro, S. Rallis. Explicit Constructions of Automorphic L-Functions. VI, 152 pages. 1987.

Vol. 1255: Differential Geometry and Differential Equations. Proceedings, 1985. Edited by C. Gu, M. Berger and R.L. Bryant. XII, 243 pages. 1987.

Vol. 1256: Pseudo-Differential Operators. Proceedings, 1986. Edited by H.O. Cordes, B. Gramsch and H. Widom. X, 479 pages. 1987.

Vol. 1257: X. Wang, On the C*-Algebras of Foliations in the Plane. V, 165 pages. 1987.

Vol. 1258: J. Weidmann, Spectral Theory of Ordinary Differential Operators. VI, 303 pages. 1987.